Internet Networks

Wired, Wireless, and Optical Technologies

Devices, Circuits, and Systems

Series Editor
Krzysztof Iniewski
CMOS Emerging Technologies Inc., Vancouver, British Columbia, Canada

Internet Networks: Wired, Wireless, and Optical Technologies
Krzysztof Iniewski

FORTHCOMING

Electronics for Radiation Detection
Krzysztof Iniewski

Radio Frequency Integrated Circuit Design
Sebastian Magierowski

Semiconductors: Integrated Circuit Design for Manufacturability
Artur Balasinki

Internet Networks

Wired, Wireless, and Optical Technologies

Edited by
Krzysztof Iniewski

CRC Press
Taylor & Francis Group
Boca Raton London New York

CRC Press is an imprint of the
Taylor & Francis Group, an **informa** business

CRC Press
Taylor & Francis Group
6000 Broken Sound Parkway NW, Suite 300
Boca Raton, FL 33487-2742

© 2010 by Taylor and Francis Group, LLC
CRC Press is an imprint of Taylor & Francis Group, an Informa business

No claim to original U.S. Government works

Printed in the United States of America on acid-free paper
10 9 8 7 6 5 4 3 2 1

International Standard Book Number: 978-1-4398-0856-6 (Hardback)

This book contains information obtained from authentic and highly regarded sources. Reasonable efforts have been made to publish reliable data and information, but the author and publisher cannot assume responsibility for the validity of all materials or the consequences of their use. The authors and publishers have attempted to trace the copyright holders of all material reproduced in this publication and apologize to copyright holders if permission to publish in this form has not been obtained. If any copyright material has not been acknowledged please write and let us know so we may rectify in any future reprint.

Except as permitted under U.S. Copyright Law, no part of this book may be reprinted, reproduced, transmitted, or utilized in any form by any electronic, mechanical, or other means, now known or hereafter invented, including photocopying, microfilming, and recording, or in any information storage or retrieval system, without written permission from the publishers.

For permission to photocopy or use material electronically from this work, please access www.copyright.com (http://www.copyright.com/) or contact the Copyright Clearance Center, Inc. (CCC), 222 Rosewood Drive, Danvers, MA 01923, 978-750-8400. CCC is a not-for-profit organization that provides licenses and registration for a variety of users. For organizations that have been granted a photocopy license by the CCC, a separate system of payment has been arranged.

Trademark Notice: Product or corporate names may be trademarks or registered trademarks, and are used only for identification and explanation without intent to infringe.

Visit the Taylor & Francis Web site at
http://www.taylorandfrancis.com

and the CRC Press Web site at
http://www.crcpress.com

Contents

About the Editor vii
Contributors ix

Chapter 1 An Exciting Future 1

  Roberto Saracco

Chapter 2 802.11n: High Throughput Enhancement to Wireless LANs 27

  Eldad Perahia

Chapter 3 Mobile WiMAX and Its Evolutions 53

  Fan Wang, Bishwarup Mondal, Amitava Ghosh,
  Chandy Sankaran, Philip J. Fleming, Frank Hsieh, and
  Stanley J. Benes

Chapter 4 Service Layer for Next-Generation Digital Media Services 79

  Abhijit Sur and Joe McIntyre

Chapter 5 Enhancing TCP Performance in Hybrid Networks:
  Wireless Access, Wired Core 99

  Milosh V. Ivanovich

Chapter 6 Femtocell—Home Base Station: Revolutionizing the
  Radio Access Network 125

  Manish Singh

Chapter 7 Satellite Communications 145

  Sooyoung Kim

Chapter 8 Next-Generation Wired Access: Architectures and Technologies ... 167

  Eugenio Iannone

Chapter 9 Broadband over Power Line Communications: Home
  Networking, Broadband Access, and Smart Power Grids 195

  Peter Sobotka, Robert Taylor, and Kris Iniewski

Chapter 10 Ethernet Optical Transport Networks 239

    Raymond Xie

Chapter 11 Optical Burst Switching: An Emerging Core Network
    Technology 265

    Yuhua Chen and Pramode K. Verma

Chapter 12 Intelligent Optical Control Planes 291

    James J. Zik

Index 315

About the Editor

Krzysztof (Kris) Iniewski manages R&D developments at Redlen Technologies, Inc., a startup company in British Columbia. He is also an Executive Director of CMOS Emerging Technologies, Inc. (http://www.cmoset.com/). His research interests are in hardware design for medical and networking applications. From 2004 to 2006 he was an Associate Professor in the Electrical Engineering and Computer Engineering Department of the University of Alberta, where he conducted research on low-power wireless circuits and systems. During his tenure in Edmonton, he edited a book for CRC Press, *Wireless Technologies: Circuits, Systems and Devices*.

From 1995 to 2003, he was with PMC-Sierra and held various technical and management positions. During his tenure he led the development of a number of very large scale integration (VLSI) chips used in optical networks. Prior to joining PMC-Sierra, from 1990 to 1994 he was an assistant professor at the University of Toronto in the Department of Electrical Engineering and Computer Engineering. Dr. Iniewski has published more than 100 research papers in international journals and conferences. He holds 18 international patents granted in the United States, Canada, France, Germany, and Japan. He received his PhD in electronics (with honors) from the Warsaw University of Technology (Warsaw, Poland) in 1988. Together with Carl McCrosky and Dan Minoli, he is an author of *Network Infrastructure and Architecture: Designing High-Availability Networks* (Wiley, 2008). He recently edited *Medical Imaging Electronics* (Wiley, 2009), *VLSI Circuits for Bio-Medical Applications* (Artech House, 2008), and *Circuits at the Nanoscale: Communications, Imaging and Sensing* (CRC Press, 2008).

Contributors

Stanley J. Benes
Networks Advanced Technologies
Motorola, Inc.
Arlington Heights, Illinois, USA

Yuhua Chen
Department of Electrical and
   Computer Engineering
University of Houston
Houston, Texas, USA

Philip J. Fleming
Networks Advanced Technologies
Motorola, Inc.
Arlington Heights, Illinois, USA

Amitava Ghosh
Networks Advanced Technologies
Motorola, Inc.
Arlington Heights, Illinois, USA

Frank Hsieh
Networks Advanced Technologies
Motorola, Inc.
Arlington Heights, Illinois, USA

Eugenio Iannone
Pirelli Labs
Milan, Italy

Kris Iniewski
CMOS Emerging Technologies
British Columbia, Canada

Milosh V. Ivanovich
Chief Technology Office
Telstra
Melbourne, Victoria, Australia

Sooyoung Kim
Division of Electronics &
   Information Engineering
College of Engineering
Chonbuk National University
Jeonju, South Korea

Joe McIntyre
IBM Corporation
Austin, Texas, USA

Bishwarup Mondal
Networks Advanced Technologies
Motorola, Inc.
Arlington Heights, Illinois, USA

Eldad Perahia
Intel Corporation
Hillsboro, Oregon, USA

Chandy Sankaran
Networks Advanced Technologies
Motorola, Inc.
Arlington Heights, Illinois, USA

Roberto Saracco
Telecom Italia Future Centre
Venice, Italy

Manish Singh
Continuous Computing
San Diego, California, USA

Peter Sobotka
Corinex Communications
Vancouver, British Columbia, Canada

Abhijit Sur
IBM Corporation
Denver, Colorado, USA

Robert Taylor
CMOS Emerging Technologies
British Columbia, Canada

Pramode K. Verma
Telecommunications Engineering
  Program
School of Electrical and
  Computer Engineering
University of Oklahoma-Tulsa
Tulsa, Oklahoma, USA

Fan Wang
Networks Advanced Technologies
Motorola, Inc.
Arlington Heights, Illinois, USA

Raymond Xie
Sycamore Networks
Chelmsford, Massachusetts, USA

James J. Zik
Ciena Corporation
Linthicum, Maryland, USA

1 An Exciting Future

Roberto Saracco

CONTENTS

1.1 Have We Reached the End of the Road? 1
1.2 Glocal Innovation 4
1.3 Digital Storage 6
1.4 Processing 8
1.5 Sensors 9
1.6 Displays 10
1.7 Statistical Data Analyses 12
1.8 Autonomic Systems 13
1.9 New Networking Paradigms 15
1.10 Business Ecosystems 18
1.11 The Internet *with* Things 21
1.12 Communications in 2020 (or Quite Sooner) 22

This introductory chapter focuses on new opportunities provided by the unrelenting technology evolution to further develop the telecommunications business well into the next decade. It takes into account the evolution of storage, processing, sensors, displays, statistical data analyses, and autonomic systems and discusses how such an evolution is going to reshape markets and business models into a new era where business ecosystems supplement value chains.

1.1 HAVE WE REACHED THE END OF THE ROAD?

The evolution we have witnessed in these last 50 years in electronics, optics, smart materials, biotech, and all fields using these technologies has been relentless. Although there is currently no sign of a plateau, we know that a physical limit to progress lies somewhere. The fact that in many fields, like electronics, this ceiling seemed to have been approaching and engineers have found ways to circumvent it does not change the fact that a physical limitation exists.

In economics we have seen what happens when we reach a ceiling, such as when we run out of liquidity: The downward spiral of stock markets in the second half of 2008 and into 2009 is a clear statement of the havoc that can happen when progress is suddenly stopped.

The technology evolution has progressed with such regularity that it no longer surprises us, we have gotten used to it, and we have actually built a world that relies on it. If the evolution of technology were to stop next year, we would need to reinvent the way we are doing business and that would cause tremendous problems.

Looking at the physical barriers, like the speed of light, the quantum of energy, the smallest dimension that exists, we can determine where the ultimate limit lies [1]. The good news is that such a limit is very far from where we are today. At the present pace of evolution we won't be reaching it for the next few centuries. This does not mean, however, that such limits will ever be reached. Actually, I feel that we will discover unsolvable issues long before getting to those physical barriers.

The investment required for chip production plants is growing exponentially and payback requires huge revenues. As we will see, this is a push toward huge volumes, with individual products costing less and less to make them sustainable by the widest possible market. The economics is already slowing down the creation of new plants but new production processes could circumvent what we see as an upper boundary today.

The continuous progress of technology has continuously increased the amount of energy being consumed (see Figure 1.1). It is estimated that the power consumed by residential households in Europe to access broadband networks in 2015 will reach 50 TWh. To give a sense of this number, 10 years ago there was no power consumption used in accessing the network; all that was needed was provided by the network itself. Networks in the period between 1980 and 2000 doubled their power consumption, and in 2008 the overall power consumption of networks in Western Europe reached 20 TWh. This means that we are expecting that a consumption level that was nonexistent 10 years ago will more than double in Europe the present consumption of all European Telecommunications Networks.

Energy is becoming a bottleneck to evolution: In 2008, China consumed as much energy in total as the United States, but the per-capita consumption is a fraction of that in the United States. Between 8 and 10 billion people are expected to populate the Earth in the next decade. This, in terms of energy, is not the 25% to 80% increase in population, but an 800% increase because that population will consume on average what is now consumed by an average U.S. due to a widespread increase in quality of life.

We simply do not have the energy available to meet those projected requirements. This means that either global wealth, in terms of energy consumption, will not be at the same level that a U.S. citizen enjoyed in 2008 or we will have found ways to dramatically decrease power consumption and increase energy production (both are indeed required to come anywhere close to meeting those requirements).

The energy issue is going to influence overall evolution in the next decade, in terms of both availability and cost. The shift toward a "greener" world, although important, is going to increase the impact of energy on evolution.

The bright side is that the energy "crunch" will force investment in alternative energy sources and decreased consumption. This, rather than slowing evolution, is likely to shift its direction, accelerating the deployment of optical networks that are much more energy savvy than copper networks, moving toward radio coverage made by smaller cells, as the energy required to cover a given surface decreases (approximately) with the square of the number of cells being used.

For the time being, and for the horizon we can reasonably consider today (not beyond 2050), we can be confident that technology evolution will continue at different paces in different sectors as it has over the last 50 years, but overall at a similar rate to what we have been experiencing in these last decades.

An Exciting Future 3

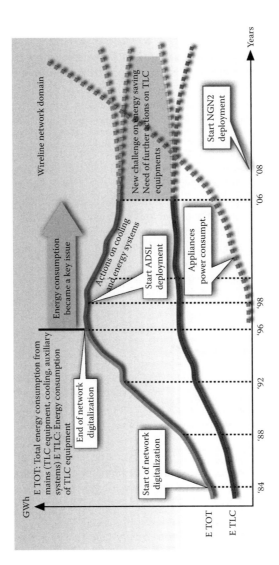

FIGURE 1.1 The growth of energy consumption.

Now, saying that the pace will be substantially unchanged does not mean that it is going to be "business as usual" for two reasons. First, as Moore's law claims, saying that in the next 18 months we are going to have an evolution that will double today's performances means that in the next 18 months we will go a distance that has taken us the last 40 years to walk. This is quite a change! Second, performance increase has a linear effect on the ecosystem until a certain threshold is reached: Beyond that point it is no longer seen as a performance increase, but rather as a change of rules.

Think about electronic watches: There was a time, in the 1970s, when owning an electronic watch was very expensive. As prices went down, more and more people could afford an electronic watch. At a certain point the cost dropped, basically, to zero (it passed the thresholds of cost perception) and there was no more market for electronic watches: The industry had to reposition itself. Long gone are ads claiming better precision of a particular watch. The marketing value of a Swiss certified chronograph dropped to zero.

A similar thing will happen with the deployment of broadband networks and their enabling optical infrastructures: Once you reach a bandwidth of 1 Gbps, and possibly before that, it will be impossible to market increased bandwidth at a premium. Bandwidth value will drop to zero and marketers will need to find new slogans.

Notice how the threshold links technology with market value. As these two factors change the business model and the rules of the game: a disruption occurs. As this happens, consolidated industries need to reinvent themselves and new ones find leverage to displace the incumbents. In discussing technology evolution, we come to this point over and over. The question to consider, therefore, is not whether a technology is reaching its evolution limit, but rather if that technology is leading to a disruption threshold.

1.2 GLOCAL INNOVATION

Innovation used to be easier to predict because only a few companies and countries were leading the way. The evolution of infrastructures was so easy to predict that the International Telecommunications Union (an international standardization body based in Geneva, Switzerland) published a table with the status of telecommunications infrastructures and telecommunications service penetration annually. Maps showed the progress made and others showed what penetration would be reached in 10 or 20 years. As an example, it would take 19 years for a country to move from 1% to 10% telecommunications penetration but only 12 years to move from 10% to 20%, 8 years to go from 20% to 30%, and so on.

The advent of wireless changed the world. Countries like India that used to have less than 3% of telecommunication density (and that is still the quota of the fixed lines) moved within 10 years from no density to 20% and are likely to reach a 50% penetration rate within the next decade. China moved from 0% to over 30% (in wireless) within 10 years.

However, more than in the infrastructure domain, where globalization has an impact in the decrease of price that in turns enables a progressively broader market thus affecting indirectly the local situation, the service domain has a distribution cost that is basically "zero." Once an application is developed, it can be made available over the network with no distribution cost hampering its marketing. The network is also playing another trick: An application that can make sense in the U.S. market can be developed and "marketed" from India.

Innovation is no longer confined to the domain of a few rich companies or countries because of huge investment barriers. The real barriers have moved from financial capital availability to educational capital availability. It is the increase in production of engineers in India and China that is placing these two countries at the forefront of innovation, not their huge (potential) market.

The market of innovation is no longer local but global. This globalization of innovation is going to continue in the next decade and it is not by chance that President Barack Obama has set the goal of increasing scientific education and the number of engineers as the way to remain at the forefront of progress, both scientific and economic.

Optical and wireless networks will further shrink the world. Distance is already irrelevant for information flow; it is also becoming irrelevant for delivery cost of a growing number of products and services, and it is certainly becoming irrelevant for innovation. Offshoring will become more and more common for big industries, whereas smaller companies will in-shore innovation from anywhere and will make a business out of their ability to localize a global world. The latter is going to stay, but the former will be viable only until a labor cost differential "plagues" the world. Eventually, it will disappear. Politics, regulations, and cultures will be the determining factors in the evolution. From a technological point of view, the Earth will be no bigger than a small village.

The Web 2.0 paradigm will evolve from being a network of services and applications made available by a plethora of (small) enterprises and made available thanks to the huge investment of few (big) enterprises to become a Web 3.0 where the interaction will take place among services and applications to serve the user's context. Questions like "When is the next train?" will become answerable because somewhere in the Web there is an understanding of my context, and it will be obvious from this understanding that I am looking for the train to Milan. It is not a small step.

Again, it is a matter of glocalization. We are moving from the syntax, from infrastructures providing physical connectivity, to semantics, to the appreciation of who I am, and this includes the understanding of who I was (the set of experiences shaping my context) and the forecasting of who I will be (my motivations and drives to act). This might seem scary, Orwellian, and definitely not the way to go.

However, the evolution, if it is to continue, has to be beneficial; otherwise it will not become entrenched. Because of this, we can expect that the balance between what is technologically possible and what the market is buying will depend on us, as individuals and as a community. Contextualization can raise many issues, from privacy to ownership, from democracy to the establishment of new communities continuously reshaping themselves.

Contextualization is not likely to result from an "intelligent, Orwellian" network, but rather from an increased intelligence of my terminal, and that is under my control. I will make decisions, most of the time unconsciously, of what to share of my context; the network will be there to enable it. My terminal, and the *my* is the crucial part, will act as an autonomous system, absorbing information from the environment, both local and, thanks to the network, global. It will let me communicate with my context, my information, my experiences, my environment, and, of course, my friends and acquaintances in the same seamless way as today I walk into a room and act according to my aims, expectations, and environment.

Therefore, telepresence, one of the holy grails of communications, will also be glocal. I will communicate locally and remotely as if both remote and local are present at the same time.

This will be possible because of technology's evolution. Although the list of technologies to consider would be very lengthy, we can examine a few of them in terms of the evolution of functionalities that such a technology evolution makes available rather than in terms of the evolution of single technologies.

1.3 DIGITAL STORAGE

Digital storage capacity has increased by leaps and bounds over the last 50 years. The original digital storage solutions have basically disappeared (magnetic cores, drums, tapes, etc.) to leave space to new technologies, like magnetic disks, solid state memory, and polymer memories (on the near-term horizon).

As of 2008, hard drives, or devices using magnetic disks for storage, reached 2 TB capacity in the consumer market, and 37.5 TB disks are expected to appear in 2010 (from Seagate). Storage capacity of 100 TB will become commonplace by the end of the next decade. The new leap in magnetic storage density is achieved through heat-assisted magnetic recording (HAMR).

Solid state memory has advanced significantly, and compact flash cards are cheap and ubiquitous these days. They were invented in 1994 and have moved from a capacity of 4 MB to 64 GB as of 2008. A capacity of 128 GB has become available in June 2009 in flash drives (solid state disks [SSD] based on flash technology appeared in 2007). The announcement at the end of 2008 of new etching processes able to reach the 22–15 nm level (down from the current 60–40 nm standard) clearly show that more progress in capacity is ahead.

This increase in capacity is placing flash memory on a collision course with magnetic disks in certain application areas, like MP3 players and portable computers. They consume only 5% of the energy required by a magnetic disk and they are shock resistant up to 2,000 Gs (corresponding to a 10-foot drop).

The bit transfer rate has already increased significantly and there is a plan to move their interface to the Serial Advanced Technology Attachment (SATA) standard, the one already used by magnetic disks, thus raising the transfer speed to 3 Gbps. By comparison, the current Parallel Advanced Technology Attachment (PATA) interface tops out at 1 Gbps.

Polymer memory has seen an increased effort by several companies to bring the technology to the market. Commercial availability is likely in 2010. Polymer memory is made by printing circuit components on plastic, a precursor to fully printed electronics. Its big advantage over other types of memory is in its extremely low cost and potential capacity. In an area the size of a credit card, one could store several terabytes (TB) of data (see Figure 1.2).

An Exciting Future

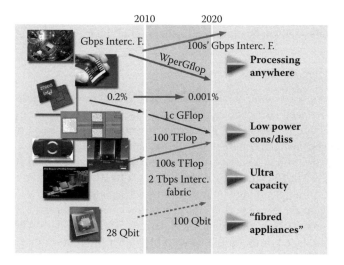

FIGURE 1.2 The evolution of digital storage.

Data will be stored both at the edges and within the network. Ericsson predicts that a 1-TB cell phone will be available in 2012, home media centers will be able to store the entire life production of a family in their multi-TB storage, exabytes (EB; a billion billion bytes) will become commonplace at data warehouses for data-based companies like Google™, Snapfish™, Flickr™, Facebook™, and those to come in the future. Institutions and governments will harvest the digital shadow of their constituencies daily to offer better services. Raw data generated by sensors will have economic value through statistical data analyses.

Storage is becoming one of the most important enablers for business in the next decade. What is the consequence of this continuous increase in storage capacity? Clearly, we can store more and more data and information; however, the real point is that this huge capacity is changing paradigms and rules of the game, affecting the value of the network and its architecture.

Because data is everywhere, the flow of data will no longer be restricted from the network toward the edges. The reverse flow will be just as important. In addition, we are going to see the emergence of local data exchange, edges to edges, terminal to terminal. The first evolution makes the uplink capacity as important as the downlink (leading to the decommission of ADSL), and the second emphasizes the importance of transaction-oriented traffic: "Updates" achieve greater importance and possibly are perceived as the real value that some provider may deliver.

Raw data (but this also applies to information as soon as one is drowning in information) makes sense only if it can be converted into perceptible chunks of information relevant to the current needs for a given user (person and machine).

As we discuss in the following, storage could disappear from sight, replaced by small "valuets," a mixture of applications and sensors or displays able to represent a meaning valuable to a user. We are starting to see this appearing as tiny apps on the iPhone™. They mask the data, the information, the transactions required, and even the specific applications being used. The new way of storage cards embedding communications and applications is a further hint of the way the future is going.

1.4 PROCESSING

Processing evolution is no longer the sole axis of increased performance. Other factors, like reduced energy consumption and ease of packaging, are growing more and more important. As in the past, when the continuous increase in processing performance expanded the market, now decreasing energy consumption and cheaper packaging of the chip in a variety of objects are opening up new markets.

Intel declared in 2005 that they were targeting a 100 times reduction of energy per GFLOP (one billion floating point instructions per second) by 2010; as of 2008, they were on target to achieve that. A decrease in power consumption enables the packaging of more processing power in handheld devices, like cell phones: The issue is not the resulting reduced drain on the battery, but rather the reduced heat dissipation. A 500-watt cell phone will burn your hand long before its battery runs out.

Second, and possibly with far-reaching consequences, very low-consuming devices could be powered using alternative power sources, such as conversion of sugar circulating in the blood into energy for tiny medical devices delivering drugs and monitoring certain parameters in the body, conversion of surface vibration into energy for sensors placed in the tarmac of roads to measure traffic, or conversion of wireless radio waves into energy using evanescent waves to power sensors in a closed environment.

As the cost of producing sensors decreases, the economics shifts to their operation and powering is a crucial factor. Because of progress in decreasing power consumption, we can be confident that the next decade will see an explosion of sensors and along with that an explosion of data. By 2005, only a tiny fraction of microprocessors produced ended up in something that could be called a "computer." Most microprocessors ended up in devices like microwave ovens, remote controls, cars, and electronic locks, to mention only a few categories. In the next decade most objects will embed a microprocessor and most of them will have the capability to be connected in a network (to The Network). This will change dramatically the way we perceive objects and the way we use them.

Part of this change will be enabled by the rising star of printed electronics. This manufacturing process, based on a derivative from inkjet printing technology, is very cheap—two to three orders of magnitude cheaper than the silicon etching currently used for chips. Additionally, printed electronics is cheaper to design (again three orders of magnitude cheaper than etching silicon) and can embed both the processing and storage and the antenna for radio communication and, if needed, a touch-based interface, avoiding the cost of packaging. In principle it will be possible to write on goods as easily as we stick labels on them today.

This described evolution is in the direction of what can be called microprocessing. We will continue to see evolution in the opposite direction, that of "supercrunchers." In this direction we are seeing a continuous increment of processing speed achieved through massive parallel computing with hundreds of thousands of chips within a single machine exceeding the PFLOPS (one million billion floating point instructions per second) today and the EFLOPS (one billion billion floating point instructions

An Exciting Future

per second) in the next decade (billions and billions of floating point operations per second). We are also going to see more diffused usage of the cloud computing paradigm in both the business and business-to-consumer environments. The consumer is unlikely to appreciate what is really going on behind the scenes, that some of the services he or she is using are actually the result of massive processing achieved through a cloud computing infrastructure.

Looking at the longer term, we can speculate that cell phones and wireless devices in general may form a sort of cloud computing for resolving interference issues, thus effectively multiplying spectrum usage efficiency. The major hurdles on this path (which has already been demonstrated as technically feasible from an algorithmic point of view) are the energy required by the computation and communications among the devices that make it practically impossible today and in the coming years.

1.5 SENSORS

Sensors are evolving rapidly, getting cheaper and more flexible. They embed the communications part and thus are ready to form local networks. Sensors open up a wide array of services. Think about the thousands of applications that are newly available on the iTouch™ and the iPhone, exploiting the accelerometer sensor.

Drug companies are studying new ways to detect proteins and other substances. What in the past required long, expensive tests executed by large, very expensive machines can now be done cheaper, quicker, and easier by one or several sensors in combination. Some of these sensors are being targeted for embedding in cell phones, like the one able to analyze breath as the person talks into the mircophone. Over time, the sensor can detect the presence of markers for lung cancer. SD-like cards containing tens, and soon hundreds, of substances will be plugged into the cell phone, enabling the detection of a variety of illnesses well before clinical signs appear.

Now, this is not just an application, although an interesting and valuable one. It is a driver to miniaturize sensors, to make them more flexible and responsive to the environment and thus able to pick up telling signs. Hundreds of sensors will be constantly producing data that will become a gold mine to derive meaning. Communications is the enabling factor because this data needs to be seen as a whole to derive meaning. We'll see this in a moment when considering statistical data analyses.

Other researchers are investigating e-textiles, special fibers that can be woven into clothing to sense a variety of conditions and the presence of special substances like sugar and proteins, thus providing data to detect several pathologies.

Printed electronics will contribute to the slashing of costs to produce and deploy sensors in any object: Pick up something and that something knows it and gets ready to interact.

Sensors are also providing what it takes to transform a collection of objects into an environment. Context awareness will make significant advances because of sensors' presence everywhere.

At the end of 2008, Intel announced a research program, Wireless Identification and Sensing Platform (WISP). They expect WISP will be available in the next decade and will be able to provide identification of any object, including our body, through a sort of miniaturized Radio Frequency Identification (RFID) forming a continuous interconnected fabric. The present RFID technology, over time, will transform itself into active components with sensing capabilities, as the price of sensors goes down. This probably won't happen before the end of the next decade. In the meantime, more and more objects will embed sensors and some of these will act as identification, thus avoiding the need for an RFID.

The transformation of an object into an entity that can communicate and can become aware of its environment leads to a change in the business space of a producer. In fact, this opens the possibility of remaining in touch with a product user, thus transforming the object into a service. In parallel this enables new business models and requires a transformation of the producer's organization. Most producers will not be prepared for this change, but it will be difficult to resist this evolution because the competition will be ready to exploit the marketing advantages provided by these new "context-aware" objects. Some producers will decide to open up their product communications and on-board flexibility to third parties to let them further increase the features and hence the perceived value of the product. This openness, in turn, will give rise to a variety of architectures, making network platforms and service platforms true service factory and delivery points.

The research work on sensors will create ripples for today's established dogmas, like the ubiquity of IP: Energy efficiency considerations are driving sensors' networks to use non-IP communications and there will be many more sensor networks using ad hoc protocols than local and backbone networks using IP; identity and authentication will need to cover objects and this might bring to the fore new approaches to assess identity. The SIM card is very effective as identification and authentication goes, but it has not satisfied the banking system and it might not be the future of identification. In fact, cell phones equipped with sensors detecting biometric parameters might provide even better authentication mechanisms and would make it possible to separate the terminal from the user (which would appease the banking system).

Finally, the need for a set of self-standing sensors within an environment coupled with the need to cut energy consumption on each sensor is pushing researchers to work out ever better autonomous systems theory and applications. This is going to have a profound effect on the network ownership and management architecture, as autonomous systems destroy the principle that one needs a central control to deliver end-to-end quality and hence the very foundations of today's telecom operators.

1.6 DISPLAYS

Display technology has brought us the wide, flat screens everybody loves. It has also populated with a growing number of devices with a screen, from digital cameras to cell phones. Digital frames have invaded our homes as well. However, some dreams have not yet come to fruition, like the holographic screen that was supposed to take center stage in our living room according to futurists in the 1960s.

There are many basic technologies available that are bound to progress, particularly in the direction of lower and lower end user cost. The improvement of production processes is the single most important factor in this progress. The lower cost makes it possible to have screens popping up everywhere, which is in sync with our perception of a world based on visual communications. The telephone has been a

An Exciting Future

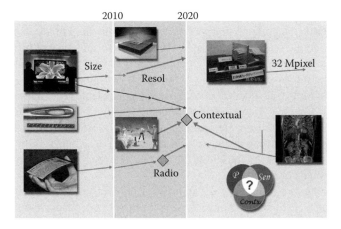

FIGURE 1.3 The evolution of the display.

compromise, but a very successful one indeed; so successful, in fact, that it created a new communication paradigm, so strong that now most people prefer talking rather than communicating over video (the latter is considered much more intrusive, as it brings you very close to the other party).

There are, however, other directions of progress that are important because of the perception impact they have. The resolution of our eye is approximately equivalent to 8 megapixels. Our brain composes the signals received from the eyes in a bigger window whose resolution is roughly equivalent to 12 megapixels (Mpixel). Present high-definition (HD) television screens have a 2 Mpixel resolution (achieved using 6 megadots, a triplet of red, green, and blue makes up one resolution pixel). Hence, although we marvel at the quality of the images, our brain is not fooled. We are looking at a screen, not at reality; we are watching a show, we are not "at the show."

The Japanese have the goal of achieving a 32 Mpixel screen (and the required production chain) by the end of the next decade. A few 4K screens are available on the Japanese market, reaching the 8 Mpixel threshold (see Figure 1.3). If we look straight at one of these screens, we cannot tell the difference from reality. We already have 8 Mpixel resolution in some products. Many digital cameras, in fact, are now capable of much higher resolutions (for example, one Nikon reflex camera that was announced at the end of 2008 has a resolution of more than 24 Mpixels).

However, most of the time we are not looking straight at something. Even without noticing, our eyes scan the environment and it is this scanning that allows the brain to create a larger image and to get the feeling of "being there." To replicate this sensation we need to have our eye scanning confined to the screen; that is, we need to be sufficiently close to the screen and the screen dimensions need to create an angle with our eyesight exceeding 160 degrees (when we are looking straight, the angle captured by the eyes is slightly less than 130 degrees).

The increasing dimension of entertainment screens and their increasing resolution will lead us into make-believe situations in the next decade. The bandwidth required to transmit that amount of information exceeds 100 Mbps, so only optical fiber connections will support this standard.

Although we will have LTE (Long Term Evolution) and LTE+ (there will always be a successor on the horizon) that will be able to handle those kinds of speed, it does not make economic sense to use all available spectra for this type of service. Within the home, the situation is different. Optical fiber may well terminate into a gateway that will beam information wirelessly at speeds up to 1 Gbps.

Smart materials will become more and more available to display images and clips. We already have special varnish that can change its colors to create images; electronic ink, on a smaller scale, can display black-and-white text and by the end of this decade it will be able to display color images. We can expect significant progress in this area that will lead by the end of the next decade to ubiquitous display capabilities on most kinds of objects.

This is going to change the look and feel of products and, as pointed out in the case of sensors, it is bound to change the relation between producer and user. As indicated for sensors, these capabilities coupled with open systems and with open service creation platforms will enable third parties to provide services on any object.

Displays are the ideal interface for human beings because we are visually oriented. The coupling of touch sensors or other kinds of "intention" detectors opens the way to new services.

The underlying assumption is that objects will be connected to the network, either directly or, more often, through a local ambient network. This connection in many instances will be based on radio waves, although another strong possibility might be the power lines within a given ambient. The fiber telecommunications infrastructure is likely to stop at the entrance of the ambient on the assumption that the fewer wires you have around your home, the better.

1.7 STATISTICAL DATA ANALYSES

The quantity and variety of data that are becoming available through the network is growing exponentially. In the next decade cell phones will be equipped with sensors and this alone will create an avalanche of data. Five billion cell phones generating data on their position, temperature, movement (direction and speed), and special sensors will provide much more information. Telecom operators will have to come to an agreement on what to do with this data and how to make it available (in a neutral way that preserves privacy).

This data can be used to monitor traffic in an urban area, to detect the emergence of epidemics, to plan and monitor the effects of public transportation, to create precise maps on environmental pollution, to understand social networks, and to gauge the interest generated by an advertisement or event. They could be used in an emergency or for urban planning. They could be used by shop managers to dynamically change their displays.

Clearly, data derived from cell phones is only a fraction of the total data that will be harvested by sensors and that can potentially be made available for analysis. As we have seen, many objects will have embeded sensors and most of them will be connected to the network, so the data they gather could be made available to third parties. Homes, department stores, schools, hospitals, parks, cars, and basically any ambient environment can generate data and make it available. The possibilities are endless.

There is a growing understanding of how to analyze massive data banks. This will further develop in the next decade to include the analyses of distributed data banks leading to the generation of metadata. Public access to metadata will stimulate the development of many services and new industries will be created. The step from raw data to metadata is crucial. Metadata should be able to capture what is of interest within a set of data, masking in a secure way any detail that can be used to trace the owner / generator of any data used to create the metadata.

The absolute guarantee of a decoupling between data and metadata is a fundamental prerequisite to enable the publication of metadata. It is likely that most of this metadata will remain invisible to the public and will be transformed into useful information through services. We are already seeing this trend in several applications available on the iPhone. As an example, the user can ask for a weather forecast in a certain area and the related information is presented without the user having to know from where the information is actually derived. It might even be that the service (the touch of a button) is actually integrating several weather forecasts based on success tracking.

This data will be distributed in several databases and their harvesting and exploitation will generate traffic with varied characteristics. On the harvesting side, it will likely be in the form of billions of tiny transactions; on the usage side it will be probably in form of bulk data transfer to feed statistical data analyzers and in the form of images and video clips to end users. Wireless and optical networks will be the supporting infrastructure. Cloud computing might offer the computational capabilities required for data analyses in several cases. Enterprises will create business ideas to leverage potential data value. They will not require massive investment in computation structures if these will be made available on demand through the network. Note how a significant portion of these enterprise-generated services might actually take the form of applets residing on users' terminals, like cell phones, which, once activated, can generate traffic on the network.

Summarizing, the availability of massive quantities of data is changing the approach to data analyses from algorithmic processing to statistical data processing. This latter implies access to distributed data banks and significant computation capabilities that will be better satisfied by cloud computing. The communication fabric is the key enabler both at the level of raw data harvesting and at the level of processing into metadata, processing of metadata versus specific applications, and customer demand and distribution of results.

1.8 AUTONOMIC SYSTEMS

The growth of independent network providers, particularly at the edges of the "old-time network," gives rise to a new paradigm for management and exploitation. In addition, what used to be "dumb terminals" connecting at the edges of the network are now more and more sophisticated devices with a level of intelligence that is comparable to, and sometimes exceeds, that of the network. These devices no longer terminate communications and services arriving from the network, but have their own local network of relationships with other devices, sometimes acting as an intelligent gateway that uses the network as a tunnel for pure connectivity.

The intelligence has moved to the edges of the network and with that has come the need for a different paradigm to understand, create, and operate services, the paradigm of autonomic systems. The basic assumption of telecom engineers has always been plan, design, engineer, control, control, and control. All of the measurements going on in the network are to provide information for control. The more complex the network, the more sophisticated the control. In the 1980s with the dissemination of computers, the dream of having fully centralized control became possible and telecom operators built their own control centers. As more equipment and technology found their way into networks, engineers and researchers developed standards to ensure the centralized control of diversity. Those in the field remember the CMIP (Common Management Information Protocol) and the SNMP (Simple Network Management Protocol), two different paradigms for controlling network elements, both based on the fact that a single comprehensive view of the network was required to ensure fault control, maintenance, and fair usage of resources.

The advent of the Internet, with the creation of a network that was the result of thousands of interconnected networks having different owners, was viewed with suspicion by most telecom engineers. The fact that the quality of the Internet was based on best effort disqualified the Internet in their eyes. As traffic kept growing, we repeatedly heard voices of imminent collapse: The Internet is not centrally managed, it will collapse, and so on. Although local parts of the Internet have, and are experiencing, outages and disservice, the traffic keeps growing and no collapse has ever occurred. On the contrary, on a number of occasions, like September 11, 2001, catastrophic events that have created havoc for the telecommunications infrastructure left the Internet unscathed. On that particular day, the only way to communicate with people in New York was through messaging via the Internet.

Now, the Internet is not a different network in the sense that it uses different wires. The wires are shared with the telecommunications network and one of the reasons why the Internet works so well is because it uses one of the most reliable infrastructures on the planet: the telecommunication infrastructure. The reason why the Internet is so resistant to local faults is because of its basic absence of a hierarchy of control. Control is local and distributed. Something might go wrong at point A, but it is not going to affect communication transiting through A because there are so many equal alternatives in the network bypassing A. The Internet is not yet an autonomic system in the full sense. Within its boundary the various components act in regard to the routing of information as autonomic systems.

A more apt example of an autonomic system is Roomba, the vacuum cleaner available on the market since the beginning of this century. It is a robot that has its own goal (which can be finely tuned by the owner) of vacuuming the spaces around it as far as it can reach. This space is cluttered with obstacles, table legs, furniture, and people (it is claimed in ads that the Roomba can avoid running into the cat, although it looks more likely that the cat will take appropriate action to step away). Over time it learns how the space is structured and works out the best vacuuming strategy. If the space changes (new furniture, reconfiguration of the present furniture, etc.) it will change its strategy accordingly. The home environment is an ideal lab for experimenting with autonomic systems. Actually, taken as a whole, it can become an autonomic system. All devices dialogue with each other and benefit from one another. As new devices are inserted in

An Exciting Future 15

the home they make themselves known in the environment and become part of it. As the external environment changes (new connections might become available, active ones might be malfunctioning), the home can react and reconfigure itself.

The evolution of wireless in the direction of more and more wireless local areas will exploit the technology of autonomic systems. Devices will have a wireless cloud around them. At any moment a device will check any overlap with other clouds and will try to establish a connection at a semantic level; that is, the device identifies itself and exposes its characteristics, objectives, and needs and expects to receive similar information from other devices sharing the cloud. This exchange of information will be followed, where appropriate, by a handshaking to bind the devices into a single system. The set of all devices will indeed behave as a single system. Because of the continuous sensing of the environment, the system will evolve in terms of participants in the system and in terms of behavior.

1.9 NEW NETWORKING PARADIGMS

The advent of autonomic systems, the multiplication of networks, the presence of huge storage capacity at the edges of the old network (more specifically in the terminals, cell phones, media centers, and so on), and the growing intelligence outside the network will change networking paradigms significantly.

Efforts in the past 30 years have focused on exploiting the progressive penetration of computers in the network to make the network more intelligent. A simple economic drive motivated this evolution: The network is a central resource, the cost of which can be split among users. It makes more sense to invest in the network to provide better services to low-cost, low-intelligence edges. The intelligent network finds economic justification in that fact.

The first dramatic shift happened with cell phones, with the mobile network. If you were to develop a network from scratch and decide to use a fixed-line network to provide services, you would have to pay almost 100% of the investment. On the other hand, if you were to deliver the same services using the mobile network approach, the overall cost would be split 30% in the network and 70% in the terminals (this latter part is likely to be sustained by customers). This reflects the shift of processing, storage, and intelligence from the network to the edges, to the terminals using it.

A possible, and likely, vision for the network of the future is very high-capacity pipes of several Tbps each, having a meshed structure to ensure high reliability and to decrease the need for maintenance, and in particular for responsive maintenance, the one that is most costly and that affects most the service quality. This network terminates with local wireless drops that will present a geographical hierarchy in the sense that we will see very small radio coverage through local wireless networks dynamically creating wireless coverage through devices; very small cells (femto and picocells); cells on the order of tens of meters (WiFi); larger cells belonging to a planned coverage like LTE, 3G, and the remnants of GSM (Group Special Mobile) or the likes; and even larger cells covering rural areas (such as WiMax when used to fill the digital divide); and even larger coverage provided by satellites (see Figure 1.4).

In this vision the crucial aspect is ensuring seamless connectivity and services across a variety of ownership domains (each drop in principle may be owned by a

FIGURE 1.4 The FEMTO solution.

different party), and vertical roaming in addition to horizontal roaming (across different hierarchy layers rather than along cells in the same layer). Authentication and identity management are crucial. This kind of evolution requires more and more transparency of the network to services.

The overall communication environment will consist of millions of data and service hubs connected by very effective (fast and cheap) links. How could it be "millions" of data and service hubs? The trend is toward shrinking the number of data centers, and the technology (storage and processing) makes theoretically possible to have just one data center for the world. Reliability requires replication of it several times in different locations, but still we can be talking of several units!

The fact is that the future will see the emergence of data pulverization in terms of storage. Basically every cell phone can be seen as a data hub and any media center in any home becomes a data hub. Put all this together and millions of data hubs actually serves as a very low estimate.

How can one dare to place on the same level 1 TB of storage in a cell phone, 10 TB in a media center, and several EB in a network (service) data center? The fact is that from an economic point of view if we do the multiplication the total storage capacity present in terminals far exceeds that present in the network service data center (TB $*$ Gterminals = 1,000 EB). The economics of value is also on the side of the terminals, as the data we have in our cell phone will be worth much more (to us) than the data in any other place. People will consider local data as "The Data" and the data in the network as very important backup. Synchronization of data will take care of the reliability but at the same time asynchronous (push) synchronization from the network and service databases (DB) to the terminals will make perceptually invisible those centralized DBs.

The same is happening for services. Services are produced everywhere, make use of other services, of data, and of connectivity, and they are perceived "locally" by the users. They are bought (or acquired for free, possibly because there is some indirect business model in place to generate revenues for the service creator and to cover its operational cost). Services can be discovered on the open Web or can be found in specific aggregator places; the aggregator usually puts some sort of markup on the service, but at the same time provides some sort of assurance to the end user (see, for example, the Apple Store). We'll come back to this in a moment.

Once we have a network that conceptually consists of interconnected data service hubs, one of which is in our hand and possibly another in our home, what are the communications paradigms used? Point-to-point communications mean that calling a specific number is going to be replaced by person-to-person or person-to-service (embedding data) communications. This represents quite a departure from today because we are no longer calling a specific termination (identified by a telephone number); rather we are connected to a particular value point (a person, a service). Conceptually we are always connected to that value point, we just decide to do something on that existing connection. The fact that such a decision might involve the setting up of a path through the network(s) is irrelevant to the user, particularly if these actions involve no cost to the user. The concept of number disappears and with it a strong asset of today's operators.

The value of contextualized personal information finds its mirror in the "sticker" communication paradigm. A single person, a machine, asks, implicitly or explicitly, to be always connected with certain information. Most of this might reside on the terminal, but a certain part can relate to the particular place the terminal is operating, or to new information being generated somewhere else. The communication operates in the background, ensuring that relevant information is at one's fingertips when needed. It is more than just pushing information; it requires continuous synchronization of user profile, presence or location, and ongoing activities. This embeds concepts like mashups of services and information, metadata, and metaservice generation. It requires value tracking and sharing. It might require shadowing (tracking data generated through that or other terminals with which that person or machine interact).

The variety of devices available for communications in any given environment, some belonging to a specific user, some shared by several users (e.g., a television), and some that might be "borrowed" for a time by someone who is not the usual owner, can be clustered to provide ambient-to-ambient communications that may be mirrored by the "cluster" paradigm. Autonomic systems will surely help in making this sort of communication possible and common. The personal interaction point a person will be using will morph into a multi-window system where one could choose the specific window(s) to use for a certain communication. Similarly, at the other end, the other user will have the possibility of choosing how to experience that particular communication. In between, there might be one or more communications links and some of these might not even be connecting the two parties, as communications can involve information that is actually available somewhere else and that is taken into play by the overall system.

This kind of communications will be at the same time more spontaneous (simple) for the parties involved and more complex to be executed by the communications manager. The communications manager can, in principle, reside anywhere, but network operators may be the ones to propose this communication service.

Contextualized communication is going to be the norm in the future and it is a significant departure from the communications model we are used to. We explore this further in the final part of this chapter, as seen by the user.

1.10 BUSINESS ECOSYSTEMS

From the previous discussion emerges the fact that the future will consist of many players loosely interconnected in the creation and exploitation of innovation. Because of this, several economists, as well as technologists, have started to wonder if the usual representation of relationships among players in a certain area still can be modeled on the basis of value chains.

There is a growing consensus that value chains modeling shall be complemented by a broader view considering business ecosystems. What characterizes a value chain is the set of contractual obligations existing among the various actors. Competition might bring one actor to discontinue the relation, with another actor connecting to a new one offering at a better quality or price the same raw product or service. Value chains tend to become more and more efficient because the competitive value of each player is to be the most efficient one in that particular place on the value chain. Innovation is pursued to increase efficiency and over time, in a competitive market, the value produced by efficiency at any point in the value chain tends to move to the end of the value chain so that the end customer is the one who really benefits. Those sustaining the cost of innovation might see increased margins for a while, but the long-lasting benefit is that of remaining part of the value chain because they remain competitive. Clearly patents may lock in innovation and preserve its value to that player. This is particularly true for manufacturers, and less so for service offerings. Services are easier to copy, circumventing any patent.

An ecosystem is characterized by the loose relationships existing among actors. Sometimes actors do not even know each other. In a way, an ecosystem is a set of autonomic systems in the sense that each player plays its own game, trying to understand how the whole ecosystem (or the part that matters to it) evolves, and it reshapes its behavior and interaction according to that. Innovation happens anywhere in an ecosystem and benefits basically anyone because it increases the perception of value of the ecosystem for the players. The party that generates the innovation basically can keep its value because there is often a direct link to the end user, and thus to the perception of that value. On the other hand, innovation is much more tumultuous, and having less inertia, the interest shifts more rapidly from one player to the other. Hence, there is much more pressure to innovate and this innovation points directly to the end users rather than at something internal. The crucial point for innovators remains unchanged: How can they take the innovation to the market? The usual tactic is to piggyback on existing connections to the end market (thus exploiting advertisement, distribution, billing, etc.). These connections are also known as control points because of their power in bringing innovation to the end market and controlling the flow of value to the end customers. iTunes is such a control point. Set-top boxes might be another example of a control point (although today they are part of a value chain and not an ecosystem, something that is likely to change in the future).

If we look at car manufacturing companies, we see the presence of very strong value chains whose effectiveness has been tuned over the years to incredible points (e.g., just in time, no more warehousing, co-design). Around these value chains we have seen the birth of an ecosystem for add-on parts (radios, seat covers, stickers, snow chains, etc.). The companies producing these add-ons simply piggyback on existing car models to offer their products. There is no contractual obligation with the car manufacturers. Hence, as shown by this example, ecosystems already existed in the industrial society. What is new is the increased flexibility and openness provided by objects with embeded microprocessors and software. It becomes possible to add features inside the object, not just on the outside. The computer industry is another example, closer to the emergence of ecosystems. Here again we have a strong value chain: from the computer components to the manufacturers, the operating systems, and the world of applications. The latter could be seen as produced by independent players who take advantage of the market created by the computer industry. Some of these applications are part of the value chain (such as Microsoft Office or the similar suite produced by Apple for its computers), and others are the result of investment by independent players. They choose to invest in one platform (e.g., Windows, OS X, Linux., Symbian), depending on their evaluation of the potential market. For some of these applications, more applications (sometimes called plugins) are developed by other players, again benefiting from the market generated by that specific application.

All plugins and applications that provide ways to refine results produced by others increase the value of the ecosystem. Sometimes, this value is so high that customers are locked in; that is, moving to a different ecosystem (e.g., from the one based on Windows to the one based on OS X) would mean the loss of a number of valuable applications, which is not acceptable. Notice that this characteristic of lock-in can be found in bio-ecosystems as well. It is not the only similarity; actually, there are many similarities between business ecosystems and bio-ecosystems resulting from a set of ground rules that apply to all complex systems where the various components are interacting on the bases of rules that are common to the ecosystem and not specific to the two parties interacting. Further, the interaction takes different paths and leads to different results on the basis of the local status of the ecosystem (local means that part of the ecosystem that is perceived by the actors involved in the interaction at that particular time).

The future will bring many more business ecosystems to the fore as result of the openness of objects, their flexible behavior, and interaction based on microprocessors, sensors, actuators, and software (they are autonomic systems). This is not enough. One single Roomba in a living room does not create an ecosystem. What it takes to reach this is meeting a certain threshold in terms of the quantity of actors. Billions of open cell phones, interacting directly or indirectly with one another, will create a huge ecosystem whose value in business terms will be enormous, and well beyond the value of the sum of each individual cell phone.

Another way of looking at the increased value produced by independent players within an ecosystem is through the concept of mashups. This is basically a seed that can aggregate an ecosystem (the typical example is Google™ Maps). Most mashups today are about the aggregation of information. In the future, mashups will be composed of services as well as information. They do not cover the whole space of ecosystems, as they do not represent something like an iPod ecosystem, where we can see actors producing external loudspeakers, pouches, and decoration, for example. Note, however, that as objects become autonomous systems, it will be progressively difficult to distinguish, from a business point of view, the atoms from the bits.

This latter observation brings us to the exploration of the future of the Internet, not in terms of its physical and architectural underpinning, but in terms of value growth. Today's Internet, the Web, consists of an endless world of information. In recent years it has become a source of services to the point that many established actors in the service business are starting to reconsider their business strategy.

The silver bullet of a killer application sought by those service providers, like mobile television on cell phones, looks more and more unlikely to happen. Telecom operators that have prospered on connectivity services and have progressively added other services (value-added services) in their closed, walled garden are now seeing a growing number of small players creating services. The sheer number of them is sufficient to guarantee a level of innovation that exceeds any resulting from massive efforts by telecom operators. They are not bound to "principles"; they just offer their wares at a low cost, leaving the broad audience of Internet surfers to try, argue, and even better their offering. The release of a service in a beta version is beyond the culture of a telecom operator. Writing along with the service the sentence that no responsibility is taken for the proper functioning of the service is unheard of in the world of telecom operators. The Internet that has brought with it the concept of "best effort" in connectivity terms is now creating a culture of "best effort" in the service area.

The mass market responds well to this offering, because one can find an interested party for basically any proposition it receives. Only a few niches might be interested in acquiring and using a service, but these niches span the planet and they are sufficient to generate a return on the (usually small) investment of those who created the services.

Although connectivity is key in enabling service access and fruition, the value from the user point of view shifts from connectivity to service. Note that the trends toward the embedding of information into services is a further drive in this direction. It is one thing to type into a browser the URL to access a site to get a weather forecast (the very fact that we have to "dial" an address brings communication value to the fore); it is quite a different thing to click an icon with the symbol of the sun on an iPhone (or an iTouch) and get the forecast for those places you care about. In this latter scenario, connectivity has disappeared. As a matter of fact, you might not know if this information is the result of the click activating a connection through the network or if the information being pushed to your device as you were recharging it. You might not know what kind of connectivity was used to bring the information to the terminal (one of the thousands of WiFi networks that the device can automatically and seamlessly connect to, or a cellular network) and you do not care. Communication is no longer on your radar screen.

The attempt of operators to charge for connectivity on a bit-by-bit basis is losing ground and competition will result in flat rates on mobile data as has become customary for fixed-line access. Once this comes to pass, and it will, operators might not expect to see an increase of revenue from connectivity, or from non-existing killer applications. In most Western markets, penetration has reached a point where

An Exciting Future

a further growth of the market is unlikely, and the progressive squeezing of margins is almost certain.

1.11 THE INTERNET *WITH* THINGS

Actually, it is the number of humans in those markets that cannot grow much further. But what about objects? We have seen how technology makes it possible to embed intelligence in objects and enable communication among them and with us. The problem with objects, from a business point of view, is that they do not have a checkbook or a bank account; that is, they cannot pay for communications or services. This is why, rather than using the standard naming of the Internet of things, I prefer to refer to it as the Internet with things (see Figure 1.5).

Does this evolution have the capacity to increase the overall value, or to free resources that can be monetized by operators and other actors? Any object can become a node for delivering services and information. The cell phone can act as a bridge between the atoms of the object and the bits of the related services and information.

The key enabler is the unique identification of the object, which can be used as an address to point to services and information, created and managed by different and independent players in the true spirit of business ecosystems. This simple mechanism will change our view of the world by bringing together both the virtual world and the real world. A cell phone may become a magnifying lens to place information available on the Web onto any object.

The connectivity and flexibility of objects will allow producers to keep in touch with their product and update or upgrade it as new features become available. If the object is "open," and most will be, the features can be enhanced by third parties, greatly improving the versatility of the object and the value perceived by the user.

Customer care will take a completely new twist as it shifts from being a way to remotely fix problems (which will become much more feasible than it is today) to

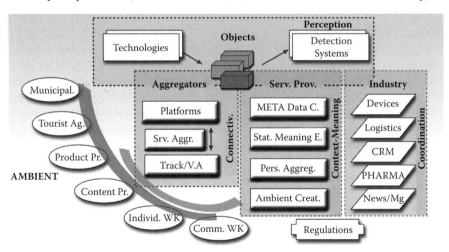

FIGURE 1.5 The Internet with think.

being a way to improve the user experience. In many instances customer care will be provided by third parties, not outsourced by the producers but completely independent, who are finding a business opportunity in providing customer care.

Objects will be able to interact and thus to create ambient environments that are aware, improving the experience of the people living in that environment.

Telecom operators will be able to act as intermediaries for these completely new businesses and reap fresh revenues from them. Basically, telecom operators have made their revenues from intermediation between people in terms of connectivity.

Now that we are running out of people, the focus will be intermediation with objects, and there are many more objects than people. We have seen how, per se, it is not the number of objects that makes the business because they will not become "subscribers," but the revenues resulting from the exploitation of the mashups of the virtual world on the physical one. Production, delivery, and postsales chains will become much more efficient, less costly, and thus will free up money that at least in part can be pocketed by operators. Clearly, this is a new market open to anybody, and competition will be fierce. Some of the assets of a telecom operator might provide some competitive advantage in this new game, but it is a new game with new rules and it will be impossible to succeed by playing the game with the old rules that were successful in the past.

The optical network and its wireless drops will be the enabling infrastructure that can sustain the traffic increase and that can economically sustain the flat rate tariff, without which the market is unlikely to take off.

1.12 COMMUNICATIONS IN 2020 (OR QUITE SOONER)

We have looked at a few technological areas, previewed their evolution, and even seen the evolution of the business framework, with the emergence of the ecosystem paradigm. However, the interesting question is this: How would all this affect our way of communicating? The terminal used in most situations will be a wireless one, a cell phone of a sort, having a nice screen of at least 3 inches that uses most of the surface, with at least VGA resolution, and likely using OLED technology (this provides a very bright screen that is also visible under sunlight with pixels of very small dimension so that high resolution can be packed onto a small surface). Some models might sport a second screen on the back, black and white, based on electronic ink technology (reflective, so that the more ambient light is available, the better the quality of vision, like paper). Actually, this terminal could be better described as a lens to discover what may be hidden at first sight; this function goes back to the 17th century, when it was fashionable to have a magnifying lens as part of one's own dress.

The terminal will be equipped with sensors, some of which will be used to identify the user handling it. When we pick it up, the screen will show the context we are in, usually the location, which could be represented as the background image. At the center it will be us, perhaps our picture. If I look at it I'll see my face; if I hand it to you, you'll see your face. This simple action, handing the device from one person to another, requires access to the identity and to the associated data. This association can happen via the network or it could be obtained by some local personal identification system (a chip embedded in the body as in the WISP initiative from Intel, or a local dingle that can be active only when in close proximity with the person to whom it refers).

Once I pick up a cell phone, it becomes my cell phone, if I am authorized by the cell phone owner; otherwise it will just remain inert and useless. On the screen I will see a number of icons, some defined by me over the years and moved from one cell phone to the next. (Changing a cell phone model does not require any work on my part to transfer data, preferences, and authorization; it just happens seamlessly on any cell phone I own as well as those I am authorized to use.) Some other icons are defined by others, described below. These icons will be the new center of communication in the next decade. A few examples of these icons are already visible on some of today's phones, although they seem very primitive just a few years from now.

One thing that is missing is the loudspeaker. The concept is to have visual communication, and to have that you need to be able to look at the screen, something that would be impossible if you placed the phone over your ear. Hence the sound will reach your ear via a "classic" Bluetooth ear plug or via more sophisticated sound-beaming devices (today we have some of these but they require several loudspeakers to focus the sound in a small area; progress in processing capacity and smart materials will make it possible to miniaturize these loudspeakers in the phone shell).

Another thing that is missing is the numeric keypad when we want to call someone or something. The point is that all communication is contextualized (with very few exceptions). It is like entering a room: you see a person, or an object, and you just get close to him or it and start interacting. Now, placing an icon to identify who you are calling might require an impossibly high number of icons and recognizing them might turn out to be more challenging than reading Chinese characters. Besides, what if I want to call something that I do not know, or that I never called before?

The present method of dialing using numbers requires typing 9 to 13 digits. This lets us reach some billion terminations. On a 3.5-inch screen, like an iPhone, one can place some 20 icons, and have the same number of choices with just seven clicks. In terms of pure clicks, therefore, we can substitute icons for numbers. This does not solve the issue of identifying the right icon, though. The problem can be solved by a tree structure where every click is basically restricting the context. Suppose I want to call my friend John, who is also my partner at tennis. I can click the Community icon, then Tennis, and there I'll see his face in an icon. I might also start by clicking Home and then Agenda, Wednesday last week, and again I'll see his icon face because we had dinner together last Wednesday at my home. You get the gist: There are many ways our brains remember things, and all of them work through association. The best interface we can think of is one based on association (and that is not the one we have been using so far by dialing numbers).

This issue of identifying something I want to call is very interesting because it is not just a matter of interface, but it has some deeper implication for data structuring and architecture and it might involve several players in an ecosystem approach. Suppose you want to call a restaurant that you were taken to by a business associate last month, and, as usual, you do not remember restaurant's name. You could just click your face and click History. Drag the arrow of time to last month and you'll start seeing a collection of memories that have been stored in your cell phone, among them that particular night at the restaurant. You did not write down the name of the restaurant because you were taken there by your business associate, but your cell phone has the location tracked and stored. Now just click the restaurant icon to view it. The location information will be sent to some directory provider (which is actually much more than a directory provider, it can be the Yellow Pages of the future; by the way, the present Yellow Pages should start rethinking their business before they discover it is no longer there) who will push onto your screen a new context containing the restaurant icon that would allow you to get inside the restaurant, browse the menu, make a reservation, call the restaurant, or create a date with several friends at that place; the transportation icon so that you know how to get there; references to comments left by other people on the place; and so on. Now, it is easy to see that an interface based on context is really the interface to our thinking and its implementation requires the involvement of several data inputs owned by different parties and located in different places, reshaped into information by some service provider that knows how "I" think, which has my profile. Part of this transformation might take place inside the terminal; remember, it is storing 1 TB of data and it is probably the most knowledgeable entity about myself. When sending questions to the outside world, it might decide to mask my identity for privacy reasons.

Personalized and contextual communications are two of the things characterizing communications in the next decade. Let's look at some more icons on our screen. One may be the shopping cart. Click it and you'll see a number of icons replacing it. One is the to-do list. This to-do list is sharing information with your family, if that is the context in which you are using it, or with your business associates if that is the context. Suppose it is the family one: When your wife gets the bread, the "bread" item disappears from the list; if you add "bulb," it appears on their to-do list to. There is no explicit action to do that. As your wife is paying for the bread with her phone at the supermarket the item disappears from the list of anyone who is connected to the list.

This list might also be shared with your car so that as you drive the car shares it (neutralizing your identity) with the stores on the way and will prompt you about the availability of the item in a particular store nearby. You might also elect to let the sharing to occur on a broader scale and accept ads for some of the items. That way you'll be able to see a variety of offers. It makes little sense to use this function for bread, of course, but if you are interested in a digital camera or renting an apartment, it would be much more useful.

The control of information sharing is going to be crucial both for its acceptance and for monetizing the ads' value. Some third party should be involved, and perhaps more than one. There are so many potential actors involved and the margins are so razor thin that most of the processing should be handled automatically and it should be up to the communications fabric to mediate among the players.

Another icon shows previous shopping sprees (also the ones of those who have decided to share with you, such as family members). How is this information captured? Clearly any time we pay with a credit card the information is potentially traceable. However, getting it might be close to impossible, given the various databases involved, the variety of security systems, and ownership domains. Life (at least in this respect) would be simpler if one were using a cell phone for any transaction.

That does not imply that the transaction will not be charged on a credit card of same sort or even paid with cash, only that the cell phone keeps track of what is going on. In the future some operators might end up developing this kind of tracer and people will be willing to use it to keep track of their lives.

Another icon might show the shops in the area. This icon, once clicked, will display several icons, one for each of the shops in the area that have pushed themselves onto your screen. Can they do that? Well, of course, some technology is required, such as a platform to manage the information and some applications to format it appropriately for the terminal visualizing it. It requires, in addition, some agreement from the owner or holder of the cell phone to allow this kind of icon on his screen. Surely, any click on one of those icons should bear no cost to the clicker. This possibility of entering the cell phone menu is very interesting for retailers who will have a way to get in touch with potential customers roaming in their area.

Let's take another icon, placed on level one, the one representing my home. By clicking it, I can activate a connection to the home (call home), but it is more interesting to stretch the icon to fill the whole screen. In this way the home becomes my context and I will have available a variety of icons (services) through which I can interact with appliances and people who at that time are part of the home environment.

I close this part, and the chapter, with one more icon: It can have the shape of an eye because its objective is to transform the cell phone into a lens to look at information and services layered on an object. This is already a reality in Japan's supermarkets and in France on billboards. A tag on the object can be read through the cell phone camera, retrieving the unique identity of the object. This identity can be used to retrieve services and information associated with that object by a number of parties, the object producer, or anyone else.

This opens up a Pandora's box for new services, as every object becomes, potentially, a distribution and access point for information and services. This is what we called before the Internet with things.

The cell phone is likely to be the chief intermediary in our communications activities and the point of aggregation of personal information that will also be used to customize services. However, we are going to have more opportunities to communicate than those offered by the cell phone. The ambient environment we are in is a communication gateway to the world. Be it our home, a hotel room, or our office, walls will display information, sensors will be present to customize the environment to our taste, and cameras will be available to bring our image to far distant places.

Any object can be overlaid by our information, in the same way as today we might want to stick a piece of paper on a box or underline a few lines of text with a crayon. Objects sent through overnight delivery or by bits, like digital photos, will have associated information that we have created for the receiver to listen to. Communications, in other ways, will make great use of wires (and wireless), but thanks to technology it will be ubiquitous, overcoming in some instances the need for a classical communications infrastructure.

However it will turn out, of one thing I am certain: Communication will be the invisible fabric connecting us and the world whenever and wherever we happen to be in a completely seamless way, connecting us so transparently, cheaply, and effortlessly that very seldom we will think about it.

REFERENCE

Lloyd, Seth, 2000. Ultimate physical limits to computation. *Nature* 406:1847–1854.

2 802.11n
High Throughput Enhancement to Wireless LANs

Eldad Perahia

CONTENTS

- 2.1 Introduction 27
- 2.2 MIMO and SDM Basics 33
- 2.3 PHY Interoperability with 11a/g Legacy OFDM Devices 37
  - 2.3.1 11a Packet Structure Review 38
  - 2.3.2 Mixed Format High Throughput Preamble 38
- 2.4 High Throughput 40
  - 2.4.1 40 MHz Channel 41
  - 2.4.2 Greenfield Preamble 42
- 2.5 Robust Performance 43
  - 2.5.1 Receive Diversity 44
  - 2.5.2 Spatial Expansion 44
  - 2.5.3 Space-Time Block Coding 44
  - 2.5.4 Transmit Beamforming 45
  - 2.5.5 Low-Density Parity Check Codes 47
- 2.6 MAC Overview 47
  - 2.6.1 Efficiency Enhancements 47
  - 2.6.2 40 MHz Coexistence 48
  - 2.6.3 Protection Mechanisms 49
- 2.7 Beyond 802.11n 50
- References 51

2.1 INTRODUCTION

In 1997 the 802.11 standard included three physical layers (PHY), up to 2 Mbps: (1) Infrared, (2) 2.4 GHz frequency hopped spread spectrum, and (3) 2.4 GHz direct sequence spread spectrum (DSSS). The 802.11b amendment in 1999 increased the data rate to 11 Mbps by enhancing DSSS with complementary code keying (CCK). The 802.11a amendment, also in 1999, introduced orthogonal frequency division multiplexing (OFDM) in 5 GHz with data rates up to

54 Mbps. The 802.11g amendment in 2003 took the 802.11a PHY and applied it to 2.4 GHz.

In January 2002, discussions began in the 802.11 community about increasing the data rates of 802.11 beyond the 54 Mbps capability of 802.11a/g. Proprietary extensions to 802.11a/g were already emerging in the marketplace. There was also interest in having wireless local area networks (LANs) keep pace with the increasing capability of wired Ethernet.

In September 2003, the High Throughput task group (802.11n) was formed with the goal of significantly increasing the throughput of the standard. Throughput was used as the key metric rather than PHY data rate to ensure that the end users experience meaningful improvements to their wireless network. Throughput was defined to be measured at the top of the medium access control (MAC) layer, which also required increased efficiency in the MAC. The requirement was set to achieve at least 100 Mbps.

Figure 2.1 demonstrates the achievable throughput when the PHY data rates are increased with an unmodified 802.11e-based MAC. The inability to achieve a throughput of 100 Mbps necessitated substantial improvements in MAC efficiency when designing the 802.11n MAC.

Considering that the typical throughput of 802.11a/g is 25 Mbps (with a 54 Mbps PHY data rate), this requirement dictated at least a fourfold increase in throughput. Defining the requirement as MAC throughput rather than PHY data rate forced developers to consider the difficult problem of improving MAC efficiency [1].

There were two other key requirements that guided the development of the 802.11n amendment. The first was robustness to achieve high throughput with reasonable range. The second was backward compatibility and coexistence with legacy devices due to the large deployment of 802.11b/g devices and hotspots.

The task group considered addressing multiple environments, including residential, enterprise, and hotspot. A residential environment typically includes only one access point (AP) and many client stations. Current applications include Internet access and streaming audio and video. The goal for 802.11n was to enhance the user experience for applications like intranetworking for local file transfer, backups, and printing with higher data rates. New applications were envisioned, such as voice over IP (VoIP) and video phones. APs could also take the form of wireless home media gateways. Such devices would distribute audio and video content throughout the home, such as DVD and standard and high-definition TV.

In an enterprise environment, the task group envisioned networking applications such as file transfer and disk backup would benefit greatly from the higher data rates of 802.11n. Higher data rates would increase network capacity, providing support for a larger number of clients. Higher throughputs would also enable new applications such as remote display via a wireless connection between a laptop and projector, simplifying presentations in conference rooms.

The developers of 802.11n also wanted to improve wireless hotspot connectivity for locations such as airport lounges, coffee shops, libraries, hotels, or convention centers. The applications typically used in a hotspot include Web browsing, Internet file transfer, and e-mail. New hotspot applications were discussed such as the ability to watch a TV program or a movie on a laptop or other display.

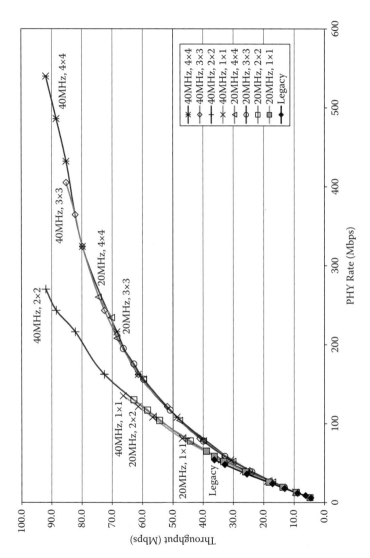

FIGURE 2.1 Throughput vs. PHY data rate assuming no MAC changes. (From Perahia, E. and R. Stacey. 2008. *Next Generation Wireless LANS: Throughput, Robustness, and Reliability in 802.11n.* New York: Cambridge University Press. Reproduced with permission.)

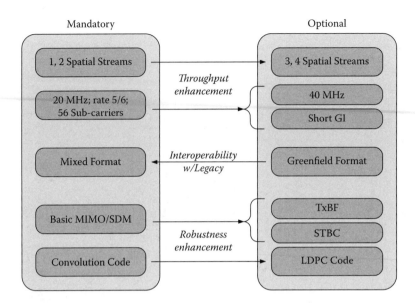

FIGURE 2.2 Mandatory and optional 802.11n PHY features. (From Perahia, E. and R. Stacey. 2008. *Next Generation Wireless LANS: Throughput, Robustness, and Reliability in 802.11n*. New York: Cambridge University Press. Reproduced with permission.)

To address the throughput requirements for the various environments and applications, two basic concepts are employed in 802.11n to increase the PHY data rates: multiple-input, multiple-output (MIMO) and 40 MHz bandwidth channels. To break the 100 Mbps throughput barrier, frame aggregation was added to the 802.11n MAC (as illustrated in Figure 2.3) as the key method of increasing efficiency.

Increasing from a single spatial stream and one transmit antenna to four spatial streams and four antennas increases the data rate by a factor of four. (The term *spatial stream* is defined in the 802.11n standard [2] as one of several bit streams that are transmitted over multiple spatial dimensions that are created by the use of multiple antennas at both ends of a communications link.) However, due to the inherent increased cost associated with increasing the number of antennas, modes that use three and four spatial streams are optional as indicated in Figure 2.2. To allow for handheld devices, the two spatial streams mode is only mandatory in an AP. As shown in Figure 2.2, 40 MHz bandwidth channel operation is optional in the standard due to concerns regarding interoperability between 20 MHz bandwidth and 40 MHz bandwidth devices, the permissibility of the use of 40 MHz bandwidth channels in the various regulatory domains, and spectral efficiency. However, the 40 MHz bandwidth channel mode has become a core feature due to the low cost of doubling the data rate from doubling the bandwidth [1].

To meet the requirement of coexistence with legacy devices, the mixed format waveform was defined that was backward compatible with 802.11a and OFDM modes of 802.11g. The requirement increased the overhead of the waveform, which reduced efficiency. In environments free from legacy devices (termed *Greenfield*), backward compatibility was not required. As illustrated in Figure 2.2, 802.11n includes an

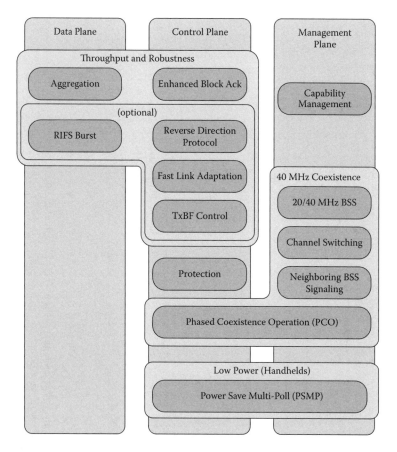

FIGURE 2.3 Summary of 802.11n MAC enhancements. (From Perahia, E. and R. Stacey. 2008. *Next Generation Wireless LANS: Throughput, Robustness, and Reliability in 802.11n.* New York: Cambridge University Press. Reproduced with permission.)

optional Greenfield format, which does not contain the components of the preamble that support backward compatibility.

To increase the data rate at a given range requires enhanced robustness of the wireless link. 802.11n defines implicit and explicit transmit beamforming (TxBF) methods and space-time block coding (STBC), which improves link performance over MIMO with basic spatial division multiplexing (SDM). The standard also defines a new optional low-density parity check (LDPC) encoding scheme, which provides better coding performance over the basic convolutional code [1]

A major issue with efficiency is that as the data rate increases, the time on air of the data portion of the packet decreases. However, the PHY and MAC overhead remain constant. This results in diminishing returns from the increase in PHY data rate as illustrated in Figure 2.1. Frame aggregation in the MAC increases the length of the data portion of the packet to increase overall efficiency.

MAC throughput with frame aggregation increases linearly with PHY data rate with traffic conducive to aggregation, as illustrated in Figure 2.4. With a PHY data

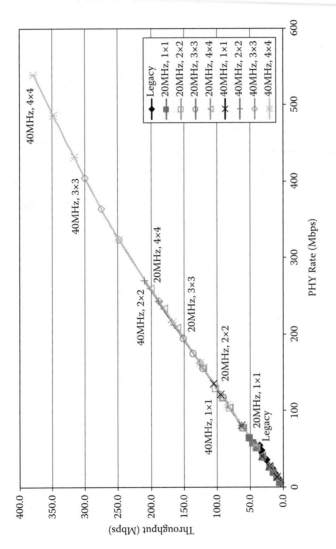

FIGURE 2.4 Throughput vs. PHY data rate with frame aggregation. (From Perahia, E. and R. Stacey. 2008. *Next Generation Wireless LANS: Throughput, Robustness, and Reliability in 802.11n.* New York: Cambridge University Press. Reproduced with permission.)

rate of 600 Mbps, a MAC throughput of over 400 Mbps is achievable with 802.11n MAC enhancements.

For further improvements to efficiency, the 802.11n MAC includes enhancements to the block acknowledgment (BA) mechanism in 802.11e. The reverse direction protocol was also incorporated, which allows a station to share its transmit opportunity (TXOP) with another station. This increases throughput with traffic patterns that are highly asymmetric, for example when transferring a large file with File Transfer Protocol (FTP) operating over Transmission Control Protocol (TCP). A simple improvement to efficiency when bursting packets is to reduce the interframe spacing (RIFS) between the packets.

Many new methods of control and management were added to 802.11n, as illustrated in Figure 2.3. To more rapidly track changes in the channel, fast link adaptation assists in the selection of the optimal modulation and coding scheme (MCS). Transmit beamforming may be considered a PHY technique, but it requires a great deal of control in the MAC for channel sounding, calibration, and the exchange of channel state information or beamforming weights. Protection mechanisms needed to be devised to ensure that legacy 802.11a/g devices were not harmed by the new modes of operation and vice versa. These new modes, which might require protection, include RIFS bursting and Greenfield format transmissions [1].

With the introduction of the 40 MHz bandwidth channel came the complexity of managing coexistence between 40 MHz bandwidth 802.11n devices and 20 MHz bandwidth 802.11n and 802.11a/g devices. With the increased interest in WiFi-enabled handheld devices, Power Save Multi-Poll (PMSP) was incorporated in the 802.11n MAC to provide a minor improvement to channel utilization and reduction in power consumption when transmitting and receiving small amounts of data periodically.

2.2 MIMO AND SDM BASICS

In a basic wireless communications system, both the transmitter and receiver are configured with one antenna, as illustrated in Figure 2.5. This is termed single-input, single-out (SISO), as in a single input to the environment and single output from the environment.

A basic communication system is described in Equation 2.1.

$$y = \sqrt{\rho} \cdot h \cdot x + z \qquad (2.1)$$

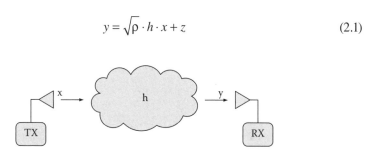

FIGURE 2.5 SISO system. (From Perahia, E. and R. Stacey. 2008. *Next Generation Wireless LANS: Throughput, Robustness, and Reliability in 802.11n*. New York: Cambridge University Press. Reproduced with permission.)

where x is the transmitted data with unity mean expected power, h is the channel fading coefficient, z is independent, complex additive white Gaussian noise (AWGN) with zero mean and unit variance, ρ is the average signal-to-noise ratio (SNR), and y is the received signal.

The capacity for a general SISO system is given by the Shannon capacity formula in Equation 2.2 [3].

$$C(bps/Hz) = \log_2\left(1 + \rho \cdot |h|^2\right) \qquad (2.2)$$

The receiver extracts the information by equalizing the received signal, as given in Equation 2.3.

$$\begin{aligned}\hat{x} &= \left(\sqrt{\rho} \cdot h\right)^{-1} \cdot y \\ &= x + \left(\sqrt{\rho} \cdot h\right)^{-1} \cdot z\end{aligned} \qquad (2.3)$$

where $\hat{x}$ is the noisy estimate of the transmitted signal x.

MIMO describes a system with a transmitter with multiple antennas that transmits through the propagation environment to a receiver with multiple receiver antennas. The transmitter may use transmit beamforming with multiple antennas. The receiver may use additional antennas for diversity combining. These systems improve robustness and increase the rate at a given range, but they do not increase the maximum data rate.

With SDM, independent data streams are transmitted on different antennas, as illustrated in Figure 2.6. With MIMO and SDM, the maximum data rate increases as a function of the number of independent data streams and transmit antennas. The number of receiver antennas must be at least equal to or greater than the number of data streams with a linear equalizer.

A more detailed graphical view of a MIMO/SDM system with two transmit antennas (and two spatial streams) and two receive antennas is given in Figure 2.7. The notation used to describe such systems is "number of Tx antennas" × "number of Rx antennas," thus Figure 2.6 and Figure 2.7 represent a 2 × 2 MIMO/SDM system.

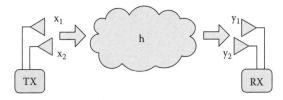

FIGURE 2.6 MIMO/SDM system. (From Perahia, E. and R. Stacey. 2008. *Next Generation Wireless LANS: Throughput, Robustness, and Reliability in 802.11n*. New York: Cambridge University Press. Reproduced with permission.)

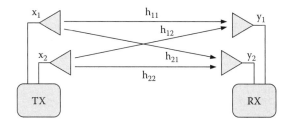

FIGURE 2.7 Mathematical model of MIMO/SDM system. (From Perahia, E. and R. Stacey. 2008. *Next Generation Wireless LANS: Throughput, Robustness, and Reliability in 802.11n.* New York: Cambridge University Press. Reproduced with permission.)

A MIMO/SDM communication system is described as follows:

$$y_1 = \sqrt{\rho/2}\, h_{11} x_1 + \sqrt{\rho/2}\, h_{12} x_2 + z_1$$
$$y_2 = \sqrt{\rho/2}\, h_{21} x_1 + \sqrt{\rho/2}\, h_{22} x_2 + z_2 \quad (2.4)$$

The MIMO/SDM system can be expressed in vector form with a similar mathematical structure as a SISO system, as given in Equation 2.5.

$$\begin{bmatrix} y_1 \\ y_2 \end{bmatrix} = \sqrt{\rho/2} \begin{bmatrix} h_{11} & h_{12} \\ h_{21} & h_{22} \end{bmatrix} \begin{bmatrix} x_1 \\ x_2 \end{bmatrix} + \begin{bmatrix} z_1 \\ z_2 \end{bmatrix}$$

$$Y = \sqrt{\rho/M}\, HX + Z \quad (2.5)$$

Conceptually, the MIMO equalizer can also be designed with a similar structure as the receiver of the SISO system in Equation 2.3, as given below.

$$\hat{X} = \left(\sqrt{\rho/M}\, H\right)^{-1} Y$$
$$= X + \left(\sqrt{\rho/M}\, H\right)^{-1} Z \quad (2.6)$$

A generalization of the Shannon capacity formula for M transmit antennas and N receive antennas is given by Equation 2.7 [4].

$$C(bps/Hz) = \log_2 \left[\det\left(I_N + \rho/M \cdot HH^*\right)\right] \quad (2.7)$$

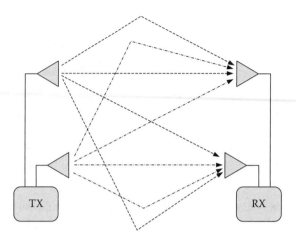

FIGURE 2.8 Multipath fading environment. (From Perahia, E. and R. Stacey. 2008. *Next Generation Wireless LANS: Throughput, Robustness, and Reliability in 802.11n*. New York: Cambridge University Press. Reproduced with permission.)

It almost seems too good to be true that with such simplicity one can increase the data rate of a system by merely adding additional transmit and receive antennas. Through the use of an example it is shown that there are bounds on this increase. Consider the SISO system of Equation 2.1, where now the antenna port of the transmitting device is directly connected to the antenna port of the receiving device, which is a typical setup for conductive testing. Mathematically, h is equal to 1, and $y = \sqrt{\rho} \cdot x + z$. With enough transmit power, the system is very reliable. Now consider a 2 × 2 MIMO/SDM system with each antenna port of the transmitter device connected to each antenna port of the receiver device. The channel H for such a system is equal to $\begin{bmatrix} 1 & 1 \\ 1 & 1 \end{bmatrix}$. To extract the transmitted data from the receive channel, we must perform an inverse of the channel matrix. In this case, the matrix is singular and cannot be inverted. The receiver fails, even in high SNR. As we see, the channel matrix must be well conditioned for the MIMO system to achieve large increases in data rate from spatial multiplexing [5].

If the taps in the channel matrix are randomly distributed and uncorrelated with each other, statistically the channel matrix will be invertible a vast majority of the time. A multipath fading creates such an environment.

In an indoor environment, rays bounce off floors, ceilings, walls, furniture, and so on, when propagating between the transmitter and the receiver [6]. The paths of the rays are different as a function of the location of the transmit and receive antennas, as illustrated in Figure 2.8. How the paths propagate determines the amount fading due to cancellation, delay, and correlation.

How do we ensure that the channel is uncorrelated enough? An antenna spacing of at least a half a wavelength should be used, in addition to low antenna coupling. Non-line-of-sight (NLOS) conditions also help by not having a highly correlated

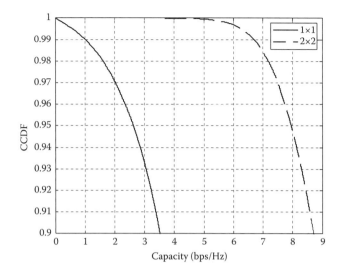

FIGURE 2.9 Capacity comparison between 2 × 2 MIMO and SISO with SNR = 20 dB. (From Perahia, E. and R. Stacey. 2008. *Next Generation Wireless LANS: Throughput, Robustness, and Reliability in 802.11n*. New York: Cambridge University Press. Reproduced with permission.)

direct path between transmit and receive devices. However, this typically means reduced signal power at the receiver.

To demonstrate the large potential gain of MIMO, in Figure 2.9 a 2 × 2 MIMO system is compared with a SISO system. The complementary cumulative distribution function (CCDF) is plotted versus capacity to show how often the capacity exceeds a certain level. The theoretical capacity of each system, in a Rayleigh fading channel, is simulated using Equation 2.7 at an SNR equal to 20 dB. For reference, the maximum data rate of 802.11a/g is 54 Mbps in a 20 MHz channel, resulting in a spectral efficiency of 2.7 bps/Hz. This is achieved 95% of the time by the SISO system in Figure 2.9. At a comparable percentage, a 2 × 2 MIMO system achieves three times the capacity. In reality, various impairments and realistic channel environments inhibit the ability to increase the capacity by that amount in all situations.

2.3 PHY INTEROPERABILITY WITH 11A/G LEGACY OFDM DEVICES

As previously mentioned, one of the requirements of the 802.11n standard was backward compatibility with 802.11a/g. This led to the definition of a mandatory mixed format (MF) preamble in 802.11n. In this section, we first review the 802.11a packet structure to fully understand the issues in creating a preamble that is interoperable between 802.11a/g and 802.11n devices. For further details regarding 802.11a/g beyond this review, refer to clause 17 and 19 in [7] and [5]. Following this overview, the MF preamble, which is part of 802.11n, is discussed.

2.3.1 11A Packet Structure Review

The 802.11a packet structure is illustrated in Figure 2.10. The short training field (STF) is used for start-of-packet detection, automatic gain control (AGC) setting, initial frequency offset estimation, and initial time synchronization. The long training field (LTF) is used for accurate frequency offset estimation, time synchronization, and channel estimation. The signal field (SIG) contains rate and length information of the packet. Example rates are 6 Mbps (BPSK, rate ½ encoding) and 54 Mbps (64-QAM, rate ¾ encoding). An issue with the SIG is that it is only protected by a single parity bit, which leads to false detects. A reserve bit in the signal field has been used by some manufacturers for more parity. This practice made it impossible for the reserve bit to be used as an indication of an 802.11n packet. The data field follows the SIG. The first 16 bits of the data field make up the service field, containing reserve bits and scrambler initialization bits.

2.3.2 Mixed Format High Throughput Preamble

The mixed format (MF) high throughput preamble begins with the legacy 11a preamble and is followed by high throughput training fields, as illustrated in Figure 2.11. The first part of the MF preamble consists of legacy (L) training identical to 11a. The legacy part of the preamble contains L-STF, L-LTF, and L-SIG fields as 11a. This enables legacy devices to detect the initial part of the MF preamble. An 802.11n device will set the rate and length of the L-SIG to an equivalent time duration of the entire MF packet. When a legacy device decodes the SIG, it computes the duration of the packet, defers access, and does not interfere with the 802.11n transmission.

As illustrated in Figure 2.11, the high throughput (HT) portion of the MF preamble consists of the high throughput signal field (HT-SIG), high throughput short training field (HT-STF), and high throughput long training field (HT-LTF). The two OFDM symbols of the HT-SIG, HT-SIG$_1$ and HT-SIG$_2$, contain new signaling information for new PHY modes of operation and are also used for autodetection between HT MF and legacy OFDM packets. The HT-STF is used to redo the AGC setting. The HT-LTFs are used for MIMO channel estimation. The HT data field follows the HT-LTFs.

A basic block diagram for a MIMO receiver is illustrated in Figure 2.12. A MIMO receiver design has multiple receive antenna chains, each consisting of an antenna, low noise amplifier (LNA), down converter, AGC, and analog-to-digital converter (ADC). To maximize the dynamic range on each antenna branch, typically each AGC in each receive chain is individually adjusted.

Because the MF packet begins with the legacy preamble, the use of the L-STF for AGC, initial frequency correction, and initial time acquisition is similar to that of 802.11a. However, 802.11n devices typically have multiple receive antennas; therefore, correlation functions for start-of-packet detection and coarse timing should be summed over all the antennas to maximize detection performance. The coarse time adjustment is applied to all receive signals. Similarly, the frequency estimate may also be summed over all receive antenna chains and applied to all received signals.

After the preamble is processed, the receiver sequentially selects 4 μsec blocks of data samples. The 0.8 μsec guard interval is removed from each block. A fast Fourier

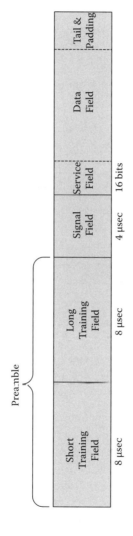

FIGURE 2.10 802.11a packet structure. (From Perahia, E. and R. Stacey. 2008. *Next Generation Wireless LANS: Throughput, Robustness, and Reliability in 802.11n*. New York: Cambridge University Press. Reproduced with permission.)

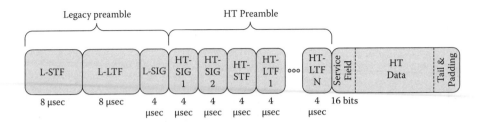

FIGURE 2.11 Mixed format preamble. (From Perahia, E. and R. Stacey. 2008. *Next Generation Wireless LANS: Throughput, Robustness, and Reliability in 802.11n*. New York: Cambridge University Press. Reproduced with permission.)

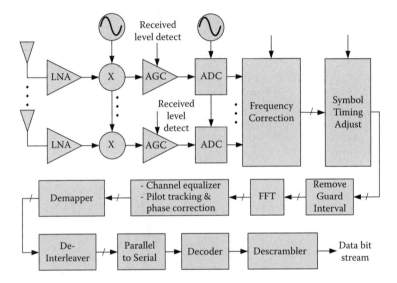

FIGURE 2.12 Receiver block diagram for MIMO. (From Perahia, E. and R. Stacey. 2008. *Next Generation Wireless LANS: Throughput, Robustness, and Reliability in 802.11n*. New York: Cambridge University Press. Reproduced with permission.)

transform (FFT) is performed on the remaining 3.2 μsec. The data subcarriers are equalized and the constellation values are extracted by the demapper. These values are deinterleaved, decoded, and then descrambled, resulting in the information bits.

2.4 HIGH THROUGHPUT

There are many PHY features in the 802.11n amendment that increase the data rate of the system. The previous two sections discussed MIMO as one of these techniques. With MIMO we achieve large data rate increases over 802.11a/g with two, three, and four spatial streams. MIMO provides a data rate increase over 802.11a/g proportional to the number of spatial streams.

In addition, in the 802.11a/g system there are 52 subcarriers—48 data subcarriers and 4 pilot subcarriers—in a 20 MHz channel. In 802.11n, this was increased to

52 data subcarriers and 4 pilot subcarriers in a 20 MHz channel. This increases the data rate by about 10%.

In the 802.11a/g system, the highest code rate is 3/4. A higher code rate of 5/6 was added to further increase the data rate by 11%. The SNR requirement with 64-QAM, rate 5/6 is higher than the highest data rate in 802.11a/g, which only uses 64-QAM, rate 3/4. However, with the added robustness features in 802.11n, the MCS's with 64-QAM, rate 5/6 are functional at a reasonable range.

The standard OFDM symbol in 802.11a/g/n is 4 μsec in length, and comprised of 0.8 μsec of guard interval (GI) and 3.2 μsec of data. With the short GI mode, the GI is reduced to 0.4 μsec, and the overall symbol length is 3.6 μsec. The data rate is increased by 12%. However, the GI is used to protect against long delay spread and filtering. Therefore, this mode is less tolerant to delay spread and should only be used in short-delay spread channels.

With the enhancements already described, the 802.11n data rate for the 20 MHz channel with one spatial stream ranges from 6.5 Mbps to 72.2 Mbps. With two spatial streams, the data rate ranges from 13 Mbps to 144.4 Mbps. With three spatial streams, the data rate ranges from 19.5 Mbps to 216.7 Mbps. With four spatial streams, the data rate ranges from 26 Mbps to 288.9 Mbps.

The other feature that dramatically increases the PHY data rate is the new 40 MHz channel in 802.11n. There is also a shorter preamble termed the Greenfield (GF) preamble to improve efficiency. These features are described in the next two sections.

2.4.1 40 MHz Channel

The 802.11 wireless local area networking (WLAN) systems predominantly operate in unlicensed bands, making spectrum free. Furthermore, there is a negligible increase in cost between a 40 MHz radio and a 20 MHz radio. Therefore, a 40 MHz waveform was added to 802.11n as an almost free way of doubling the data rate. Two issues arise when we increase the bandwidth of the channel. First, using wider channels leaves fewer distinct channels, possibly resulting in an increase in frequency reuse. This could result in an increase in interference if not properly managed. Second, having two channel widths requires management of coexistence and interoperability between 20 MHz and 40 MHz capable devices.

The typical channels used for 802.11a systems were four channels in the 5.15–5.25 GHz band and four channels in the 5.25–5.35 GHz band. With the use of dynamic frequency selection, WLAN systems are now permitted to operate in the 5.47–5.725 GHz band in many regulatory domains, including the Federal Communications Commission (FCC) and the European Telecommunications Standards Institute (ETSI). This adds ten new channels in 5 GHz. Previously there were typically eight 20 MHz channels available for 802.11a systems. Now there are eighteen 20 MHz channels or nine 40 MHz channels available for 802.11n. Therefore, the frequency reuse with 802.11n 40 MHz channels is no more stringent than it was with 802.11a 20 MHz channels.

The 40 MHz channel is composed of two adjacent 20 MHz channels, one comprising the primary channel and the other comprising the secondary channel. Any 20 MHz traffic in a 40 MHz basic service set (BSS) will be sent on the primary

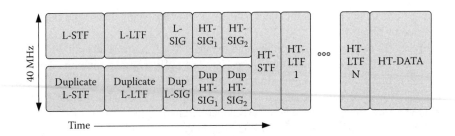

FIGURE 2.13 40 MHz mixed mode preamble. (From Perahia, E. and R. Stacey. 2008. *Next Generation Wireless LANS: Throughput, Robustness, and Reliability in 802.11n*. New York: Cambridge University Press. Reproduced with permission.)

channel. In the 5 GHz band, the frequency plan is defined such that there are no overlapping 40 MHz channels. In the 2.4 GHz band, the 5 MHz channel numbering is still maintained from 802.11b/g, making possible partial overlapping 40 MHz channels. This is restricted by other rules.

The 40 MHz waveform design includes 114 populated subcarriers, of which 108 are data and 6 are pilot. This is more than double the number of data subcarriers for the 20 MHz waveform. With one spatial stream, 40 MHz, and short GI, the maximum data rate is 150 Mbps. With two and three spatial streams, the maximum data rates are 300 Mbps and 450 Mbps, respectively. The maximum data rate in 802.11n is 600 Mbps, achieved with four spatial streams, 40 MHz, short GI, 64-QAM modulation, and rate 5/6 coding.

To interoperate and coexist with legacy 802.11a/g OFDM devices, the legacy portion of the 20 MHz MF preamble is replicated over both 20 MHz portions of the 40 MHz band. This is illustrated in Figure 2.13. Duplication of the L-STF, L-LTF, and L-SIG fields enables a legacy OFDM receiver to process the signal field and properly defer transmission. Duplication of the HT-SIG transmitted by a 40 MHz device allows 20 MHz 802.11n devices on either half of the 40 MHz channel to decode the HT-SIG and also properly defer transmission. Furthermore, a 20/40 bit in the HT-SIG indicates whether the signal is a 20 MHz or 40 MHz transmission. As a power-saving mechanism, the 20 MHz 802.11n device can terminate reception in the presence of a 40 MHz transmission.

New waveforms are defined for the 40 MHz HT-STF, HT-LTFs, and HT-Data, filling in the subcarriers in the null space that results from bonding two adjacent 20 MHz channels. This is why the 40 MHz data rates are more than double their 20 MHz counterparts.

2.4.2 GREENFIELD PREAMBLE

The MF preamble trades throughput efficiency for coexistence between HT and legacy devices. The optional GF preamble provides a little better efficiency than MF by eliminating the legacy portion of the preamble and reducing the preamble by 12 μsec. The fields in the GF preamble are illustrated in Figure 2.14 and compared

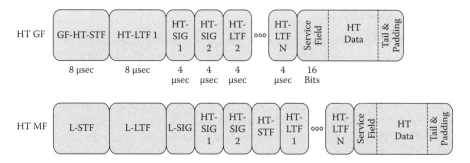

FIGURE 2.14 Greenfield format preamble. (From Perahia, E. and R. Stacey. 2008. *Next Generation Wireless LANS: Throughput, Robustness, and Reliability in 802.11n*. New York: Cambridge University Press. Reproduced with permission.)

to the MF preamble. The GF HT-STF is the same as the MF HT-STF, except double in length. Similar to L-STF, it is used for acquisition and AGC setting. Similarly for the GF HT-LTF1, it is the same as the MF HT-LTF, except double in length to allow for frequency and timing estimation. The GF HT-SIGs and HT-LTFs are same as the MF HT-SIGs and HT-LTF field.

The GF preamble can be used when no legacy devices are associated with the AP. However, legacy devices associated with neighboring APs might still experience interference. In the presence of legacy devices, the GF preamble may be used during a MAC protected transmission period. The efficiency improvement of the GF preamble over MF preamble occurs mainly with short packet lengths in conjunction with higher data rates. In this situation, the reduction of 12 μsec of preamble time is more noticeable. This could occur with VoIP packets on the order of 100 bytes. The time on air is shorter, which means the system can support more callers. The shorter time on air reduces the time that the transmitter and receiver are active, which decreases power consumption and increases battery life of handheld devices.

There are some issues with the GF preamble. Obviously without the legacy portion of the preamble, the GF preamble does not appear as an 802.11 packet format to legacy OFDM devices. Furthermore, with two HT preamble formats it is now necessary for 11n devices to autodetect between multiple HT preamble formats. And because GF is optional, not all HT devices support this preamble type. With longer packet lengths and typical MAC overhead, there is negligible network efficiency improvement.

2.5 ROBUST PERFORMANCE

Several options are included in 802.11n to maintain comparable robustness to 802.11a/g, typically measured in range. With multiple antennas, receive diversity and transmit spatial expansion are achieved naturally. STBC is employed to achieve larger gains on transmission with minimal complexity. A closed-loop transmit beamforming mechanism is defined for even more gain than STBC. LDPC coding is included as an enhancement over the binary convolutional coding (BCC).

2.5.1 Receive Diversity

A basic MIMO/SDM system with a linear receiver requires at least as many receive antennas as transmit data streams, as illustrated in Figure 2.7. In a multipath fading channel, higher two-stream data rates might not be achievable with, for example, only two receive antennas. Extra receive antennas provide receive diversity gain and reduce required SNR. For example, with two streams, 64-QAM, rate 5/6 coding, the required SNR is reduced by 9 dB with an extra receive diversity antenna. Receive diversity enables signal reception of the peak two-stream data rate at a feasible SNR. This extends the range of the highest data rates. With multiple spatial streams the diversity benefit is significant with higher modulation orders as these are sensitive to fading.

Even 802.11n devices with only two receive antennas provide diversity gain over single-antenna 802.11a/g devices. A single stream transmission received by a two-antenna receiver has performance comparable to a 1 × 2 diversity system. For example, in a multipath fading environment, a two-antenna 802.11n device achieves in excess of 7 dB receive diversity gain over a single-antenna 802.11a/g device. This gain increases the range at the higher single-stream data rates. More important, however, this increases the range of the lowest rate (BPSK, rate ½), which significantly increases the absolute range of the BSS. This translates into larger coverage area for 802.11n devices.

2.5.2 Spatial Expansion

Spatial expansion provides a simple means to transmit fewer spatial streams with more antennas. For example, if the device has two transmit antennas, but the channel only supports one spatial stream, it is desirable to still transmit with both antennas to utilize both power amplifiers and maximize the output transmit power. With the appropriate selection of cyclic shifts, spatial expansion provides a small transmit diversity benefit in channels with more pronounced flat fading. Typically a gain on the order of 1 dB is achievable.

2.5.3 Space-Time Block Coding

STBC combines signals over two OFDM symbols between multiple antennas for transmit diversity gain. With the Alamouti scheme [8], transmit gain can be theoretically achieved equivalent to received diversity. However, assuming normalized output transmit power, STBC will have a transmit power penalty with respect to receive diversity. This is due to splitting the transmit power over multiple transmit antennas.

Two transmit antennas and one receive antenna (2 × 1) is the basic system configuration for the Alamouti scheme. In a wide-band frequency selective fading channel, a gain of over 5 dB is achieved with single stream, 64-QAM, rate 5/6 coding relative to single transmit antenna performance of 802.11a/g. STBC has a clear benefit in extending the range of the higher data rates of single stream, single receive antenna devices. This enables robust performance for size and power-limited devices, such as mobile handheld devices.

Unfortunately at lower order modulation, STBC does not provide much gain as a result of the power penalty and implementation impairments. So, unlike receive diversity, STBC does not provide a significant improvement in absolute coverage.

The 802.11n standard amendment includes STBC configurations of 2×1, 3×2, 4×2, and 4×3. The basic STBC building block is transmitting one spatial stream with two antennas using the 2×1 Alamouti scheme. With three transmit antennas, one spatial stream is transmitted with the 2×1 STBC configuration, and the second spatial stream is transmitted without STBC. With four antennas and two spatial streams, each spatial stream is separated coded with the 2×1 STBC mapping. With four transmit antennas and three spatial streams, one spatial stream is transmitted with the standard 2×1 configuration, and the second and third spatial streams are transmitted without STBC.

The most gain is achieved with 2×1 and 4×2, as 3×2 and 4×3 systems include streams without STBC.

2.5.4 Transmit Beamforming

With transmit beamforming, weights are applied to the transmitted signal to improve reception. The weights for each subcarrier are adapted from channel state information (CSI) for every subcarrier. The key advantage is the ability to improve link performance to low-cost, low-complexity devices. For example, an AP with four transmit antennas can transmit two streams to a two-antenna client device with 4×2 transmit beamforming. Then when the client transmits to the AP, the link is balanced with 2×4 receive diversity.

The most common approach for generating transmitter beamforming weights is performing singular value decomposition on the CSI. With this approach, the output SNR is a function of the singular values of the CSI matrix. In addition, the singular values are by definition ordered in decreasing order. This results in two dramatic effects. Because transmit beamforming is performed on a subcarrier basis, the first effect is that the SNR of the spatial streams becomes separated across the band. The first few streams will have higher output SNR than without transmit beamforming. Therefore, each stream can be matched to the appropriate modulation and coding scheme. The second effect is that the first few spatial streams are relatively flat across the band and will exhibit improved coding performance similar to AWGN rather than a frequency-selective fading channel.

Transmit beamforming may be used with "square" system configurations; that is, two transmit, two receive antennas, or three transmit, three receive antennas, or four transmit, four receive antennas. In such configurations the gain is limited, especially at the higher order modulations.

The true benefit of transmit beamforming is achieved when the number of transmit antennas exceeds the number of receive antennas and the number of spatial streams. In such a configuration, stronger eigen modes are utilized, whereas the weaker ones are disregarded. This leads to gains up to 11 dB with high-order modulation and coding rate.

As described, calculating the weights for transmit beamforming depends on having the CSI. Therefore, the channel needs to be sounded between two devices to measure the CSI. This basic concept is as follows: Device A transmits a packet to device B. Device B estimates the CSI from the LTF of the preamble of the packet. However, consider a poor channel that will only support one spatial stream prior to transmit beamforming, but will support many spatial streams with transmit beamforming. In a normal packet, the preamble contains one LTF for each spatial stream. When device A transmits a single stream packet to device B, the packet will only have one LTF. However, the full dimensionality of the channel is equivalent to the number of transmit antennas at device A and number of receive antennas at device B. With only one LTF, the CSI for only one spatial stream can be estimated.

In the 802.11n standard amendment, there are two approaches to channel sounding. The first approach is the Null Data Packet. The number of LTFs in the preamble is selected to cover the full dimensionality of the channel. This type of packet contains no data; therefore, there is no issue with a poor channel. The second approach is the Extension LTFs. This type of packet contains data. However, extra LTFs are included in the preamble beyond those necessary for channel estimation of the data. The extra LTFs are used to measure full dimensionality of the channel.

Once the channel is sounded, it is necessary to feed back the information to the device applying the transmit beamforming weights. There are two feedback mechanisms in 802.11, implicit and explicit feedback. Implicit feedback is based on the reciprocity relationships of electromagnetism. Because 802.11 employs time division duplexing, the same frequency carrier is used for both link directions. Therefore, the CSI measured at either end of the link would ideally be equivalent.

The exchange of feedback and beamforming with implicit feedback is as follows. Device B transmits a sounding packet to device A. Device A estimates the CSI from the LTFs. Device A computes beamforming weights from the CSI. Then device A applies beamforming weights to the next transmission to device B. Radio frequency distortions in the transmitter and receiver of the devices at both ends of the link will degrade reciprocity. Calibration is necessary to minimize the degradation.

The second approach is explicit feedback. With explicit feedback, the device performing the transmit beamforming is the same device that transmits the sounding packet. The packet flow with explicit feedback is as follows. Device A transmits a sounding packet to device B. Device B estimates the CSI from the LTFs. Then device B transmits the CSI to device A, or device B computes the beamforming weights from the CSI and transmits that to device A. Finally, device A applies beamforming weights to the next transmission to device B. Ideally, the channel remains the same between sounding and beamforming.

With explicit feedback, the format of the feedback must be defined. 802.11n contains three types of explicit feedback: CSI, noncompressed beamforming weights, and compressed beamforming weights. In addition, quantization and subcarrier grouping techniques are utilized to reduce overhead.

Whether to choose to implement explicit or implicit feedback depends on many factors. Implicit beamforming will work with as many transmit antennas as contained in the device performing beamforming, whereas explicit beamforming is limited to four transmit antennas. In addition, explicit has overhead from the feedback of CSI or beamforming weights. With explicit, the feedback delay could be lengthy due to the time to compute the weights and gain channel access for transmission of the feedback. Implicit will work with all recipients, whereas explicit requires the recipient to compute beamforming weights. However, implicit requires calibration and there could be issues with calibration stability over many conditions. This makes the hardware design of an implicit beamforming implementation involve both an analog and digital design. Explicit does not require calibration, and can be implemented completely digitally.

2.5.5 Low-Density Parity Check Codes

LDPC coding is an advanced, capacity-approaching forward error correction scheme. They are iteratively decoded like Turbo codes. However, LDPC codes are made up of binary block codes. These constituent codes have simple parity check relationships, resulting in the low-density name. LDPC was originally developed by Bob Gallager in his PhD thesis in 1962. They were more recently rediscovered when Turbo codes made iterative decoding popular.

Because BCC is the mandatory coding scheme of 802.11n, a device is required to have a BCC encoder and decoder. Using optional LDPC coding means implementing both the BCC and LDPC encoders and decoders. A designer needs to ask several basic questions regarding LDPC coding to decide whether to implement LDPC. How much additional gain does it provide to the system? Is it worth having a second encoder or decoder type in the device? With a wide-band 802.11n MIMO system in a frequency-selective fading channel, the gain of LDPC over BCC ranges from 1.5 dB to 3 dB, depending on the modulation and coding rate.

2.6 MAC OVERVIEW

To break the 100 Mbps throughput barrier, many efficiency enhancement techniques were added into the 802.11n standard amendment. These included RIFS, aggregation of data, aggregating data with control, and implicit BA request. With the inclusion of a 40 MHz channel came the necessity to manage coexistence between 40 MHz and 20 MHz devices. Furthermore, the adoption of all the new modes in 802.11n requires protection mechanisms to ensure that legacy 802.11a/g devices are not harmed by the new modes of operation and vice versa.

2.6.1 Efficiency Enhancements

There were two forms of aggregation adopted in the standard: MAC protocol data unit aggregation (A-MPDU) and MAC service data unit aggregation (A-MSDU). Logically, A-MSDU resides at the top of the MAC and aggregates multiple MSDUs into a single MPDU. Each MSDU is prepended with a subframe header consisting of the destination address, source address, and a length field giving the length of the SDU in bytes. This is then padded with 0 to 3 bytes to round the subframe to a 32-bit word boundary. Multiple such subframes are concatenated together to form a single MPDU. An advantage of A-MSDU is that it can be implemented in software. A-MPDU resides at the bottom of the MAC and aggregates multiple MPDUs. Each MPDU is prepended with a header consisting of a length field, an 8-bit cyclical redundancy check (CRC), and an 8-bit signature field. These subframes are similarly padded to 32-bit word boundaries. Each subframe is concatenated together. An advantage of A-MPDU is that if an individual MPDU is corrupt, the receiver can scan forward to the next MPDU by detecting the signature field in the header of the next MPDU. With A-MSDU, any bit error causes all the aggregates to fail [1].

To improve on BA, 802.11n makes three improvements: compressed BA, implicit BA request, and "No Ack" delayed BA. The original BA format from 802.11e included a 128-byte MSDU bitmapping to handle fragmentation of MSDUs. At higher data rates, fragmentation is no longer effective. The compressed BA format eliminates the fragmentation bitmap and reduces the MSDU bitmap to 8 bytes, creating a shorter frame.

With the 802.11e BA mechanism, a BA request is necessary to solicit a BA after data bursting. In a normal acknowledgment policy, a data packet is always immediately acknowledged by the recipient. To improve efficiency of acknowledgments, 802.11n defines that the normal acknowledgment policy of an aggregate of data frames solicits a BA. This eliminates the overhead of the BA request. In 802.11e the delayed BA mechanism allows a BA response to BA request to be sent in a later TXOP. However, both the BA request and the BA require acknowledgments. The No Ack Delayed BA mechanism in 802.11n allows for a delayed BA response without the overhead of acknowledging both the BA request and the BA. With this, the BA can be efficiently aggregated with data in separate TXOPs.

2.6.2 40 MHz Coexistence

Clear channel assessment (CCA) is a basic mechanism in 802.11 to determine whether the medium is free. The CCA rules are extended to 40 MHz channels in 802.11n. These rules are used in both 5 GHz and 2.4 GHz bands.

The 40 MHz CCA requirements are a simple extension of the 20 MHz CCA rules when the secondary channel is idle. In this case the CCA on the primary channel must go busy with a sensitivity of at least −82 dBm for a valid 20 MHz physical layer convergence procedure (PLCP) protocol data unit (PPDU) transmission (signal detection). Furthermore, CCA for the primary channel must go busy with a sensitivity of at least −62 dBm for any signal on the primary channel (energy detection). These rules are equivalent to the 20 MHz CCA rules.

Additional rules exist when a signal is present on both channels. The CCA for both the primary and secondary channel must go busy with a sensitivity of at least −79 dBm to a valid 40 MHz PPDU transmission (signal detection). The CCA for both the primary and secondary channels must go busy with a sensitivity of at least −59 dBm to any signal present on both channels (energy detection).

The 40 MHz rules conceptually differ from 20 MHz when the primary channel is idle. In this case, CCA for the secondary channel must go busy with a sensitivity of only at least −62 dBm to any signal on the secondary channel. This allows a 40 MHz device to only perform energy detection on the secondary channel. Because a 40 MHz device may only employ energy detect on the secondary channel, in the 5 GHz band it is recommended that an AP not establish a 20/40 MHz BSS where the secondary channel is occupied by a 20 MHz BSS. Conversely, it is recommended that an AP not establish a 20 MHz BSS in the secondary channel of an existing 20/40 MHz BSS.

If there is an existing 20/40 MHz BSS on the desired channels then the AP must ensure that the primary and secondary channels of the new BSS match the primary and secondary channels of the existing BSS. In the 5 GHz band, once the 20/40 MHz BSS has been established the AP may periodically scan the secondary channel for 20 MHz overlapping BSSs.

The coexistence rules in 2.4 GHz are much more onerous because channel numbering is incremented by 5 MHz, causing complicated partial overlapping channel conditions between neighboring APs. Rules were put in place mandating an AP scan for neighboring BSSs prior to establishing a 20/40 MHz BSS and preventing the establishment of a 20/40 MHz BSS when neighboring BSSs are detected in partially overlapping channels. Furthermore, during the operation of a 20/40 MHz BSS in the 2.4 GHz band, active 20/40 MHz bandwidth stations must periodically scan overlapping channels. If conditions change, disallowing 40 MHz operation (e.g., a new 20 MHz BSS appears in an overlapping channel), the AP must switch the BSS to 20 MHz bandwidth channel operation. As a note, complete overlap of a 20 MHz BSS with the primary channel of a 20/40 MHz BSS is permitted because this is equivalent to an overlap between an 802.11g BSS and an 802.11b BSS [9].

Participants in the 802.11n standards development were also concerned that 40 MHz operation would cause too much interference for non-802.11 devices in the 2.4 GHz band (i.e., Bluetooth). A 40 MHz Intolerant bit was created to allow a station to indicate to an AP that it may not operate in 40 MHz. Receipt of a packet with the 40 MHz Intolerant bit set requires the AP to stop operating in 40 MHz. A station may also broadcast the 40 MHz Intolerant bit to overlapping BSSs to force them to stop 40 MHz operation. A system with both 802.11 and Bluetooth radios could use this bit to protect ongoing Bluetooth transmissions.

2.6.3 PROTECTION MECHANISMS

Mechanisms are necessary to protect HT transmissions and certain HT sequences from stations that might not recognize these formats and thus not defer correctly. However, protection mechanisms add significant overhead and an attempt should be made to use these mechanisms only when needed. The 802.11n standard amendment includes many optional features that might require protection in mixed environments.

The first environment is when 802.11b stations are present. These devices will not recognize new 802.11n formats and will not defer access to the medium correctly. Therefore, when 802.11b devices are present in the BSS, the use of request-to-send (RTS)/clear-to-send (CTS) or CTS-to-Self with DSSS/CCK format is required. When 802.11b devices are present in an overlapping BSS, protection is only recommended, just as in 802.11g.

When 802.11g or 802.11a stations are present, the use of the HT mixed format frame inherently provides protection. However, some HT transmissions might not be interpreted correctly by non-HT stations and require protection when legacy devices are present in the BSS. These include RIFS bursting and the Greenfield format. To provide protection for overlapping BSS legacy stations, the AP may indicate that protection should be used when it detects the existence of overlapping BSS legacy stations.

One further requirement exists to protect legacy devices in an overlapping BSS from Greenfield transmissions when operating in a regulatory class that is subject to dynamic frequency selection (DFS) rules with 50 to 100 μsec radar pulses. Prior to establishing a BSS, scanning must be performed for the presence of an overlapping BSS with legacy devices. If such a condition exists, transmission of the Greenfield format is not permitted.

The last environment is when HT stations that do not support Greenfield format are present in a BSS. In this condition, Greenfield format protection is also required.

2.7 BEYOND 802.11N

In mid-2007, the 802.11 community formed the Very High Throughput Study Group (VHTSG) to initiate the investigation of throughput enhancements beyond 802.11n. Initial discussion revolved around the possibility of accessing new microwave spectrum via the International Telecommunications Union's new International Mobile Telecommunications-Advanced initiative. However, it was immediately unclear whether any of the new spectrum under consideration would be available for unlicensed use by systems like 802.11. Due to this uncertainty, discussions in the study group emphasized the 5 GHz bands currently used by 802.11.

To assist in the discussions, 802.11 requested input from the Wi-Fi Alliance (WFA) on future usage models. The report from WFA included usage models in six categories: wireless display, in-home distribution of HDTV and other content, rapid upload and download of large files to and from a server, back haul traffic, campus or auditorium deployments, and manufacturing floor automation [10]. The wireless display category included uncompressed video driving wireless requirements to 3 Gbps video rate, 5 msec of jitter, and 5 msec of delay.

The group agreed on a minimum throughput requirement of 1 Gbps. However, there was lengthy discussion as to how this was to be achieved in the 5 GHz band. As we have seen, 802.11n can achieve approximately 400 Mbps throughput with four antennas and 40 MHz. Requiring enough antennas to achieve 1 Gbps would most likely eliminate mobile devices from being VHT capable. The discussions led to enhancing network capacity to at least 1 Gbps by emphasizing multistation capability [11]. Throughput would be measured as a sum of the throughput across all devices within the BSS. Techniques discussed in the study group included orthogonal frequency division multiple access (OFDMA) and spatial division multiple access (SDMA).

To increase the link throughput of 802.11 in the bands below 6 GHz, the group decided on a single link requirement of at least 500 Mbps. Although theoretically possible with 802.11n, it would require four antennas, 40 MHz, and 83.3% efficiency. 80 MHz channels were postulated, which would enable 500 Mbps with only two antennas for handheld devices.

Because the 5 GHz band would be the primary band of operation, a requirement was levied to ensure backward compatibility and coexistence with legacy 802.11a/n devices. Furthermore, due to the overpopulation of devices in the 2.4 GHz band, operation in the 2.4 GHz band was explicitly excluded from consideration in VHT. This will avoid the difficulties that 802.11n had to face with coexisting with 802.11b/g devices and non-802.11 devices like Bluetooth.

Early on in the VHTSG, strong interest was expressed in considering the 60 GHz unlicensed band. With uncompressed video requiring single-link throughput of 3 Gbps or higher, many felt that the large swath of bandwidth (up to 9 GHz in some regulatory bands) was the only way to achieve such requirements. Much of the difficulty in establishing requirements for 60 GHz was the apparent overlap with other personal area networking (PAN) groups already creating standards for this band.

A primary objection was that range would be very limited in the 60 GHz band, more suited to the PAN standardization group rather than 802.11. However, 802.11 is used quite often in short ranges, especially in dense enterprise deployments or home offices. It was felt that the application and not necessarily just the range should be considered when deciding on what group, PAN or LAN, should address the usage. The other objection to 802.11 creating a 60 GHz standard was potential interference with other 60 GHz standards currently under development. To alleviate this concern, a requirement was created to "provide mechanisms that enable coexistence with other systems in the band" [13].

As an outcome of the study group, two tasks were formed: one to create a VHT amendment to the 802.11 in the <6 GHz band and one in the 60 GHz band. The scope and purpose (which are the guiding requirements for a task group) for the <6 GHz band task group (802.11ac) are given in [12]. The scope and purpose for the 60 GHz band task group (802.11ad) are given in [13]. The goal of each of these task groups is to create an amendment to the 802.11 standard by the end of 2012 addressing their respective requirements.

REFERENCES

1. E. Perahia, "IEEE 802.11n development: History, process, and technology," *IEEE Communications Magazine,* pp. 48–55, July 2008.
2. IEEE P802.11n™/D7.0—Draft Amendment to STANDARD for Information Technology—Telecommunications and Information Exchange Between Systems—Local and Metropolitan Networks—Specific Requirements, Part 11: Wireless LAN Medium Access Control (MAC) and Physical Layer (PHY), Amendment 5: Enhancements for Higher Throughput.
3. G. J. Foschini and M. J. Gans, "On the limits of wireless communications in a fading environment when using multiple antennas," *Wireless Personal Communications,* Vol. 6, pp. 311–335, 1998.
4. G. J. Foschini, "Layered space-time architecture for wireless communication in a fading environment when using multi-element antennas," *Bell Labs Technical Journal,* pp. 41–59, Autumn 1996.
5. E. Perahia and R. Stacey *Next Generation Wireless LANs: Throughput, Robustness, and Reliability in 802.11n*. New York: Cambridge University Press, 2008.
6. T. S. Rappaport, *Wireless Communications. Principles and Practice.* Upper Saddle River, NJ: Prentice Hall, 1996.

7. IEEE Std 802.11™-2007—IEEE Standard for Information Technology—Telecommunications and Information Exchange Between Systems—Local and Metropolitan Area Networks—Specific Requirements, Part 11: Wireless LAN Medium Access Control (MAC) and Physical Layer (PHY) Specifications (Revision of IEEE Std 802.11–1999).
8. S. Alamouti, "A simple transmit diversity technique for wireless communications," *IEEE JSAC*, Vol. 16, pp. 1451–1458, Oct. 1998.
9. E. Perahia, M. Fischer, H. Ramamurthy, et al., "40 MHz coexistence in 2.4 GHz tutorial," IEEE 802.11–08/1360r0, Nov. 2008.
10. A. Myles, and R. de Vegt, "Wi-Fi Alliance VHT Study Group usage models," IEEE 802.11–07/2988r4, March 2008.
11. B. Kraemer and E. Perahia, "VHT SG Report to EC," IEEE 802.11–08/0813r1, July 2008.
12. E. Perahia, "VHT below 6 GHz PAR Plus 5C's," IEEE 802.11–08/0807r4, Sept. 2008.
13. E. Perahia, "VHT 60 GHz PAR Plus 5C's," IEEE 802.11–08/0806r7, Nov. 2008.

3 Mobile WiMAX and Its Evolutions*

Fan Wang, Bishwarup Mondal, Amitava Ghosh, Chandy Sankaran, Philip J. Fleming, Frank Hsieh, and Stanley J. Benes

CONTENTS

3.1 Mobile WiMAX and IEEE 802.16 Standards 54
3.2 Mobile WiMAX Frame Structure 56
3.3 WiMAX Control Channel Coverage 58
3.4 Fractional Frequency Reuse 59
3.5 Multiple-Antenna Technologies in Mobile WiMAX 60
    3.5.1 Open Loop Multiple-Antenna Technologies 60
        3.5.1.1 Space Time Block Coding 60
        3.5.1.2 Cyclic Shift Transmit Diversity 61
        3.5.1.3 Open Loop Spatial Multiplexing MIMO 61
        3.5.1.4 Adaptive Mode Selection Between STBC and Open Loop MIMO 61
    3.5.2 Closed Loop Multiple Antenna Technologies 62
        3.5.2.1 Maximum Ratio Transmission 62
        3.5.2.2 Statistical Eigen Beamforming 63
3.6 Mobile WiMAX System Performance 63
    3.6.1 Mobile WiMAX Link Performance Summary 64
    3.6.2 Link-to-System Mapping in System Performance Evaluation 65
    3.6.3 Mobile WiMAX Downlink System Performance Summary 67
    3.6.4 Mobile WiMAX Uplink System Performance with Multiple Receive Antennas 69
    3.6.5 Mobile WiMAX with Fractional Frequency Reuse 69
    3.6.6 VoIP over Mobile WiMAX 71
3.7 Mobile WiMAX Evolution 71
    3.7.1 802.16m Frame Structure 73
    3.7.2 802.16m Physical Resource Allocation 73
    3.7.3 802.16m Control Channels 74

---

* Part of this chapter was published in the IEEE Communications Magazine [1].

  3.7.4 802.16m MIMO Technologies 75
  3.7.5 Other Advanced Technologies in 802.16m 76
3.8 Summary 76
Acknowledgment 77
References 77

3.1 MOBILE WIMAX AND IEEE 802.16 STANDARDS

The WiMAX forum is an industry consortium promoting the Institute of Electrical and Electronic Engineers (IEEE) 802.16 family of standards for broadband wireless access systems. Historically, the first IEEE 802.16 standard (and associated 802.16c profile definitions) addressed primarily line-of-sight (LOS) environments at high-frequency bands (10–66 GHz) via conventional QAM modulated single-carrier techniques. The limited market potential for mm-wave LOS systems resulted in the development of the IEEE 802.16a amendment to support non-line-of-sight (NLOS) modes in radio bands between 2 and 11 GHz. The 802.16–2004 standard [2] (also known as 802.16d) made more radical changes to 802.16 physical layer (PHY) operations for low-frequency (2–11 GHz) bands by adding two orthogonal frequency division multiplexing (OFDM)-based PHY modes:

- A 256-point FFT (Fast Fourier Transform) OFDM PHY mode.
- A 2,048-point FFT OFDMA (Orthogonal Frequency Division Multiple Access) PHY mode.

The well-understood goal of these developments was to use OFDM to enable relatively simple, high-performance receiver structures in the presence of frequency-selective fading channels.

This new PHY capability was further augmented by several additional features:

1. Frequency-diverse and frequency-specific subchannelization schemes where respective groups of physically distributed and physically adjacent subcarriers are used to construct subchannels. These schemes enable both frequency-diverse and frequency-selective scheduling and resource allocation methods.
2. Adaptive modulation and coding based on Hybrid Automatic Repeat Request (ARQ) (HARQ) techniques (previously used in Third Generation Partnership Project (3GPP)/3GPP2 systems such as Enhanced Data Rate for GSM Evolution (EDGE), High Speed Packet Access (HSPA), Evolution Data Only (EVDO), etc.) along with support for Chase Combining (CC) and Incremental Redundancy (IR).
3. Fast scheduling based on Channel Quality Indication (CQI).
4. New forward error correction schemes including Convolution Turbo Code (CTC) and low-density parity check (LDPC) codes.
5. Support for multiantenna operations including optional Advanced Antenna Subsystem (AAS) modes, open loop Space Time Coding (STC) modes (supporting between two and four transmit antennas), closed loop multiple-input,

multiple-output (MIMO) modes, and uplink coordinated spatial division multiple access (SDMA).
6. Efficient multicast-broadcast transmission schemes using single frequency network (SFN) concepts.
7. Variable frame sizes (e.g., 2 ms, 2.5 ms, 5 ms).

The mobility enhancements provided by the later 802.16e amendment [3] further enhanced operation of nomadic, portable, and mobile wireless access and were published by IEEE at the beginning of 2006. The 802.16e specification (a.k.a. 802.16–2005) provides improved support for intercell handoff, directed adjacent-cell measurement, and sleep modes to support low-power mobile station operation. Another important addition is the introduction into the 802.16e OFDMA PHY of FFT sizes of 128, 256, 512, and 1024 in addition to the original length of 2048. This permits so-called scalable deployment, wherein the OFDM symbol duration and intersubcarrier separation are constant regardless of the carrier bandwidth.

Table 3.1 shows an example of such a carrier bandwidth-scaling process for a 5-ms frame duration, where the cyclic prefix (CP) duration is one-eighth of the useful symbol duration.

The IEEE 802.16 family of standards contains many optional features that are not implemented by the vendors and operators. Two tasks of the WiMAX forum are to decide on a commonly agreed system profile by reducing the number of options in the 802.16 specifications and to promote interoperability among equipment vendors and system operators.

The first release of mobile WiMAX system profile [4] supports time division duplex (TDD) downlink and uplink transmissions. Besides scalable OFDMA, this profile also supports various frequency permutation schemes and multiple antenna technologies. Next, we introduce the mobile WiMAX frame structure followed by WiMAX control channel coverage discussion. We then describe the fractional frequency reuse configuration of mobile WiMAX. Next, we describe multiple-antenna technologies supported in mobile WiMAX. The section after that presents the mobile WiMAX system performance under various configurations and channel conditions and for different data traffics. Finally, we provide an overview of mobile WiMAX evolution.

TABLE 3.1
IEEE 802.16e Scalable OFDMA Parameters

System bandwidth (MHz)	1.25	2.5	5	10	20
Sampling frequency (MHz)	1.4	2.8	5.6	11.2	22.4
FFT size	128	256	512	1,024	2,048
Subcarrier spacing (KHz)	10.94				
OFDM symbol duration (μs)	102.86				
Useful symbol time[a] (μs)	91.43				
Cyclic prefix[a] (μs)	11.43				

[a] Cyclic prefix is 1/8 of useful symbol time according to the mobile WiMAX profile.

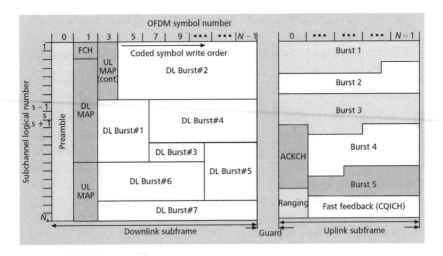

FIGURE 3.1 Mobile WiMAX TDD frame structure.

3.2 MOBILE WIMAX FRAME STRUCTURE

Figure 3.1 shows the mobile WiMAX TDD frame structure. Each frame is configured to be 5 ms long and is time division duplexed into downlink and uplink subframes. There are time gaps between downlink and uplink subframes, including mobile transceiver radio turnaround time and a guard time avoiding interference between downlink and uplink signals. The time gap for the transition from downlink subframe to uplink subframe is called the Transmit Time Gap (TTG). The time gap for the transition from uplink subframe to downlink subframe is called the Receive Time Gap (RTG). The time gaps in the mobile WiMAX profile [4] support a maximum cell size of approximately 20.7 Km for 3.5 MHz or 7 MHz bandwidth, and 8.4 Km for 5 MHz or 10 MHz bandwidth mobile WiMAX systems. The 802.16e standard also specifies several options for mapping OFDM subcarriers to data subchannels so that both frequency-diverse and frequency-selective scheduling can be supported. For frequency-diverse mode of subchannelization, IEEE 802.16e supports full usage subchannelization (FUSC) and partial usage subchannelization (PUSC). Other options include tile usage subchannelization (TUSC) and optional full usage subchannelization (O-FUSC) for downlink, and optional partial usage subchannelization (O-PUSC) for uplink. For frequency-selective mode of subchannelization, IEEE 802.16e supports downlink and uplink band AMC (B-AMC). The minimum resource allocation unit in mobile WiMAX is a time-frequency slot containing 48 modulated data symbols.

At the beginning of each frame, downlink control information is transmitted, consisting of a preamble, a frame control header (FCH) downlink and MAP messages. The first symbol of every downlink subframe is the preamble, which is used for synchronization, downlink channel estimation, and so on. The subcarriers allocated to the preamble are uniformly distributed throughout the spectrum and occupy every third subcarrier.

TABLE 3.2
Mobile WiMAX MAP IEs and Sizes

MAP IEs	Size in Bits
Fixed compressed MAP (DL + UL + CRC)	152
Ranging region allocation IE	
(3 IEs: initial, periodic, and bandwidth request IEs)	168
Fast feedback allocation IE	32
HARQ ACK region allocation IE	56
UL interference and noise level IE	28
Fixed overhead in HARQ DL MAP IE	72
Fixed overhead in HARQ UL MAP IE	64
UL HARQ per scheduled user	40
DL HARQ per scheduled user	44

The first message at the beginning of the second downlink OFDM symbol is the FCH, which provides the information required to decode the subsequent DL-MAP message. The FCH carries information like the subchannels being used by the sector in the current frame, coding, and size of the DL-MAP.

The MAP message indicates the resource allocation for downlink and uplink data and control signal transmission. Each MAP message includes several information elements (IEs) and has a fixed part and a variable part. The size of the variable part is proportional to the number of downlink and uplink users scheduled in that frame. The number of OFDM symbols required for transmitting MAP messages depends on system bandwidth, number of users scheduled in the frame, frame size, IEs included in the MAP, and so on.

Table 3.2 lists several essential IEs that are included in a compressed MAP message. It should be noted that the sizes of control messages in Table 3.2 might be different for different implementations. The compressed MAP messages in Table 3.2 are broadcast to all mobiles in the sector and thus need to be coded properly so that the mobiles at the edge of cells can decode the message correctly. The total MAP overhead in OFDM symbols and the number of OFDM symbols available for transmitting bearer data with 5 or 10 users scheduled per frame for each downlink and uplink transmissions is shown in Table 3.3 for a system with 10 MHz bandwidth allocation. With five scheduled users per frame (e.g., optimal for delay-insensitive data traffics such as FTP and HTTP), the number of symbols available for both downlink and uplink bearer data is 37 if 95% cell coverage is required for downlink MAP (achieved using 4 Tx cyclic shift transmit diversity [CSTD] and a repetition rate of 4; see Section 3.3 for control channel coverage details). This translates into an overhead of approximately 23% without accounting for cyclic prefix and the pilot overhead for bearer data. It may be noted that for Voice over IP (VoIP) traffic, more users will be scheduled per TDD frame (e.g., 15 to 20 users per frame for each downlink and uplink transmission).

Besides the control messages in the compressed MAP shown in Table 3.2 that are broadcast as one packet, the mobile WiMAX system also supports another

TABLE 3.3
Mobile WiMAX Number of OFDM Symbols Available for Bearer Data (10 MHz)

	5 Users Scheduled Per Frame	10 Users Scheduled Per Frame
MAP overhead symbols with Rep = 6	10	12
Other overhead symbols including guard time	5	5
Symbols for (DL + UL) bearer for Rep = 6	33	31
MAP overhead symbols with Rep = 4	6	8
Symbols for (DL + UL) bearer for Rep = 4	37	35

MAP control message transmission mode known as SUB-DL-UL-MAP. Using SUB-DL-UL-MAP, different modulation and coding schemes can be adaptively applied for user-specific information within the MAP message. For users that are scheduled within the frame and having good channel conditions, user-specific control messages can be transmitted using a more efficient modulation and coding scheme and thus the MAP overhead can be reduced [5].

Uplink control channels support ranging, channel quality information channel (CQICH), and ACK/NACK transmissions. Ranging channel provides the random access capability for initial entry, timing adjustment, periodic synchronization, bandwidth request, and handover entry. The uplink CQI or fast-feedback channel is used by a mobile station to report the measured downlink carrier to interference and noise ratio (CINR) back to the access point. This information is used for functions such as selecting the downlink modulation and coding rates. The CQI information is quantized into a 6-bit format and carried over on one uplink slot consisting of 24 subcarriers by three OFDM symbols. Another uplink control channel, the uplink ACK channel, transports ACK/NACK feedback for the downlink hybrid ARQ data transmission and occupies half an uplink slot.

3.3 WIMAX CONTROL CHANNEL COVERAGE

WiMAX system coverage is limited by the control channel in the downlink and data channels in the uplink. The downlink control MAP message is coded with rate 1/2 CTC and modulated using quadriture phase-shift keying (QPSK). Because this message is broadcast to all mobile stations in the system, the CTC coded symbols are repeated one, two, four, or six times so that it can cover the mobile stations at cell edge.

The MAP control channel system coverage is shown in Figure 3.2 with and without CSTD. CSTD is a space-time coding scheme used to achieve spatial diversity without explicit signaling to the mobiles. For the coverage performance shown in Figure 3.2, every mobile station has a maximum ratio combining (MRC) receiver with two diversity receive antennas. Details of the simulation configurations are listed in [6].

Mobile WiMAX and Its Evolutions

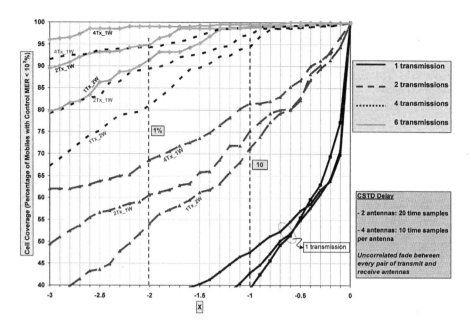

FIGURE 3.2 Cell coverage of mobile WiMAX MAP control channel.

In Figure 3.2, the cumulative distribution function (CDF) of the control channel coverage for various repetition rates using one, two, or four transmit antennas is shown under Typical Urban (TU) channel with 50% of the users at 3 km/hr and 50% at 30 km/hr. It may be observed that more than 95% cell coverage is achieved at an frame error rate (FER) operating point of 1% using CSTD with four transmit and two receive antennas, $R = \frac{1}{2}$ CTC code with a repetition of four. Using more advanced mobile receivers, such as a receiver with interference rejection, the coverage of the MAP control channel can be significantly improved, especially for mobiles with interference that is from one or two dominant interfering signals.

3.4 FRACTIONAL FREQUENCY REUSE

Under the full frequency reuse system configuration, mobiles in different sectors use the same frequency and result in low cell edge throughput due to interference from all their neighboring sectors. By taking advantage of flexible resource allocation on frequencies utilizing OFDMA, mobile WiMAX supports full frequency reuse, partial frequency reuse, and even a mixture of full and partial frequency reuse within one TDD frame. This feature of mobile WiMAX is known as fractional frequency reuse and can provide high cell edge throughput.

In partial frequency reuse, each neighboring sector takes a disjoint part of the frequency band. A typical example is a system with three sectors per cell and each sector takes one third of the frequency band. There is no cross-sector interference between the three neighboring sectors with the partial frequency reuse configuration.

The disadvantage of partial frequency reuse, however, is a reduction in effective bandwidth for each sector.

Fractional frequency reuse takes advantage of the partial frequency reuse benefit of reduction in cross-sector interference for users at the cell edge while avoiding the disadvantage of reduced effective bandwidth in partial frequency reuse. In fractional frequency reuse, mobiles close to cell sites are configured to operate with full frequency reuse for part of the time (known as "zone"), whereas mobiles at the cell edge are configured to operate with partial frequency reuse in another zone for the remaining portion of the subframe.

As a comparison, 3G code division multiple access (CDMA)-based systems typically only operate in a full frequency reuse configuration.

3.5 MULTIPLE-ANTENNA TECHNOLOGIES IN MOBILE WIMAX

Mobile WiMAX supports various multiple-antenna system (MAS) technologies. The MAS technologies can in general be classified into two categories:

- Open loop MIMO.
- Closed loop MIMO.

Open loop MIMO includes space time block coding (STBC or MIMO-A), open loop spatial multiplexing (SM-MIMO or MIMO-B), and adaptive mode selection between the two. Transmitters using open loop MIMO do not require explicit knowledge of the fading channels.

For closed loop MIMO, a transmitter forms antenna beams adaptively based on channel-side information. These technologies are commonly referred to as transmitter adaptive antenna array (Tx-AA) techniques.

3.5.1 Open Loop Multiple-Antenna Technologies

3.5.1.1 Space Time Block Coding

In mobile WiMAX systems, STBC encoding (known as Matrix-A, or Alamouti scheme) is performed on pairs of encoded modulated symbols and occurs immediately after constellation mapping,

$$A = \begin{bmatrix} s_1 & -s_2^* \\ s_2 & s_1^* \end{bmatrix}$$

where each row corresponds to a TX antenna, and each column corresponds to one time interval. Each modulated symbol s_1 or s_2 is transmitted twice, once per antenna, so that the overall space-time coding rate is one. Subcarrier mapping is performed independently for each transmit antenna signal. In addition, the signal from each transmit antenna has orthogonal dedicated pilots.

3.5.1.2 Cyclic Shift Transmit Diversity

CSTD takes advantage of the OFDM receiver by adding spatial diversity from each transmit antenna. With CSTD, each antenna element in a transmit array sends a circularly shifted version of the same OFDM time-domain signal. For example if Antenna 1 sends an unshifted version of the OFDM symbol, then antenna m transmits the same OFDM symbol, but circularly shifted by $(m - 1)D$ time-domain samples. An equivalent operation can be done in the frequency domain by multiplying a phase shifting sequence. Note that each antenna adds a cyclic prefix after circularly shifting the OFDM symbol and thus the interblock interference protection offered by the cyclic prefix is unaffected by CSTD.

Compared to STBC, a receiver for CSTD with short circular delay does not need knowledge of the transmission scheme. Thus, dedicated pilots corresponding to each individual transmit antenna are not required. Because the WiMAX system is based on OFDM, the added multipath interference due to cyclic shift delay from different transmit antennas does not increase the complexity of the mobile receiver but provides additional spatial diversity. However, the spatial diversity gain of CSTD is usually smaller than that of STBC.

3.5.1.3 Open Loop Spatial Multiplexing MIMO

The mobile WiMAX system profile supports allocating two downlink data streams on the same time-frequency resource to one mobile receiver. This scheme (known as Matrix B) consists of encoding a single stream of data and then splitting each pair of modulation symbols between the two antennas. SM-MIMO doubles the peak data rate and improves the system performance when the system is bandwidth limited. However, the achievable system performance improvement of SM-MIMO over single stream STBC depends on system configuration issues such as frequency reuse factor, mobile geometry, traffic type, and so on.

A similar spatial multiplexing scheme is supported in mobile WiMAX on uplink. In this uplink coordinated SDMA scheme, an access point scheduler allocates two uplink data streams on the same time-frequency resource from two mobile transmitters. The access point receiver separates the two data streams utilizing the receiver antenna array. Similar to downlink SM-MIMO, uplink SDMA doubles the peak sector data rate and improves system performance when the system is bandwidth limited.

In detecting the received signal of SM-MIMO, the cross-interference between the two spatially multiplexed signals needs to be mitigated. A simple linear receiver is based Minimum Mean Square Error (MMSE) design. A more advanced nonlinear receiver is based on maximizing log-likelihood detection (MLD) receiver. An MLD receiver provides significantly better performance than an MMSE receiver, but with a higher cost of complexity. Simplified MLD receivers, such as a pseudo-MLD receiver based on list sphere decoding, are typically applied in practice. Later we provide a link performance comparison among MMSE, MLD, and pseudo-MLD receivers.

3.5.1.4 Adaptive Mode Selection Between STBC and Open Loop MIMO

Both STBC and SM-MIMO are downlink open loop MIMO schemes. The criterion to switch between Matrix-A (STBC) and Matrix-B (SM-MIMO) should be based on

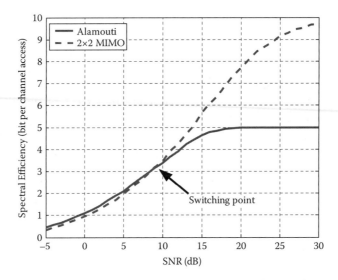

FIGURE 3.3 Downlink sector spectral efficiency comparison with full buffer traffic.

picking the scheme that gives the highest spectral efficiency. Using MMSE receivers, Figure 3.3 shows a comparison of channel capacity of single data stream STBC and a two-stream SM-MIMO over a fading channel. At low SINR, the single data stream STBC outperforms SM-MIMO. More important, STBC is more robust than SM-MIMO with realistic receivers. At high SINR, however, the spatial multiplexing scheme outperforms STBC. This is because at high SINR, the system becomes bandwidth limited instead of power limited.

3.5.2 Closed Loop Multiple Antenna Technologies

The TDD mobile WiMAX system supports closed loop transmission by taking advantage of the reciprocity of the downlink and uplink transmissions. In this chapter, two types of closed loop MIMO schemes are described. In the first scheme, known as maximum ratio transmission (MRT), an antenna beam is formed on each OFDM subcarrier. The second scheme is known as statistical Eigen beamforming (EBF) where only one antenna beam is formed across multiple allocated subcarriers.

3.5.2.1 Maximum Ratio Transmission

MRT is similar to adaptive beamsteering. While adaptive beamsteering phases the transmit array to point the main beam toward the subscriber using knowledge of the subscriber's location, MRT attempts to track the channel response between the transmit array and the receive array on each subcarrier in frequency. MRT tries to maximize the received signal-to-noise ratio (SNR) on each subcarrier at the receiver's antenna array, and generally provides better performance than adaptive beamsteering because it is robust to high angular spread channels. For a mobile station with one receive antenna, the MRT weight $w(i,k)$ on transmitter antenna i and subcarrier k

at the serving base station is obtained from the conjugate of the estimated channel between the transmitter antenna i and the mobile station receive antenna. This may be expressed as $w(i,k) = H^*(i,k)$ where $H(i,k)$ is the estimated channel. The MRT antenna weights can also be normalized so that the transmit power on each antenna is guaranteed to be identical. In this case $w(i,k) = \dfrac{H^*(i,k)}{|H^*(i,k)|}$.

One difficulty of implementing MRT in practice lies in the difficulty of obtaining accurate channel knowledge $H(i,k)$. In mobile WiMAX, the channel knowledge can be obtained through the uplink sounding channel for base stations with calibrated antenna arrays. The calibrated antenna array at base stations guarantees the reciprocity of uplink and downlink transmission channels in a TDD mobile WiMAX system.

One benefit of MRT is that it poses no strict constraints on the characteristics of the TX antenna array. It makes no assumption about antenna placement, polarization, orientation, or even antenna patterns. Although some of these factors can limit ultimate performance, the algorithm attempts to take full advantage of the given physical hardware.

3.5.2.2 Statistical Eigen Beamforming

In Eigen beamforming (EBF), a single set of transmit weights are computed and applied to all subcarriers within a desired band, instead of calculating a separate set of weights for each subcarrier as in MRT. A covariance matrix of the channel, $\mathbf{R}_H$ defined as $\mathbf{R}_H = \dfrac{1}{N}\sum_{k=1}^{N} \mathbf{H}(k)\mathbf{H}^*(k)$ is first computed over the band of interest and the transmit antenna weights are computed from the largest eigenvector of this covariance matrix $\mathbf{R}_H$, and applying it to all the subcarriers within the band.

MRT requires accurate knowledge of the channel and is better for high SINR and low velocity. Statistical EBF is more robust and outperforms MRT for low SINR and for high-velocity mobiles with a large delay between channel measurement and beamforming. Additionally, statistical EBF simplifies downlink channel estimation because the same beamforming weights are applied for all subcarriers within the band of interest.

3.6 MOBILE WIMAX SYSTEM PERFORMANCE

A system simulator based on the mobile WiMAX system profile [4] and system evaluation methodology [7] is used to study the mobile WiMAX system performance with different MIMO technologies. This simulator uses 19 hexagonal three-sector cells. The number of mobile stations in the simulator is adjusted to meet the target outage probability requirement. The cell layout, path loss model, antenna configurations at access points and mobile stations, and traffic models are listed in Table 3.4. More detailed assumptions for system simulations can be found in [8] through [10].

To simplify the system simulations, mobile WiMAX link performance is studied first using link simulators [11]. Then a link-to-system mapping is applied in a system

TABLE 3.4
System Simulation Parameters

Parameter	Value
Number of sectors	19 cells, 3 sectors/cell
Carrier frequency and bandwidth	2.5 GHz, 10MHz
Frequency reuse	PUSC 1/1. PUSC 1/3 and FFR
Propagation model	Path loss (dB): L= 126.2 + 36 log (d), d in km
Lognormal shadowing	8 dB std, 50 m correlation distribution
Number of AP transmit antennas	1/2/4
Number of AP receive antennas	2/8
Number of MS TX/RX antennas	½
MS antenna gain	–2 dBi
Traffic models	Full buffer, Web browsing (HTTP), and VoIP

simulator that replaces detailed modulation/demodulation and coding/decoding processes (see later for more details).

3.6.1 Mobile WiMAX Link Performance Summary

Two types of fading channels are used in link simulations. The 12-ray Typical-Urban (TU) model in [12] is used for baseline performance study for a mobile WiMAX system with a single Tx antenna and two uncorrelated diversity receive antennas. The Spatial Channel Models (SCMs) described in [13] are used to model the multipath channels for antenna arrays used in downlink (DL) beamforming and uplink (UL) spatial division multiplexing.

For downlink channel estimation, a channel estimator based on two one-dimensional MMSE [11] is used with common pilots. For UL transmission, the number of pilot subcarriers for each user is proportional to the number of tiles allocated to that user and can vary dramatically depending on user data allocation. Furthermore, there is subchannel rotation or frequency hopping at every three-symbol interval. These preconditions make accurate channel estimation difficult for the base station receiver. On the other hand, all subcarriers within one tile are contiguous in time and in frequency. Thus, a simple channel estimator is used to measure the frequency response on the four pilot subcarriers of a tile and take the average value to be the estimated frequency response for all data subcarriers within the same tile.

Figure 3.4 shows the block error rate (BLER) performance as a function of Es/No in TU 30 kmph channel using ideal and actual channel estimation. The degradation attributed to channel estimation error is 0.2–0.3 dB. With HARQ using CC, where the retransmitted encoded blocks are identical to the first transmission, Figure 3.5 shows the spectral efficiency (known as hull curves) of different MCS.

Similar link performance curves can be derived for mobile WiMAX uplinks. For a mobile WiMAX base station with 2-Rx antennas and one wavelength spacing between antennas, and carrier frequency of 2.5 GHz, the uplink link performance using two different types of receivers is compared in Figure 3.6.

Mobile WiMAX and Its Evolutions

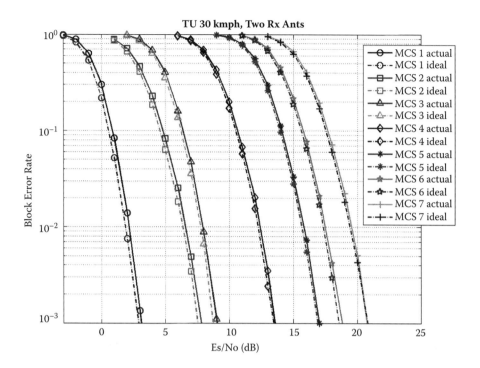

FIGURE 3.4 BLER performance of various MCSs in TU30 channel using ideal and actual channel estimation.

3.6.2 Link-to-System Mapping in System Performance Evaluation

To simplify the system-level performance evaluation, transmitter and receiver details such as modulation/demodulation and coding/decoding are typically not modeled exactly in the system simulation. Instead, a link-to-system mapping is applied to evaluate the link performance in a system simulation. The Exponential Effective SIR Mapping (EESM) is a widely used link-to-system performance mapping method for OFDM systems. In this approach, a mapping is constructed using an exponential function to approximate the constraint capacity. This approach can be derived from the Union-Chernoff bound for BPSK modulation. For higher order modulations, an empirical factor β is introduced to match the link simulation results.

Figure 3.7 describes the basic methodology. In the system simulator, the effective SINR γ_{eff} is calculated from

$$\gamma_{\mathit{eff}} = -\beta \ln \left(\frac{1}{N_u} \sum_{k=1}^{N_u} \exp\left\{ \frac{-\gamma_k}{\beta} \right\} \right) \quad (1)$$

where β is an empirical factor depending on encoding rate and modulation order and N_u is the number of allocated subcarriers. γ_k is the instantaneous received SINR at the kth subcarrier. In the case of single-input, single-output (SISO), γ_k is computed from

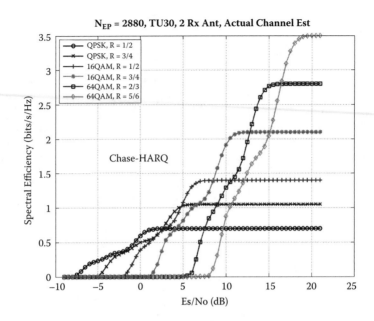

FIGURE 3.5 Downlink spectral efficiency of various MCSs (based on 2,880-bit block size) in Chase HARQ mode in TU30 channel. Spectral efficiency includes the overhead due to pilots and OFDM symbol cyclic prefix.

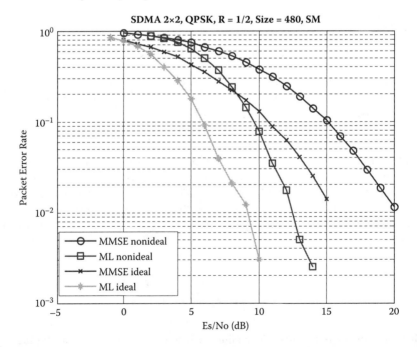

FIGURE 3.6 Mobile WiMAX uplink performance with SDMA, MLD, and MMSE receivers with ideal/nonideal channel estimations (suburban macro SCM channel).

Mobile WiMAX and Its Evolutions

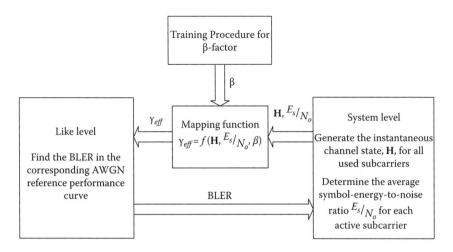

FIGURE 3.7 Basic link-system mapping methodology.

$$\gamma_k = \frac{\hat{I}_{or}}{I_{oc} + N_o} \cdot \frac{N_{used}}{N_d + PDR \cdot N_p} \cdot H_k \quad (3.2)$$

In Equation 3.2, H_k is the channel gain of the kth subcarrier, N_{used} is the total number of subcarriers, PDR is the pilot-to-data subcarrier power ratio, N_d is the number of data subcarriers per OFDM symbol, N_p is the number of pilot subcarriers per OFDM symbol, N_o is the receiver thermal noise power, and I_{oc} is the other-cell noise power density (assumed spatially and temporally uncorrelated) and $\hat{I}_{or}$ is the desired signa power density. Link simulation results for different modulation and coding rates and for different fading channels were used in determining the empirical factor β in Equation 3.1 and to minimize the difference between the predicted BLER and the simulated BLER.

In deriving the BLER in system simulations, the effective SINR γ_{eff} is calculated according to Equation 3.1 for each OFDM symbol. Then the effective SINR γ_{eff} is applied on a reference curve, generated by link simulations using an additive white Gaussian noise (AWGN) channel. The corresponding BLER is the instantaneous BLER for the packet.

Besides EESM, other link-to-system mapping methods can also be used in system evaluations. For example, mutual information-based methods such as Mutual Information Effective SINR Mapping (MIESM) have been proposed to evaluate the 802.16m (see [14] for more details).

3.6.3 Mobile WiMAX Downlink System Performance Summary

The spectral efficiency improvement using open loop MIMO and closed loop beamforming over SIMO depends on factors such as traffic model (full buffer or HTTP), frequency reuse pattern, and so on. Figure 3.8 and Figure 3.9 show the comparison of sector spectral efficiency for various multiple-antenna technologies including

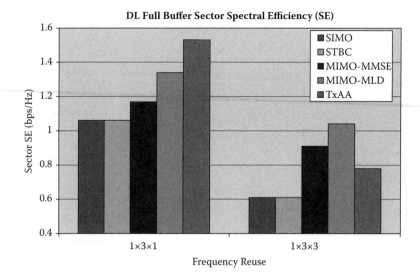

FIGURE 3.8 Downlink sector spectral efficiency comparison with full buffer traffic.

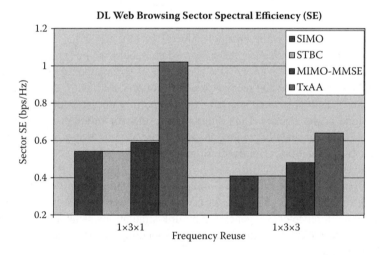

FIGURE 3.9 Downlink sector spectral efficiency comparison with HTTP traffic.

SIMO, STBC, adaptively switching between STBC and MIMO (denoted as MIMO in Figure 3.8 and Figure 3.9), and closed loop beamforming (denoted as Tx-AA in Figure 3.8 and Figure 3.9).

As shown in Figure 3.8 and Figure 3.9, STBC does not improve the system throughput. Although STBC improves the link performance and system coverage, the system throughput does not improve with the additional spatial diversity due to other diversities in mobile WiMAX systems (e.g., receive diversity, multiuser diversity, etc.).

By adaptively switching between single data stream STBC and two data streams SM-MIMO, the WiMAX system performance can be improved over SIMO by about

Mobile WiMAX and Its Evolutions

50% for 1 × 3 × 3 frequency reuse and by 10% for 1 × 3 × 1 frequency reuse using MMSE receivers. With more complex pseudo-MLD receivers, the system performance can be further improved by an additional 15% to 20%.

For 1 × 3 × 1 frequency reuse, the system performance can be further improved using closed loop beamforming. For 1 × 3 × 3 frequency reuse, however, open loop MIMO outperforms closed loop beamforming for full buffer data traffic.

Figure 3.9 shows the performance comparison for HTTP data traffic. It can be observed that the relative performance of various multiple-antenna technologies for HTTP is different from that for full buffer. Closed loop beamforming provides higher system performance improvement for equal data rate types of traffic such as HTTP than with full buffer. Open loop MIMO is more effective for data traffic with large payloads such as full buffer because the scheduler gives preference to users with good channel conditions. These full buffer users are bandwidth limited and can use the extra bandwidth through spatial multiplexing.

3.6.4 Mobile WiMAX Uplink System Performance with Multiple Receive Antennas

In this section, the performance of MU-MIMO and SU-MIMO is analyzed for two types of access points with two and eight receive antennas, respectively. For the first type of access points, the two receive antennas at the access point are four wavelengths apart. For the second type of access points, each one has a linear array that contains eight antenna elements and half of a wavelength between two adjacent antenna elements. In the case of SDMA, the scheduler assigns one or two mobiles to transmit on each time-frequency resource at each scheduling instance. Similar to downlink MIMO, the peak sector data rate of uplink transmission can be doubled using SDMA.

Uplink sector spectral efficiencies using MRC and SDMA with MMSE receivers are shown in Figure 3.10. We note that for 1 × 3 × 3 frequency reuse, the system is bandwidth limited but not power limited. In this case, even though the combined signal power from multiple receive antennas is increased by increasing the number of receive antennas, the uplink system performance is only slightly improved in the SIMO mode. However, by allocating two users to transmit simultaneously using SDMA, the system performance can be improved by more than 80% using the second type of access points with eight receive antennas. In the case of 1 × 3 × 1 frequency reuse, the system performance improvement of SDMA over MRC using the second type of access points is around 20%.

3.6.5 Mobile WiMAX with Fractional Frequency Reuse

Figure 3.11 shows the relative improvement with fractional frequency reuse when compared to full frequency reuse (PUSC 1/1) configuration and partial frequency reuse (PUSC 1/3) configuration. The result is for mobile WiMAX systems with 10 MHz bandwidth, a load of 150 users per sector and HTTP traffic. Cell edge throughput is defined as the minimum data rate achieved by 95% of users in the

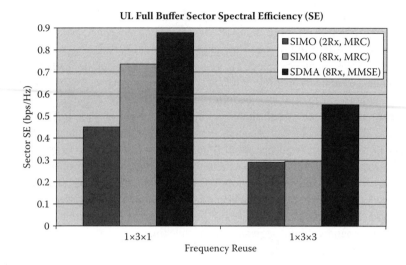

FIGURE 3.10 Uplink sector spectral efficiency comparison with full buffer traffic.

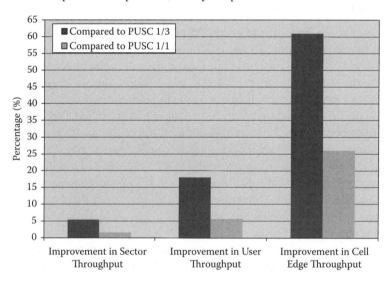

FIGURE 3.11 Relative improvements of fractional frequency reuse over PUSC 1/3 and PUSC 1/1.

sector. The results show that fractional frequency reuse outperforms the full and partial frequency reuse configurations, especially for cell edge throughput.

To get good cell edge reliability, it is preferable to send the MAP message in a PUSC 1/3 zone. The fractional frequency reuse configuration would be a good choice in this kind of deployment because it can combine both the PUSC 1/3 zone and a PUSC 1/1 zone in the same downlink subframe to provide a good cell edge coverage and cell edge throughput performance, while not compromising on the overall sector throughput and average user throughput.

Mobile WiMAX and Its Evolutions

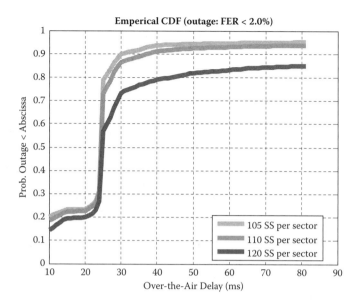

FIGURE 3.12 VoIP performance of mobile WiMAX systems.

3.6.6 VoIP over Mobile WiMAX

VoIP capacity of mobile WiMAX is studied using the system simulator, and is shown in Figure 3.12. Because the system capacity for VoIP-only service is uplink limited, the simulation results in Figure 3.12 are for uplink system performance with MRC receiver and two diversity receive antennas at each access point. The VoIP model in Figure 3.12 is 12.2 Kbps adaptive multi-rate (AMR) vocoder with 50% voice activity factor (see [7]). A VoIP user is in outage if less than 98% of its packets are delivered successfully within a given delay bound. System VoIP capacity is defined as the number of users that can be supported in each sector such that the percentage of users in outage is less than a certain criterion. From Figure 3.12, it can be observed that with a load of 105 active VoIP users in one sector and an over-the-air delay request of 50 ms, the system outage is less than 5%. With a higher system load of 110 active VoIP users in one sector, the system outage increases to 8%. With an even higher system load of 120 active VoIP users in one sector, the system outage increases to almost 20%.

An alternative analytic methodology described in [15] is also used to study the packet queuing delay. The result from [15] substantiates the VoIP capacity from system simulations. Further, the result from [15] shows that to meet the delay request of VoIP service, the system utilization is around 80%. This observation implies that more delay-insensitive services (such as HTTP or FTP) can be added without affecting the VoIP capacity shown in Figure 3.12.

3.7 MOBILE WIMAX EVOLUTION

Since early 2007, the WiMAX forum and the IEEE 802.16 working group have started separate evolution projects to improve the performance of the current release of mobile WiMAX and to keep the momentum of evolving mobile WiMAX as a leading mobile broadband wireless communication solution.

In the WiMAX forum, the evolution project is known as mobile WiMAX release 1.5, whereas the current mobile WiMAX release is 1.0. The focus of this project includes adding frequency division duplexing (FDD) configuration in the mobile WiMAX profile, enabling MIMO/BF under FDD mobile WiMAX, adding persistent allocation to improve the efficiency for VoIP data traffic, reducing MAP control message overhead, and reducing handoff latency. The target of this project is to provide performance enhancement to current mobile WiMAX systems within a short time period (targeting deployment in late 2009) and without introducing significant changes to the IEEE 802.16–2005 specification.

The project that introduces more fundamental performance enhancements to mobile WiMAX is led by the IEEE 802.16 working group. This project and the associated future standard are known as 802.16m. The target for 802.16m is to meet the requirements of IMT-Advanced, the 4G successor of IMT-2000. In other words, 802.16m will be the 4G mobile WiMAX evolution. According to the system requirement document (SRD) of 802.16m [16] the key performance targets for 802.16m include the following:

- 802.16m system should be backward compatible with current mobile WiMAX systems.
- 802.16m system should provide more than twice the spectral efficiency compared to the current mobile WiMAX release 1.0 for both downlink and uplink for metrics like average sector throughput, average user throughput, and cell edge user throughput.
- 802.16m system should provide more than 1.5 times the VoIP capacity of the current mobile WiMAX release 1.0.
- 802.16m system should provide lower latency than the current mobile WiMAX release 1.0.

The 802.16m evolved mobile WiMAX profile is targeted to finish during the fourth quarter of 2009, and to be deployed in 2011. More importantly, 802.16m is expected to be one of the candidate technologies (along with LTE-Advanced from 3GPP) for the IMT-Advanced radio interface and will be submitted to the International Telecommunication Union (ITU) in 2009. The IEEE 802.16 Task Group m (TGm) is working actively to develop the new 802.16m standard with a focus in the following areas:

- Improved and backward-compatible frame structure and system protocol.
- Smaller frame size to reduce latency.
- New multiantenna technologies.
- Improved interference coordination and management schemes for both downlink and uplink.
- New control channel design with better system coverage and reduced overhead.
- Persistent scheduling for VoIP and real-time video services. Optimized handover, more efficient paging and random access, and so on.

Mobile WiMAX and Its Evolutions

Because 802.16m is still being actively developed by the IEEE 802.16 working group, we describe only several key technology improvements of the 802.16m over the current mobile WiMAX system in this section. These technologies may be evaluated and further improved with the development of the 802.16m standards.

3.7.1 802.16M FRAME STRUCTURE

To reduce the latency* of current mobile WiMAX system and also the control channel overhead, the 802.16m frame structure is based on superframe/frame/subframe hierarchical structure. 802.16m frame duration is still 5 ms and thus keeps the backward compatibility with the current mobile WiMAX system. 820.16m introduces two additional frame structure units: a superframe and a subframe. A superframe is defined as a time interval of 20 ms made up of four consecutive 5-ms frames. A subframe duration is typically one eighth of a frame.

Deployment-wide and sector-specific system information like system bandwidth, TDD ratio, and subframe configuration that do not change rapidly but are essential for network access are broadcast at the beginning of a superframe in the form of a superframe header. The broadcast information is protected by strong modulation and coding so that its coverage does not limit the system coverage. More time-sensitive information related to synchronization, cell ID, and measurement pilots are broadcast every frame. Further, user-specific control information such as resource allocation information, acknowledgments, and power control commands can be transmitted in a subframe and through unicast to specific users.

Usage of subframes and multiple downlink/uplink switching points within a 5-ms frame are introduced to reduce the latency of an 802.16m system compared to the current mobile WiMAX system. The overhead (control information) for supporting subframes may be reduced by subframe bundling or transmitting control information in only selected subframes.

The 802.16m frame structure is backward compatible with legacy mobile WiMAX systems. The downlink transmissions to legacy mobiles and to 802.16m mobiles can be time division multiplexed (TDM), and uplink transmissions from legacy mobiles and from 802.16m mobiles can be either TDM or frequency division multiplexed (FDM). The frame structure of 802.16m supports coexistence with other cellular communications standards including EUTRA (TDD-LTE) and UTRA LCR TDD (TD-SCDMA). The efficient coexistence is supported by (1) a configurable time offset that may be introduced between the radio frames of 802.16m and another standard and (2) puncturing of OFDM symbols within the 802.16m frame for minimizing the interference between the two systems. The basic 802.16m frame structure is shown in Figure 3.13.

3.7.2 802.16M PHYSICAL RESOURCE ALLOCATION

Resource structure or subchannelization defines a mapping of logical data subchannels or control channels for allocation to physical resource (OFDM symbols and

---

* Latency specifically refers to user-plane latency and is loosely defined as the time interval between data being available at the IP layer in the BS (or MS) for transmission and being available at the IP layer in the MS (or BS) after reception.

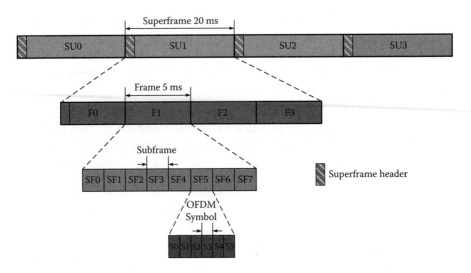

FIGURE 3.13 802.16m basic frame structure. (From IEEE 802.16 Broadband Wireless Access Working Group. 2009. IEEE 802.16n system description document. Used with permission.)

frequency subcarriers). 802.16m supports both frequency-diverse and frequency-selective subchannelization schemes similar to the current mobile WiMAX. A primary difference, however, is that 802.16m supports both frequency-diverse and frequency-selective subchannelization within a same OFDM symbol, whereas the current mobile WiMAX partitions a frame into multiple zones in time and applies different subchannelization schemes in each partition. Additionally, 802.16m allows the usage of any combination of multiple-antenna techniques within an OFDM symbol, whereas the current mobile WiMAX does not allow the usage of open loop (Matrix-A/B) and closed loop (TxAA) antenna techniques in the same OFDM symbol.

While a resource allocation unit (a slot) in the current mobile WiMAX contains 48 modulation symbols, the resource unit of 802.16m spans over 18 subcarriers in frequency and five, six, or seven OFDM symbols. The exact number of modulation symbols within a resource unit depends on the pilot format and the length of the subframe. One important improvement of 802.16m over the current mobile WiMAX is the reduction of pilot overhead (see Table 3.5).

3.7.3 802.16m Control Channels

An important goal for 802.16m is to improve downlink control channel coverage and to reduce uplink control channel overhead in current mobile WiMAX systems. In 802.16m downlink, user-specific control information (analogous to MAP of mobile WiMAX earlier) is coded separately and transmitted using dedicated control channels. This is similar to the LTE control channel design and is in contrast to the current mobile WiMAX, where the downlink control information for multiple users is transmitted using a broadcast control channel. This, along with user-specific power control and MCS, is designed to extend the 802.16m control channel coverage without excessive overhead. In addition to this improvement, the downlink control information is structured hierarchically to

TABLE 3.5
Comparison of Pilot Overhead of Current Mobile WiMAX and 802.16m

# Stream (DL/UL)	Resource Allocation	Pilot Overhead Mobile WiMAX (%)	Pilot Overhead 802.16m (%)
1-stream DL	Distributed	14.28	5.55
	Localized	11.11	5.55
2-stream DL	Distributed	14.28	11.11
	Localized	11.11	11.11
3-stream DL	Distributed	—	11.11
	Localized	22.22	11.11
4-stream DL	Distributed	28.57	14.81
	Localized	22.22	14.81
1-stream UL	Distributed	33.33	11.11
	Localized	11.11	5.55
2-stream UL	Distributed	33.33	22.22
	Localized	11.11	11.11
3-stream UL	Distributed	—	—
	Localized	—	11.11
4-stream UL	Distributed	—	—
	Localized	—	14.81

fit physical control channels at different time scales (superframe, frame, and subframe level) and to provide efficient network entry and coverage (see earlier).

In the uplink, 802.16m control channels extend a control channel physical resource unit (control channel minitile) to six OFDM symbols and have less pilot overheads than current mobile WiMAX.

3.7.4 802.16M MIMO TECHNOLOGIES

The multiple-antenna transmission options in 802.16m can be described by

$$Z = WS$$

where Z represents the transmitted symbols ($M_t \times M_s$), W is a precoding matrix ($M_t \times M_s$), and S is a symbol vector ($M_s \times 1$). M_t is the number of transmit antenna elements and M_s is the number of data streams. The precoding matrix W is chosen arbitrarily at the transmitter in the case of open loop and is determined based on user feedback in the case of closed loop. In addition, a 2×2 Alamouti-matrix-based space frequency block coding (SFBC) (a transmit diversity mode) is also supported in 802.16m.

1. Control channel transmission: The downlink broadcast control channels may be transmitted using 2-Tx SFBC. This is in contrast to current mobile WiMAX, which only supports 1-Tx transmission or CSTD for downlink control channels.

2. Single-user data transmission: 802.16m supports up to eight data streams in the downlink and up to four data streams in the uplink for both single-user open loop and closed loop transmissions. This is targetted to improve the peak spectral efficiency. In downlink, 802.16m is expected to support open loop MIMO and closed loop MIMO with sounding feedback or codebook-based feedback. In uplink, 802.16m also supports open loop MIMO and codebook-based closed loop MIMO.
3. Multiuser data transmission: Multiuser MIMO is supported in both uplink and downlink with up to four mobiles sharing a time-frequency resource. This feature is targeted to boost the throughput performance of the system. In the case of downlink, the transmit weights for MU-MIMO can be determined from sounding signals or from codebook feedback from the mobiles. Each mobile is restricted to a single data stream and multiple mobiles are allocated to share the multiple data streams over the same time-frequency physical resource. A mobile can use dedicated pilots to estimate the cross-interference among data streams and employ interference supression receivers such as MMSE or MLD. In the case of uplink, both open loop and codebook-based closed loop MU-MIMO are supported.

3.7.5 Other Advanced Technologies in 802.16m

802.16m supports several additional advanced features. Details of these advanced features are still being discussed in the IEEE 802.16 working group. These advanced features include the following:

- Multicarrier operation where a base station utilizes multiple carrier frequencies for resource allocation.
- Coordination among multiple base stations through higher layer messages for downlink MIMO transmissions.
- Enhanced multicast and broadcast services by multiple coordinated base stations transmitting the same information using open loop MIMO techniques.
- Efficient estimation and reporting of a mobile location for location-based services (LBS).
- Multihop relays to improve the coverage of a system.
- Femtocell base station (a low-power, low-cost base station) for providing local wireless service.

3.8 SUMMARY

Mobile WiMAX has spurred tremendous interest from operators seeking to deploy high-performing yet cost-effective broadband wireless networks. This chapter provided a detailed overview of the mobile WiMAX system and summarized its coverage and performance capabilities for various features, including different types of MIMO schemes, fractional frequency reuse schemes, receiver structures, under different workloads, and different channel conditions.

With the ongoing phased deployment of the first release of mobile WiMAX systems in the U.S. and globally, the wireless industry has started the evolution process toward more advanced mobile WiMAX systems. One evolution is known as mobile WiMAX 1.5, which was developed by the WiMAX forum, and the corresponding standards evolution at IEEE is known as the 802.16Rev2. A more important evolution is the 802.16m that is being developed by the IEEE 802.16 working group. This chapter highlighted several key requirements and techniques of the 802.16m. It is expected that 802.16m will be a candidate for the IMT-Advanced technology that will be evaluated by the ITU.

ACKNOWLEDGMENT

We would like to acknowledge Dr. R. Ratasuk, Dr. W. Xiao, Lisa Whitelock, and Rick Keith for their comments and suggestions.

REFERENCES

1. F. Wang, A. Ghosh, C. Sankaran, P. J. Fleming, F. Hsieh, and S. J. Benes, "Mobile WiMAX systems: Performance and evolution," *IEEE Communications Magazine*, Vol. 46, pp. 41–49, Oct. 2008.
2. IEEE 802.16–2004, "IEEE Standard for Local and Metropolitan Area Networks—Part 16: Air Interface for Fixed Broadband Wireless Access Systems," Oct. 2004.
3. IEEE 802.16–2005, "Part 16: Air Interface for Fixed and Mobile Broadband Wireless Access Systems Amendment 2: Physical and Medium Access Control Layers for Combined Fixed and Mobile Operation in Licensed Bands and Corrigendum 1," Feb. 2006.
4. WiMAX Forum, WiMAX Forum Mobile System Profile Release 1.0, revision 1.5.0, Nov. 2007.
5. WiMAX Forum, Mobile WiMAX—Part I/II: A Technical Overview and Performance Evaluation, Feb. 2006.
6. F. Wang, A. Ghosh, C. Sankaran, and P. Fleming, "WiMAX overview and system performance," presented at the IEEE Vehicular Technology Conf. (VTC), Montreal, Canada, Sept. 2006.
7. IEEE 802.16 Broadband Wireless Access Working Group, IEEE 802.16m Evaluation Methodology, Dec. 2007.
8. F. Wang, A. Ghosh, R. Love, K. Stewart, R. Ratasuk, R. Bachu, and Y. Sun, "IEEE 802.16e System performance: Analysis and simulations," presented at the 16th Annual IEEE International Symposium on Personal Indoor and Mobile Radio Communications (PIMRC), Berlin, Germany, 2005.
9. F. Wang, A. Ghosh, C. Sankaran, and S. Benes, "WiMAX system performance with multiple transmit and multiple receive antennas," presented at IEEE Vehicular Technology Conf. (VTC), Dublin, Ireland, April 2007.
10. C. Sankaran, A. Ghosh, and F. Wang, "Performance Study of a Fractional Frequency Reuse (FFR) Scheme for an 802.16e Based WiMAX System," preprint, 2008.
11. F. Hsieh, F. Wang, and A. Ghosh, "Link performance of WiMAX PUSC," presented at the IEEE Wireless Communication & Networking Conf. (WCNC), Las Vegas, Nevada, April 2008.
12. 3GPP TS 45.005, "Radio transmission and reception," 3GPP Technical Specification Group GSM/EDGE.

13. 3GPP/3GPP2 SCM Ad-Hoc Group, "Spatial channel model text description," v6.0, April 2003.
14. IEEE 802.16 Broadband Wireless Access Working Group, IEEE 802.16m Evaluation Methodology Document (EMD), Jan. 2009.
15. P. J. Fleming and B. Simon, "A fluid model of VoIP over wireless broadband," in *Proc. 45th Allerton Conf. on Communications, Control and Computing,* Urbana-Champaign, Illinois, 2007.
16. IEEE 802.16 Broadband Wireless Access Working Group, IEEE 802.16m system requirements, Jan. 2009.
17. IEEE 802.16 Broadband Wireless Access Working Group, IEEE 802.16m system description document, Feb. 2009.

4 Service Layer for Next-Generation Digital Media Services

Abhijit Sur and Joe McIntyre

CONTENTS

4.1 Introduction 79
4.2 Changing Trends of Consumer Services: Consumer Picture 80
    4.2.1 Context-Aware Services..81
    4.2.2 User-Generated Content 82
    4.2.3 Social Networking 83
4.3 Service Delivery Platform for Communication Service Provider:
CSP Picture...83
    4.3.1 Service-Oriented Architecture 84
        4.3.1.1 How to Incorporate Digital Media in a Service-Oriented
Architecture 85
        4.3.1.2 Two Technologies for SOA: SOAP and REST 87
    4.3.2 Service Delivery Platform 88
    4.3.3 Web 2.0 and Telecom..89
4.4 Standards 90
    4.4.1 Major Standards Activities 91
4.5 Authentication and Authorization Considerations.......................................91
4.6 Illustrative Call Flows: Putting It All Together 92
    4.5.1 Case Study 1: Upload Content into Your Digital Locker Account
through MMS 93
    4.5.2 Case Study 2: Example of Mashup Telco Service 94
4.7 Business Model Considerations for Service Layer 96
4.8 Conclusion 97
References..97

4.1 INTRODUCTION

Over the last several decades, the telecommunications industry has experienced dynamic expansion in its scope and reach. Telecommunication service providers have seen their competitive landscape evolve in both technology and business competition. Until recently, an average telco simply competed with other telcos. With changing regulation in several countries, they began to compete with cable service providers that also provide communication services. Now they have to compete with Internet service providers that can also provide voice communication. *Convergence* has played a major role in changing the competitive landscape for telecom carriers. Convergence does not just refer to consolidation among companies, but it also includes convergence between different technologies or industries, such as traditional telecom, media and entertainment, digital media, and Internet-based service providers. Even though the objective of this convergence was to provide better service to the consumer, the reverse is equally true as well. Demand for better services is further driving this convergence.

In other words, the role of an average telecom service provider has evolved from the mere bit-pipe provider to more of a service provider. Hence this question is presented: How does a telco stay competitive in this changing landscape and generate revenue? The landscape gets even more complicated because telecommunication service providers can be further classified into mobile service providers and those that focus solely on wireline or landline services. If that is not enough, we now have companies that are known for leadership in Internet technologies (e.g., Google™) venturing into providing wireless access to their customers. The last mile that provides the final access to the customer is often considered the weakest link of the end-to-end chain. If that is the case, the service layer that resides on top of the access layer can actually make it the strongest one as well—as the deciding factor for today's customer on which service provider should be chosen over the other.

This chapter discusses the service layer that resides on top of the access layer. This layer deals with services that are finally delivered to customers. We discuss the services, service enablers, and service delivery platforms that need to be executed at the edge. We illustrate how a service delivery platform at the edge can leverage the capabilities of the core network and deliver rich services to consumers. Even though this section focuses on wireless technology, the concepts discussed in this chapter can be extended to nonwireless technologies as well. Finally, as a convention, we use the term *communication service provider* (CSP) when referring to a telco service provider that is providing the access technology (or the last mile) to the end customers for next-generation services.

4.2 CHANGING TRENDS OF CONSUMER SERVICES: CONSUMER PICTURE

Voice communication has been the basic form of communication, and it will continue to remain so. However, today's average consumer is aware of the pervasive nature of data communication, so much so that, Tim O'Reilly [1] refers to data as "the Next Intel Inside." It's no wonder that we have seen a growing trend toward datacentric services. A datacentric service is more than just a simple short message service (SMS) or a multimedia message service (MMS). To understand the role of data in the rich services of the next-generation network, let us first define the hierarchy of services.

At the most basic level, we have the *foundation service* (as shown in Figure 4.1), a basic service that can be sold in its own right. Examples are basic telephony, instant messaging, and streaming video. *Service enablers* are deployed as common reusable

Service Layer for Next-Generation Digital Media Services

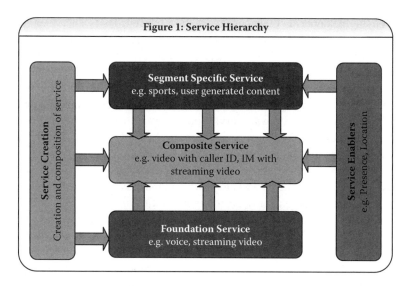

FIGURE 4.1 Service hierarchy.

elements that can be deployed across multiple services. They ensure a common ground between services and ensure cost reduction for service providers through reuse. Examples are presence, location, and the network address book. Finally, *composite service* refers to a combination of two or more foundation services, and/or service enablers. It is important to note that there has to be at least one foundation service in a composite service. Examples include watching streaming video with caller ID information or instant messaging with friends who are watching the same video. Composite services can be developed to meet segment-specific requirements (for interest group such as online gaming) as well. Based on the preceding classification, data services such as SMS and MMS can be considered foundation data services. However, we are interested in composite services wherein data plays a role.

Before we delve into discussing the architecture framework that will be useful for the delivery of such composite services, let us look into three growing trends that will shape the type of services a CSP would be expected to deliver:

- Context-aware applications.
- Consumers becoming producers of content.
- Social networking.

We discuss each one of these three trends in detail and the role that they are playing in changing the service layer for a CSP.

4.2.1 Context-Aware Services

With the rise in consumption of digital media, there is an increasing need to personalize the content to the customers' needs and requirements. This trend is not entirely new. For instance, many e-commerce Web sites recommend certain products based on the buying behavior of their customers. In a telecom landscape, a service can be personalized based on what we refer to as context of the current situation of the customer. As mentioned in the paper "SOA-Based Context Aware Services Infrastructure" [2], context can be associated with a user profile (such as teenager), session (type of session, such as instant messaging, or type of content used in the session), type of content, or a service-enabling capability such as location and presence information. A context-aware service incorporates one or more of the core contextual capabilities mentioned earlier as part of a business service. For instance, a user wants to download a 30-minute episode of her favorite TV sitcom. Let's say the user subscribes to the IPTV (Internet Protocol Television) and mobile service from the telecom carrier. When she wants to download the episode (say using her mobile portal), the content automatically gets downloaded to the IPTV set-top box. This example demonstrates how we can use the content characteristics for a service.

Context-aware capabilities need not be restricted to services alone. They can be used for mobile advertisements. For instance, if a user prefers to receive ads and promotions on golf resorts, the user will receive promotional ads from neighboring golf resorts whenever he is in a new neighborhood. This example uses location as the context attribute. We discuss how a user can build (using available tools) situational or contextual applications in order to meet specific requirements in a later section of this chapter.

4.2.2 User-Generated Content

Traditionally, entertainment companies have been the sole producers of digital media content meant for consumption by the general public. The recent rise of YouTube™ indicates the extent to which users are sharing their personal content (photographs and video). Of course this is related to the growing trend of social networking (which we discuss in the next subsection), but the point to note here is to assess the opportunities and challenges of allowing consumers to produce content that is meant for public consumption.

User-generated content (UGC) is typically offered through a digital locker or vault service that can be offered to customers. There are several application service providers who provide this service for free or for a nominal fee, allowing their customers to upload and store, share, and download their personal digital content. A telecom carrier can also offer a similar service to customers or collaborate with other external service providers providing similar services. The trusted relationship between a telco and its customers is a very key element here. A telco can leverage this aspect to provide a safe digital vault for storing customers' personal documents as well (say as a backup service). An average mobile handset today can take decent photographs and short video clips. A mobile carrier can provide a capability to upload such user content directly to a personal storage locker through an MMS (either through a short code or an e-mail address). The paper "Extending the Service Bus for Successful and Sustainable IPTV Services" [3] discusses certain aspects of a digital locker service when offered by an IPTV carrier—for example, using UGC of local communities to be broadcast (through a video on demand server) for local news.

4.2.3 SOCIAL NETWORKING

There has been a recent shift in our communication trends. Traditionally, communication meant a two-way, point-to-point voice conversation. In more recent times we have seen communication becoming more collaborative in nature and becoming a many-to-many style of conversation. Additionally the many-to-many communication is not what we would think of a voice teleconference, but rather a rich multimedia experience that involves voice and exchange of digital media. All these trends are reflective in the rise in popularity of the new trend, commonly referred to as social networking (SN). SN can be done in several modes—through multiperson chat or with the use of blogs and wikis, wherein users share their content and collaborate with their inputs and feedback. Some figures reported in the "The Changing Face of Communication" [4] are worthy of mention here. Before 2005 not a single SN site was ranked among the world's top 20 English-language Web sites. However, by June 2008 SN sites (e.g., YouTube, Facebook™, MySpace™) made up half of the same list.

We discuss the architecture platform in more detail in the next section; however, an appropriate question at this juncture is this: What does SN mean for a telecom carrier? How do all these evolving trends impact a telecom carrier, and more specifically a mobile carrier? Telecom carriers can either compete with the external SN sites (by providing similar sites to their own customers) or collaborate with those external SN sites that are already popular with their customers. A most likely answer would perhaps be a happy medium between the two. A telecom carrier can also add value in their partnership in the following ways:

1. Provide context-aware capabilities (presence, location), including context-aware advertisements.
2. Provide messaging (SMS, MMS) capabilities among the users. Together with these features we have the evolving trend of mobile SN.
3. Leverage their distribution network to provide higher performance for distributing digital content by caching the content in the edge of the network (VHO [Video Hub Office] and CO [Central Office]).

These three factors add value in a joint collaboration between a telecom carrier and the external SN site. However, a telco can use social networking for its operations as well. SN can be used as a means to employ "viral marketing," allowing users or groups of users to spread the good word about a recently introduced service by a carrier. Similarly, if users do not like a service, their feedback can be useful to a service provider in modifying or removing the service completely.

4.3 SERVICE DELIVERY PLATFORM FOR COMMUNICATION SERVICE PROVIDER: CSP PICTURE

There has been a gradual shift in how CSPs have provided services to their customers. Initially they provided different network-level services through their proprietary TDM-based infrastructure. Next came a walled garden approach wherein a CSP collaborated with application service providers and hosted those applications for their customers on their network. In the next stage a CSP exposed its core network capabilities to its partners, which used these capabilities to provide rich composite services. These services were no longer needed to be hosted on the CSP's network (although they could be). In the next stage we are seeing how a CSP is going to expose these services to its users to develop mashup applications themselves. This evolution has led to a few significant changes in how a CSP provides services:

- Time to deploy services has been reduced significantly.
- A flexible open-standards-based architecture can expose network capabilities as loosely coupled reusable services.

In this section we discuss the architecture framework that needs to be deployed by a CSP for providing next-generation services. We discuss a few key concepts such as service-oriented architecture (SOA), Web services, and service delivery platform (SDP).

4.3.1 Service-Oriented Architecture

The competitive landscape of the CSPs has broadened significantly as the scope of services has reached into the Internet, data services, and personalization. There is an increasing demand for CSPs to be able to differentiate in these areas to achieve greater revenue and customer loyalty together—while reducing operation costs. CSPs need to be able to offer new services faster than ever—both to attract new customers and keep existing customers happy. These factors drive the need for better alignment of the business, IT, and network organizations to enable greater flexibility and coordination in the definition, deployment, and delivery of new services. SOA represents an integration technology that allows business logic to be encapsulated into atomic units of logic, which we refer to as services. Each service is reusable and loosely coupled with other services, while maintaining autonomy and control over the business logic that it encapsulates. Adopting SOA principles and supporting an SOA framework for integration enables CSPs to deploy a loosely coupled and federated architecture relying on reusable services and assets. Detailed concepts of designing an SOA-based model can be found in [5].

The concept of Web services is closely tied with SOA and often SOA is considered to be "intrinsically reliant" [5] on Web services; however, it is important to understand that Web services are a means to implement the service orientation within an SOA model. Equally important is to note that simply using Web services does not imply that that an architecture platform is following the SOA model. The concept of Web service stemmed from a drive to steer away from proprietary or difficult-to-deploy protocols, and to enable a common approach for service definitions, protocol integration, and use of Internet standard capabilities for routing, security, and transport. Thus a Web service would rely on a standards-based open interface for communication. This would enable it to get identified and invoked during the service execution process. A designer has to decide the granularity of a Web service. A simple example might make this clear. For instance, a CSP wants to offer MMS service to its prepaid customers. Instead of embedding the business logic of a *balance check* within the MMS service, it can design the balance check part as a separate Web service that could be invoked by the MMS service. Doing so will enable the CSP to use this feature in other services as well. Web services can communicate with one another in two ways—Simple Object Access Protocol (SOAP) and Representational State Transfer (REST). We briefly discuss these two communication mechanisms later.

A key component in the implementation of SOA is the service bus that glues the different services together. At the most basic level, a service bus connects service requestors and providers of that service. A service bus also enables publishing and discovering available services (registry lookup) during mediations of messages. This is essential for dynamic routing of messages. To provide the mediation and interaction of messages, a service bus needs to perform conversion (message transformation) between protocols (SOAP/HTTP, XML/MQ) and interaction patterns. The role of a service bus is crucial for choreographing a composite service. The combination of these three features—registry lookup, dynamic routing, and message transformation—enables another crucial feature of SOA—*service virtualization*. Service virtualization provides the capability of selecting the actual service provider at runtime based on the static service characteristics (e.g., quality of service, cost) and real-time performance data (e.g., CPU utilization, disk space, queue length for one service provider). In a nutshell, while executing a composite service (service choreography), a service bus has to ensure that messaging properties—persistence, transformation, and routing—are strictly enforced.

In conclusion of our introduction to SOA, we would like to summarize why we need a Web-services-based SOA approach as our integration platform:

1. Promotes *interoperability* due to open-standards-based Web services for SOA, thus reducing vendor lock-in for CSPs.
2. Promotes a *federated architecture* for service layer, which can allow a CSP to offer composite services that are composed of loosely coupled basic services offered by the CSP and its business partners.
3. Dynamic service discovery and invocation (with the help of a service bus) enable easier service creation/design and service execution.

In the remainder of the section on SOA, we discuss two key topics of interest. First we introduce the concept of media-enabled service bus and see how it can enable a CSP to offer a composite service that relies on the execution of a workflow involving digital media. We conclude this discussion with a brief comparison of two technologies that can be used for executing SOA—REST and SOAP.

4.3.1.1 How to Incorporate Digital Media in a Service-Oriented Architecture

We have discussed how digital media will play a dominant role in all next-generation services for a CSP. At this juncture a crucial question is how SOA deals with large contents of digital media such as video and pictures. We have seen how a service bus plays a crucial role for implementing SOA. Does this mean that just like messages,

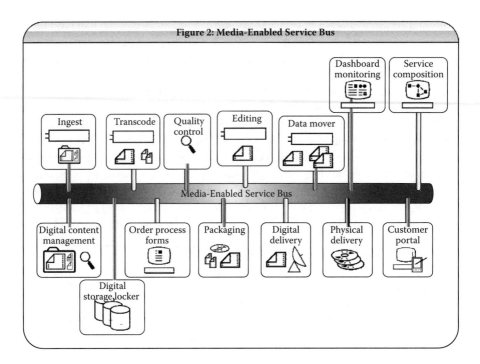

FIGURE 4.2 Media-enabled service bus.

digital media would also be sent across different endpoints of a service bus as it gets processed—such as transcoding, watermarking, and preinsertion of advertisements? That is definitely not the right approach for handling digital media in an SOA-based solution. To ensure persistence, transformation, and routing of digital media (just as we do for messages) in an SOA we need a *media-enabled* service bus. Figure 4.2 shows how a media-enabled service bus fits in an overall scheme of a digital media solution.

Fundamentally a media-enabled service bus is an extension of the service bus concept that we discussed earlier. It is not agnostic to the digital content that needs to be passed across several endpoints in the service while choreographing a composite service. A basic criterion to media enable a service bus is to include and manage the metadata of the digital content. An IBM implementation of a media-enabled service bus is explained in [3]. It relies on MPEG-21 Digital Item Declaration (DID) to provide an abstract model for declaring digital items [6, 7]. MPEG-21 DID also provides an Extensible Markup Language (XML) schema definition to represent the model. It is important to note that the MPEG-21 DID exchanged between services contains only the references to the media and not the actual media resource. In addition to this we also need a parallel content transport infrastructure (a data mover service) that would physically move the content across the different service endpoints of a service bus. The data mover service can be represented as a virtual content bus controlled by the service selection and service invocation process of the service bus. The data mover service can be a simple implementation based on standard network protocols such as File Transfer Protocol (FTP) or a commercial system.

Putting it all together, we can say a media-enabled service bus supports the exchange of messages containing control information for the transport of content and metadata information describing the content. This is very similar to what telecom carriers refer to as control network. For the transfer of the actual digital media, we rely on a parallel content bus, and the actual move of the digital content is based on the service selection and service invocation processes of the service bus. With this we have extended the concept of SOA to digital media and can ensure the persistence, transformation, and routing of digital media, in the same way as we ensure these attributes for messages passing through a service bus.

4.3.1.2 Two Technologies for SOA: SOAP and REST

As Web services gained prominence together with SOA, SOAP [8] became the de facto protocol for communication between Web services. SOAP allows both message-style communications and remote procedure call (RPC)-style communications, using a small set of patterns to define interactions between endpoints. Using XML Schema to define data structures, SOAP provides the message format for both the data and service invocation information for communications between a service requestor and a service provider. Bindings enable use of SOAP over a variety of transports, both HTTP based and non-HTTP based (such as XML/MQ for communications with existing message-orientated middleware [MOM]-based systems). SOAP also enables definition of information within headers, enabling extensibility to be defined independent of the service definition, enabling substitution of facilities such as security and routing to be defined at deployment time. This rich set of functions across many areas provides significant flexibility for Web services using SOAP, but also incurs an inherent amount of complexity.

As the Web evolved to allow richer interactions (commonly referred to as Web 2.0), including embedded logic within client-side functions with rich server interactions, the value of Web services has been broadened to incorporate applications that are much more client oriented, rather than the application-to-application model common to EAI (Enterprise Application Integration)- and ESB (Enterprise Service Bus)-based systems. REST is a minimalist architectural style that utilizes HTTP concepts to allow Web services to access resources through HTTP using existing Uniform Resource Identifier (URI) and HTTP operation capabilities. Resources, both data and function, are made available using HTTP operations such as GET and POST using unique identifiers to indicate the resource to be accessed and the item within the resource. REST-based services are often called RESTful. For example, a CSP might expose an address book application using a REST interface that enables HTTP GET to be used to retrieve a group (e.g., www.example.com/addressbook/johndoe/group/BowlingTeam), which will return an XML document with the list of people in this group.

SOAP and REST might often be used to provide the same function, and there are cases where a CSP might wish to provide both types of interfaces for the same services. SOAP is typically used when more operational type operations are being exposed, and where the additional flexibility offered addresses the requirements for a particular application. REST is a natural choice for interactive services that are focused on the exchange of data and invocation of services in a presentation context, and where ease of integration at the client is a key differentiator. A good comparison of the two technologies can be found in [9].

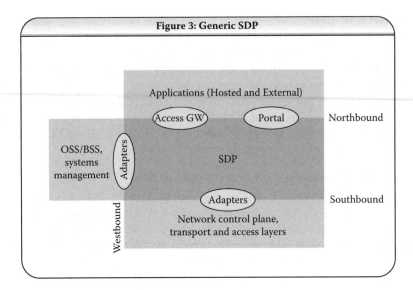

FIGURE 4.3 Generic service delivery platform.

4.3.2 Service Delivery Platform

The SDP provides the infrastructure and framework on which services are executed, managed, and exposed. The SDP has three primary logical interfaces: toward the network, toward the operational and business systems, and toward the users and applications that access the services. Figure 4.3 shows a simplified SDP architecture, highlighting the boundary interfaces with which SDP-hosted services and enablers interact.

The interface toward the network, often referred to as a southbound interface, provides the interactions to and from the network, and across networks. Following the layered architecture model common in both standards such as IP Multimedia Subsystem (IMS) and in many existing deployments, the services do not interact directly with the transport and access layers; rather, they utilize protocols and interfaces that abstract these layers—enabling services to be portable across networks, to be deployed in multinetwork environments, and to remain operational while networks migrate to new generations of technologies.

The interfaces toward the operational and business systems, often referred to as eastbound and westbound interfaces, enable services to be deployed without requiring changes to the operational and business systems that support them. Processes such as service activation may be utilized across services, and integration toward systems for fault management, billing mediation, and other functions utilizes common interfaces. Many of these processes and interfaces are defined by the TeleManagement Forum (TMF) in the enhanced Telecom Operations Map (eTOM) and Next Generation Operational Support Systems (NGOSS) activities.

The interfaces toward users and applications, often referred to as northbound interfaces, provide the access functions enabling the services to be utilized by hosted applications, external applications, and devices. These interfaces can be programmatic, enabling interaction by applications using Web services, for example, or they can be presentation based, where the user is presented with a Web- or device-based application with which to interact. Programmatic interfaces include Parlay X Web Services published by 3GPP and Open Mobile Alliance (OMA), and presentation interfaces often utilize HTML and HTTP to present services in a browser, or are provided as downloadable device applications. Combinations are also common, where presentation function is provided that invokes SDP-hosted services, such as a device application that invokes a messaging service from the address book to send a text message to a group of friends.

Within the standards activities of the telecommunications industry, there are a number of areas in which standards are related to SDP.

- Southbound interfaces utilize a wide variety of protocol standards ranging from traditional networks (SS7—Signaling System #7) to Session Initiation Protocol (SIP) to HTTP to Short Messaging Peer-to-Peer Protocol (SMPP). Proprietary protocols can also be utilized. The use of the protocols may be specified (e.g., IMS defines the ISC interface), or may follow de facto or deployment-specific criteria.
- Service enablers define a functional component, with a set of interfaces that could include a combination of southbound, eastbound/westbound, and northbound interfaces. A service enabler is usually not meant to be a service itself, but rather a common component that is used by many services. Examples include the Presence Server definitions from 3GPP and OMA, and the Push-to-Talk over Cellular enabler from OMA.
- SDP interfaces often include presentation and access technologies from the Internet space, and as such, many standards activities at IETF, W3C, OASIS, and other bodies are incorporated into telecom standards and the definition of SDP interfaces.

As one considers the spectrum of interacting elements, and the richness of new services that will drive more varied interactions, the use of network resources will grow not only in numbers (number of requests, amount of data) but also in the number of interaction points that need to be served. Where the SDP provides the flexibility to support new services and new interaction models, providing a common platform to access these, the underlying network will also have the opportunity to support richer functions and interactions that can be optimized by the SDP to enhance the user experience. Likewise, the SDP platform can improve services as the network evolves, providing longevity and value to services without requiring replacement.

4.3.3 Web 2.0 and Telecom

The term Web 2.0 was first coined during a conference between O'Reilly and MediaLive International in 2004. The term has evolved in defining a new ecosystem of application service providers for the Web. A good description of Web 2.0 can be found in Tim O'Reilly's paper "What Is Web 2.0" [1]. Fundamentally, Web 2.0 relies on collaboration between users. The underlying principle of Web 2.0 is very similar to the terminology that is popular in a traditional telecommunication industry—*network effects*. Similarly, a service based on Web 2.0 (such as community blogs, social networking sites) gets better as more users avail the service.

Web 2.0 is disruptive in nature for a traditional telecom carrier in many ways. A significant shift in the mindset of a telco is the phenomenon of "perpetual beta," wherein a service and its features are constantly being evolved and changed with the feedback of users. Such concepts are not common in a traditional telco landscape. Moreover, traditional telco has been very protective about its services and customer data. Many telecom carriers provide rich digital media services through a "walled garden" approach. For example, telecom carriers expose their network capabilities (such as call control features, messaging) as abstracted Web services that can be used by their trusted application service providers for providing third-party applications. On the other hand, in the Web 2.0 landscape a CSP relies on an open model.

A common meeting ground between these two ends is perhaps the *mashup platform* that provides users to create applications by assembling multiple services and content sources. These services could very well be the abstracted services exposed by the telco such as user location and presence information. Mashups can be created by users to create situational or personalized applications that in turn can be shared within a community. A user can build a mashup to fetch his buddy list from his address book and plot their location on a map (with the help of a map service) and send an SMS to his friends within a 10-mile radius to meet for a drink. The paper "Web 2.0 Meets Telecom" within the Moriana report [10] discusses how an operator's SOA that resides on an SDP framework can actually enable a carrier's service platform to collaborate with partner ecosystems that are based on Web 2.0 technology.

4.4 STANDARDS

The SDP provided a number of references to standards in describing the interfaces provided in the context of the SDP. Given the many standards activities, and the dependencies among them, getting a good grasp on how the different activities fit together is not always obvious. The following provides some background information.

3GPP (www.3gpp.org) and 3GPP2 (www.3gpp2.org) provide a large set of specifications for wireless networks (GSM [Global System for Mobile Communications] and CDMA [Code Division Multiple Access]-based, respectively). Some of the specifications have been further adopted for use with nonwireless networks, such as use of some IMS specifications for wireline and cable networking standards. 3GPP collaborates with other bodies to reference many standards, and to provide requirements to enable standards to become usable by 3GPP. An example of this collaboration is the referencing of SIP specifications from IETF (www.ietf.org) in many 3GPP specifications.

Where 3GPP generally focused on the transport, access, and control functions, OMA (www.openmobilealliance.org) has a significant amount of work that focused on service enablers and service and application-level interfaces. Over the last few years, 3GPP and OMA have collaborated to identify work areas for this complementary relationship, and migrated work as appropriate to maintain these focus areas.

The TMF (www.tmforum.org) has traditionally focused on the architecture models for the support systems, with the intersection between the services and the support systems coming together in deployments at the CSP. TMF has been progressing toward providing more implementation materials in its deliverables, enabling more implementation functions to be packaged. However, it is still an integration activity to define the service integration toward the support systems—although it is eased with the user of SOA and Web services technologies.

The collaborative nature of the standards activities has been effective in enabling the expertise of each body to be utilized across a wide range of activities. Although not a simple hierarchical model, navigation across bodies has continued to improve and the complementary nature of many of the activities has reduced conflicting work across activities.

4.4.1 Major Standards Activities

In the voice services area, the IMS standards have been a primary effort to define the evolution of networks to IP. The IMS architecture has been utilized in deriving implementations for GSM and CDMA networks, and has been subsequently utilized in other networks, including wireline and broadband within both traditional voice and cable networking standards. The IMS standards reference many existing standards from IETF for protocols, enabling use of the same specifications across networks and industries. In addition, as data services and Internet integration have been introduced within solutions utilizing IMS, existing standards from W3C, OASIS, and others have been utilized, enabling ease of integration for these solutions.

OMA has a variety of work areas, including enablers for mobile browsing, messaging, location, presence, group list and document management, and Web services interfaces for network services (recently migrated from 3GPP and The Parlay Group). Like IMS, many of the OMA activities utilize existing standards, and define service-specific protocols, functions, and operational specifications for these services and enablers.

TMF defines process and data standards for operational systems, enabling the operations of CSPs to be architected in a consistent manner. The TMF specifications are often realized in vendor products utilizing standards including process languages, such as BPEL (Business Process Execution Language) and communications standards (e.g., Web services).

Whereas previous generations of telecom network services utilized many technologies that were specific to the telecommunications industry itself, the next generation of services is being delivered on an infrastructure that shares its technology definitions, utilizing many of the same technologies as Internet services, while specifying their use to meet the functional and nonfunctional requirements of the CSPs' specific set of services.

4.5 AUTHENTICATION AND AUTHORIZATION CONSIDERATIONS

Authentication enables the CSP to determine the identity conveyed by a requester, and to verify the identity conveyed. Authentication can also include identity management functions, where the identity conveyed represents an access to a service on behalf of another provider. Authentication can occur at different points of entry to the network—whether through a device interaction initiating contact through an access network interface or from an application interaction through an application-facing interface.

Authorization can occur in multiple places, and is usually complementary in nature—with each element applying its context to the request being made to determine whether to authorize its continuance and behavior. To illustrate this, consider a request by an application to send a video clip to a mobile device.

- The first authorization is for the application to gain access to the network. This requires only knowledge of the identity of the requester (from its authentication) and whether access through the network entry point accessed is permitted.
- The second authorization is for the service: Is the requester permitted to use this particular service? It can also apply the granularity to the operation requested. This requires access to the subscription information of the requester, including the policy regarding use of the service and operation requested.
- The third authorization is related to the work requested by the request: Is the requester permitted to consume the resources requested by the service (at the service level, not yet at the network level)? For example, a requester might be constrained in the size of the video that can be sent or the number of users to whom it can be sent, or an aggregate of the two.
- The fourth authorization is related to privacy: Does the requester have the right to send a video to the destination user? This authorization uses the context available from a privacy management subsystem.
- The fifth authorization is related to content: Is the specific content allowed for the destination user? Rated content may be restricted to delivery to specific users, or excluded from other users. This authorization requires both profile information for the user and content rating knowledge of the video to be delivered.
- The sixth authorization is user acceptance, enabling the user to indicate whether he or she is willing to accept the content or not. This might be an interactive authorization or might be automatically retrieved from a profile.
- The seventh authorization is the acceptance of the request by the network: Is the requester–user combination permitted to utilize the functions and bandwidth necessary to deliver the content?

Not all of these authorization steps might be required for all services, but consider the overall message flows and the processing related to messages at different nodes in their processing, and the ability to reject messages early in their processing when authorization is not permitted to deliver the services.

4.6 ILLUSTRATIVE CALL FLOWS: PUTTING IT ALL TOGETHER

Based on the concepts that we have discussed so far, we next illustrate, with a few call flow examples, how an open architecture can execute composite services for a next-generation CSP.

Service Layer for Next-Generation Digital Media Services

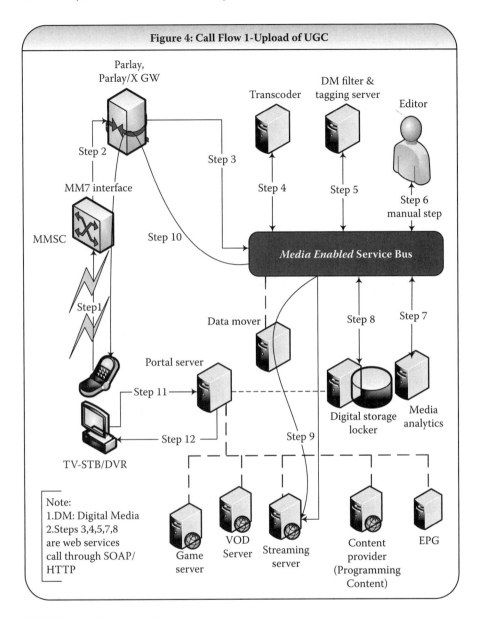

FIGURE 4.4 Call flow 1: Upload of user-generated content.

4.6.1 Case Study 1: Upload Content into Your Digital Locker Account through MMS

Our first call flow (Figure 4.4) is a modified version of a call flow that we had published [3]. This use case pertains to UGC. For simplicity we assume that our CSP offers mobile, IPTV, and digital locker services and our user, Jenny, has subscribed to all of them. The use case demonstrates how Jenny uploads video of a community event (e.g., Sunday Farmer's Market) from her mobile and receives a notification.

In Step 1, Jenny's feed is taken from a video-camera-enabled mobile phone (as a mobile-originating MMS) over the wireless network with short code for her digital locker (which is a service offered by the CSP). The Parlay X Web Services gateway (GW) receives short code MMS from the MMSC GW (Step 2), and infers from the short code that the message is intended for the digital locker and subsequent distribution. So in Step 3, the Parlay X Web Services GW forwards the message to the service bus (after transforming the message to a SOAP message). Once the service bus receives the SOAP message, it creates a representation of the content in an open standard format, MPEG 21, and gets the content transcoded by a transcoder to (say) MPEG 4. Because this is personal video, the IPTV provider invokes the digital media filter and tagging server (Step 5) [11]. This server would perform a video context search for objectionable frames and also tag the video content. A provider would still like to ensure quality control by having human eyes inspect the content (Step 6). The editor inspects the results from Step 5 and approves (or rejects) the content for distribution. Assuming it is approved, the service bus invokes the media analytics Web service to perform analytics of the video content (adds context information) in Step 7. In Step 8 the service bus uploads the content in Jenny's personal storage locker account with the provider. After successful upload, the storage locker sends a confirmation to the service bus. In Step 9, the service bus forwards the content to the streaming server (with the help of a data mover service). Once complete, the service bus service sends a confirmation via MMS with the URI of the location where the content is loaded (Step 10). At a later time Jenny opens her STB browser. The STB browser queries for her UGC files from the portal server (as an HTTP request) in Step 11. (Actually, Jenny could perform a *unified search,* a service from her provider that would let her search for contents from multiple servers—Electronic Program Guide (EPG), video on demand [VoD], and game servers—in addition to her storage locker.) Jenny retrieves a list of her uploaded (Step 12) videos and ensures that her recent video is marked for distribution. At a later point in time Jenny can also check (from the STB browser portal) the number of times her video has been watched and any feedback on her video.

This scenario demonstrates the need to have built-in flexibility for developing workflows so that workflows can be built (or changed) rapidly from loosely coupled and reusable services. We saw how a Web-services-based SOA infrastructure, together with the media awareness of the service bus, can help a CSP provide composite services that rely on multiple technologies—mobile and IPTV.

4.6.2 Case Study 2: Example of Mashup Telco Service

Our second call flow (Figure 4.5) is based on SN and Web 2.0 principles that we have just discussed. It demonstrates how a service provider, a content provider, and a mashup assembler (could be a business partner or an experienced user who is familiar with mashups) come together to build a situational application that can actually be used by an end user. For simplicity we take the mashup assembler as an individual (say Jenny) who is also a customer of the wireless company Best Wireless. For our example, we assume that Jenny is the user of this application. This example can be

Service Layer for Next-Generation Digital Media Services

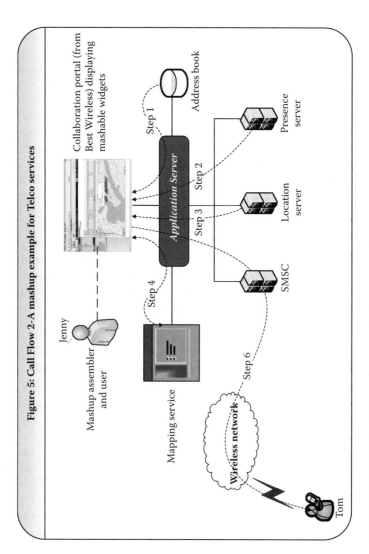

FIGURE 4.5 Call flow 2: A mashup example for telco services.

broken into two—the design and build phase (mashup assembly) where Jenny builds the mash-up, and the execution phase when Jenny wants to execute the call flow.

Let's first discuss the mashup assembly stage. Jenny logs on to the Collaboration Portal that is provided by Best Wireless. She searches for widgets that are available to her for building a mashup. After doing a search, she finds two categories of widgets that are available to her:

1. Telco services exposed as widgets by Best Wireless: Presence, Location, and SMS.
2. Application widgets: Weather (from AccuWeather), Mapping Service (from Google), and an address book (provided by Best Wireless).

Jenny decides to build a mashup that will display the location of her friends on a map along with their presence status. She would use this to see which of her friends are nearby (using the mapping service) and check if any of her friends would like to meet with her at a nearby shopping mall.

Once the assembly stage is complete, she is ready to execute the use case. The application polls her buddy list (Step 1) and fetches the presence status of each one (Step 2). The application next fetches the location of each of her friends who are available from the location service (Step 3), and then provides the location data to the mapping service for getting the location plotted on a map. The mapping service returns a map with the location plotted (Step 4). Jenny finds Tom to be closest to where she is located and his presence status is displayed as "available on mobile" (Step 5, not shown in Figure 4.5). So now she uses the Telco widget to send him an SMS (Step 6), "Would you like to meet at the Promenade Mall?"

This scenario demonstrates how a service provider can provide services for the "long tail" by making services and content or data "mashable." Nonprogrammers can create situational content by "wiring" exposed services and content, enabling them to collaborate with others. This broad collaboration ultimately helps all the parties involved—end users, business partners, and the CSPs themselves.

4.7 BUSINESS MODEL CONSIDERATIONS FOR SERVICE LAYER

We have discussed how collaboration between telecom carriers and different business partners (content producers, application service providers) can be the key factor to foster growth and revenue for a telco. A telecom carrier brings three valuable assets to a collaborative relationship with business partners:

1. A reliable distribution network.
2. A trusted relationship with their customers.
3. Network-level services that can be exposed as abstract services, such as call handling, presence, and telecommunication.

Such collaborative relationships between application service providers (say ones that are Web 2.0 based) and telecom carriers present an opportunity to provide multiple services to common customers. However, merely offering multiple services will not suffice. A key aspect, as we all know, is how innovative a carrier can get in bundling these services that would appeal to a customer and pricing these bundles. With the transition of network backbone to a common IP layer, such bundling becomes even more meaningful across multiple services, and yet maintaining the service level agreements (SLAs) across all these services is crucial. Carriers also need to be creative in adopting different billing models for services. For instance, a family tracking service, wherein a parent can be alerted by an SMS whenever his or her child leaves a geo-triangle of three coordinates, need not be charged by network resource consumption (bits and bytes) for every SMS and location notification, but perhaps by a monthly fee that would cover a certain number of alerts per month.

In addition to forming alliances with application service providers and content producers, alliances between a telecom carrier and other enterprises will also shape how the service layer of telecom carriers can be used to increase the productivity of these enterprises. Telecom carriers have always provided communication infrastructure to enterprises; however, earlier it was limited to the bit pipe. Services exposed by a next-gen CSP, such as presence and location, can be used to enhance the communication within an enterprise. An example of such collaboration is the health care industry. For example, health care providers can view presence and location information of a patient and other care providers of the patient and execute different messaging options for communicating with them.

4.8 CONCLUSION

The evolution of networks will have a profound and complementary relationship to the evolution of the services infrastructure that interacts with them. Today's service platforms tackle many legacy integration challenges, enabling services to operate across multiple networks and to provide a consistent user experience regardless of delivery network. They also enable services to be decoupled from their legacy charging systems—supporting new business models such as indirect charging (e.g., advertising-based services) and charging based on the content provided by the service instead of just resources consumed.

As the evolution to optical networks continues, and the value of the service becomes the primary consideration for competing, and for pricing services, the role of the service platform will evolve to reflect the need for service functions to utilize the new capabilities offered by the network—replacing the legacy adapter model of today with a network exploitive set of capabilities. This parallel evolution of service and network will be enhanced by the flexibility offered by the network evolution and the ease of integration with peer networks at both the network and service levels.

REFERENCES

1. T. O'Reilly. (2005, Sept.). What Is Web 2.0. O'Reilly Network [Online]. Available: http://www.oreillynet.com/pub/a/oreilly/tim/news/2005/09/30/what-is-web-20.html
2. A. Sur, A. Arsanjani, and S. Ramanathan, "SOA-based context aware services infrastructure: An approach to enable telecom service provider transformation in a Web 2.0 environment," presented at the IEEE SOA Industry Summit, Salt Lake City, Utah, 2007.

3. A. Sur, F. Schaffa, J. McIntyre, et al., "Extending the service bus for successful and sustainable IPTV services," *IEEE Communications Magazine,* pp. 96–103, Aug. 2008.
4. IBM. The Changing Face of Communication [Online]. Available: http://www.ibm.com/iibv, Dec, 2008.
5. T. Erl, *Service Oriented Architecture: Concepts, Technology and Design.* Upper Saddle River, NJ: Prentice Hall, 2005.
6. J. Bekaert, E. De Kooning, and H. De Sompel, "Representing digital assets using MPEG-21 Digital Item Declaration," *International Journal on Digital Libraries,* Vol. 6, pp. 159–173, April 2006.
7. I. Burnett, R. Van de Walle, K. Hill, J. Bormans, and F. Pereira, "MPEG-21: Goals and achievements," *IEEE Multimedia,* pp. 60–70, Oct.–Dec. 2003.
8. F. Curbera, M. Duftler, R. Khalaf, W. Nagy, N. Mukhi, and S. Weerawarana, "Unraveling the Web services web: An introduction to SOAP, WSDL, and UDDI," *IEEE Internet Computing,* Vol. 6, pp. 86–93, March 2002.
9. E. List, D. Ayers, E. Bruchez, et al., *Professional Web 2.0 Programming.* Indianapolis, IN: Wiley, 2007.
10. Moriana. (2008, Sept.). Moriana SDP 2.0 operator guide [Online]. Available: http://www.morianagroup.com/index.php?option=com_content&view=article&id=148&Itemid=129
11. J. R. Smith, et al., "Video indexing using model vectors," presented at IS&T/SPIE Electronic Imaging 2004: Storage and Retrieval for Media Databases, San Jose, California, Jan. 2004.

5 Enhancing TCP Performance in Hybrid Networks
Wireless Access, Wired Core

Milosh V. Ivanovich

CONTENTS

5.1	Introduction	100
5.2	Introductory TCP Concepts	100
5.3	The Problem: TCP-Impacting Characteristics of Widely Used Wireless Access Networks	101
	5.3.1 Cellular Radio	101
	5.3.1.1 cdma2000 1xRTT	101
	5.3.1.2 GPRS/GSM	103
	5.3.1.3 WCDMA/UMTS	104
	5.3.1.4 Packet-Switched Shared Channel Technologies: HSDPA and 1xEVDO	104
	5.3.2 Wireless Local Area Networks	105
5.4	The Possible Solutions: "Tweaking" TCP, Replacing TCP, or Inserting a Proxy	106
	5.4.1 Use of the TCP Selective Acknowledgments (SACK) Option	106
	5.4.2 Implementing IETF Recommendations: The PILC Working Group	107
	5.4.3 Modifying TCP	108
	5.4.4 Introducing an Intermediary Between Sender and Receiver: Split Connection	110
	5.4.4.1 Nontransparent Approaches	110
	5.4.4.2 Transparent Approaches	111
	5.4.5 Replacing TCP: Introducing Proprietary Protocols	112
	5.4.6 Replacing the Wireless: Making New Technologies TCP Friendly	113
5.5	Case Study: A Novel TCP Performance-Enhancing Proxy	114
	5.5.1 Architectural Considerations	114
	5.5.2 Performance Enhancement Techniques Employed in the TRL-PEP	115
	5.5.2.1 Aggressive Link Utilization	115
	5.5.2.2 Additional Data Transmission on DUPACKs	115
	5.5.2.3 Retransmission on Partial Acknowledgments	115
	5.5.2.4 Intelligent Use of Selective Acknowledgments	116

                5.5.2.5   TCP State Prediction... 118
        5.5.3   Simulation Results 118
        5.5.4   Test Network Trial Results 119
Acknowledgment 122
References.. 123

5.1   INTRODUCTION

With the ever-increasing convergence of the Internet and wireless modes of access, pressure has been steadily increasing on the networking community to improve the performance of Transmission Control Protocol (TCP), the Internet's most widely used transport protocol, when being carried by wireless networks. For this reason, TCP performance enhancement in wireless access networks with a wireline core is an important ongoing area of research. The hostile nature of the wireless channel and the mobile nature of wireless users interact adversely with standard TCP congestion control mechanisms, often causing a drastic reduction in throughput.

This chapter introduces the problems associated with the use of TCP over diverse cellular radio access technologies based on the Global System Mobile (GSM), Code Division Multiple Access (cdma2000), and Universal Mobile Telecommunications System (UMTS) standards, as well as over wireless local area networks (WLANs). A selection of approaches for solving, or at least partially mitigating these problems, is surveyed.

A novel TCP performance-enhancing proxy (PEP) is treated as a case study, for which the results of simulation and field trials are presented. The case study highlights key metrics used in judging the success of TCP enhancement approaches, including the level of transparency to both TCP endpoints, the ability to adapt to different access technologies, and the ability to maximize the wireless access technology's available throughput.

5.2   INTRODUCTORY TCP CONCEPTS

Historically, TCP's main area of application has been in wired networks, where user traffic congestion is the main cause of packet loss, and errors due to the transmission media are rare. Subsequently, TCP behavior has been optimized to deal with these issues. However, when the transmission medium is a wireless link, such "traditional" TCP behavior is inappropriate, because radio-induced packet errors will often overtake congestion as the dominant source of packet loss.

Consider Figure 5.1, where the vertical axis represents the amount of unacknowledged data in TCP segments (called the congestion window [CWND]) that the TCP sender is permitted to transmit to the receiver, and the horizontal axis is time. The dotted red line represents the optimal bandwidth delay product for the transmission link—the amount of sender unacknowledged data required to fully utilize the TCP pipe. For this reason, we seek to keep the blue line (effectively a measure of achieved throughput) consistently close to the red dotted line (the maximum possible link throughput). Figure 5.1 shows TCP's normal reaction to a packet loss (signaled by a triple duplicate acknowledgment) and to a retransmission timeout; in the former

Enhancing TCP Performance in Hybrid Networks

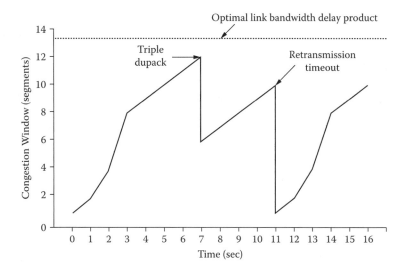

FIGURE 5.1 Standard TCP reaction to a random packet loss (signaled by a "triple dupack") and a retransmission timeout.

case, the congestion window is halved, whereas in the latter it is reduced to one TCP segment. In a wireless network where IP packets and hence TCP segments are lost due to random radio events, such substantial throughput reduction is often unnecessary. In the following, we refer to NewReno [1], the most widely deployed TCP variant in the Internet [2], as the de facto standard TCP.

Assuming that the reader has a basic level of familiarity with TCP protocol fundamentals, we turn our attention to reviewing the key characteristics of several widely used wireless access technologies, and how these characteristics may cause problems when the given access technology carries standard TCP.

5.3 THE PROBLEM: TCP-IMPACTING CHARACTERISTICS OF WIDELY USED WIRELESS ACCESS NETWORKS

Many wireless access networks in widespread use today suffer from a range of characteristics that are undesirable from the perspective of standard TCP. The unpredictability of the physical radio channel or the operational characteristics of the media access control (MAC) and link layers often negatively impact observed TCP throughput and latency through increased packet loss and delay variation. We now turn to exploring these characteristics in more detail for popular wireless access technologies in use around the world today.

5.3.1 CELLULAR RADIO

5.3.1.1 cdma2000 1xRTT

In the 1xRTT system [3], data is transmitted to a mobile station (MS) on two channels. An MS has (1) continuous access to a low-rate fundamental channel (9.6 kbits/sec at the link layer), and (2) shared access to a variable-rate supplemental channel based on both the MS's own connection requirements and the number of competing MSs it is sharing with (up to a peak rate of 153.6 kbits/sec at the link layer). Ignoring the sharing aspect, the amount of supplemental channel resources obtained by an MS is determined by the amount of data residing at the 1xRTT Base Station (BS) link layer buffer. Lower supplemental channel rates are assigned to MSs that have less data to receive. As the amount of data destined for a particular MS changes, the result is a "channel switching" behavior similar to that of Wideband Code Division Multiple Access (WCDM)A, described later, which results in TCP seeing a variable bandwidth presented to it by the lower layers.

A more significant source of link layer bandwidth variation is bandwidth oscillation, a well-studied effect initially identified in the cdma2000 1xRTT cellular access technology [4]. The oscillation results from the fact that the supplemental channel rates to mobile stations are renegotiated periodically (e.g., ~5 seconds in a typical network setting) with the actual time for renegotiation lasting a nontrivial amount of time (e.g., ~1 second). During this time, all mobile stations have access only to their exclusive-use but much slower fundamental channel. This results in a scenario where the bandwidth available to an MS at the link layer may be as high as 163.2 kbits/sec for a period of time (assuming no competing stations), only to be reduced by a significant factor as high as ×17 for a subsequent and nonnegligible period of time.

As the authors of [4] point out, such behavior can be very detrimental to standard TCP due to the spurious TCP retransmission problem it causes. Spurious retransmissions will typically occur in situations where an acknowledgment for outstanding data is delayed. Reasons for this delay include excessive queuing in the end-to-end path, congestion or loss of ACKs on the return path, or, as in the case described here, a sudden change (i.e., reduction) in the link bandwidth of the forward path (i.e., sender-to-receiver bandwidth). Whatever the reason for the delayed ACK packet, it causes TCP to register a retransmission timeout—a very serious outcome from the point of view of throughput degradation—due to the resetting of the congestion window to a single TCP segment. This is especially true when the TCP timeout is known to be spurious, in the sense that no data was actually lost (it has only been delayed, instead), and hence no retransmission was required. Figure 5.2 shows an example of such a timeout followed by the subsequent spurious retransmission, based on actual measurements of bandwidth oscillation in a cdma2000 1xRTT test network. Other than the obvious bandwidth wastage incurred by the sending of needless data, a secondary problem caused by such spurious retransmissions is that they can cause problems when interacting with other TCP options or configuration settings.

One example is that the use of aggressive Fast Retransmission [5] becomes compromised. That is, the early triggering of the Fast Retransmission mechanism (on reception of only two instead of three duplicate ACK packets) will most often do more harm than good in circumstances where spurious TCP retransmissions are likely to generate many instances of two duplicate packets. In such a scenario, the negative effect of spurious TCP retransmissions would be greatly magnified: Not only are we needlessly resending data assumed to be lost instead of delayed, but we also incorrectly assume the same packet is (again) lost when the original delayed ACK packet arrives.

Enhancing TCP Performance in Hybrid Networks

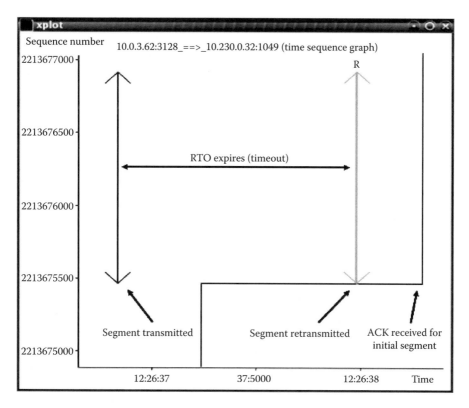

FIGURE 5.2 Time-sequence diagram showing a spurious TCP retransmission caused by the bandwidth oscillation effect in a cdma2000 1xRTT test network.

5.3.1.2 GPRS/GSM

The General Packet Radio Service (GPRS) [6], a part of the wider GSM second-generation (2G) mobile telephony standard, is designed to transmit low-rate packet data by utilizing radio resources (i.e., GSM radio channels) that are not used by GSM voice services. Therefore, the number of channels allocated to GPRS is a random variable depending on the voice traffic, which has the ability to preempt data users. Preemption involves the ability of an incoming voice user to take radio resources away from ongoing GPRS data connections, down to some operator-configurable minimum (which can be set to zero, in theory). Radio layer bandwidth variation is thus also present in GPRS, but occurs over much longer time scales than the cdma2000 1xRTT oscillation described earlier, as voice call holding times typically average many tens of seconds. As a result, spurious TCP retransmissions are less likely, and GPRS poses less of a problem to standard TCP in this regard.

From the point of view of TCP performance, a more detrimental characteristic of GPRS is its very high round-trip time (RTT), on the order of magnitude of around 1 second, found to be generally due to a combination of relatively slow and variable transmission rate and significant system processing latencies resulting from heavy signaling overheads [7]. In the most general sense, TCP running over links with such long latencies suffers from "sluggishness," both in terms of its packet loss recovery mechanisms and transmission rate management mechanisms (i.e., low throughput particularly for short- to medium-length TCP flows, due to the inordinately long time spent in the TCP slow start state). This is perhaps unsurprising, given that the long link layer latencies introduced by GPRS effectively act to slow down the rate at which TCP can perform its signaling functions. Namely, TCP's basic unit of signaling, the ACK packet, takes a very long time to return to the sender.

5.3.1.3 WCDMA/UMTS

WCDMA for UMTS is a third-generation (3G) cellular access technology that is often viewed as the direct successor to the GSM/GPRS 2G standard 0. WCDMA largely mitigates the high RTT problems encountered in the older GPRS technology by reducing the round-trip latency to around 200 ms [9].

However, the problem of TCP seeing a variable bandwidth at link layer is still present in WCDMA through a phenomenon known as *channel switching*. As described in [10], fluctuations in packet-switched WCDMA data traffic mean that reserving resources for a dedicated channel without regard to the actual load is not efficient. To optimize the use of scarce radio resources, channel switching is proposed as a technique to dynamically change the resources allocated during a packet data session, according to the amount of data that need to be transmitted in the uplink and downlink. When large amounts of data are being sent or received, a dedicated transport channel is used. The supported uplink/downlink bit rates are 64/64, 64/128, and 64/384 kbit/sec, respectively. Similar to the mechanism described for cdma2000 1xRTT earlier, a WCDMA user will switch between the different rate channels during a session, dependent on fluctuations in (1) the amount of available radio resources and (2) the amount of data ready for transmission in its link layer buffers.

Note that the amount of available radio resources is influenced not only by the number of competing WCDMA users, but also by the radio channel conditions. Channelization code and downlink power utilization are monitored, along with the uplink and downlink "noise rise" of the system, which are related to the user load. If any of these monitored physical layer quantities increase above their designated thresholds, a switch to a lower rate dedicated channel or even the very low rate shared channel (32 kbit/sec) may be required. This behavior is dictated by the congestion control algorithms of the WCDMA system.

The link layer bandwidth variability in WCDMA is not as drastic as that observed in cdma2000 1xRTT, and hence does not cause as severe a problem to the operation of standard TCP. Coupled with WCDMA's significant latency improvement over GPRS, the overall result is more "TCP-friendly" behavior and less need for complex TCP performance enhancement solutions when comparing WCDMA to technologies like GPRS and cdma2000 1xRTT.

5.3.1.4 Packet-Switched Shared Channel Technologies: HSDPA and 1xEVDO

Technologies such as cdma2000 Evolution Data-Only (1xEVDO) [11] and High Speed Downlink Packet Access (HSDPA) [12] represent the next evolutionary step beyond original 3G standards such as WCDMA/UMTS and are based on the concept of a high-speed packet-switched radio channel that is dynamically shared between many MSs.

Time is slotted into very short transmission intervals (1.67 ms in 1xEVDO and 2 ms in HSDPA), which are used as the basis of dynamically sharing the radio channel among many MSs. In 1xEVDO the shared downlink can only serve a single MS in one transmission interval, because the system dedicates its entire resources to that MS during that interval. In HSDPA, on the other hand, it is acknowledged that there may be intervals during which a single MS cannot use all of the system resources, so it is permitted for up to four MSs to be served during a single interval using a technique known as *code multiplexing*. In both access technologies, the scheduling algorithm at the BS takes advantage of the paradigm of multiuser diversity. In broad terms this means scheduling for transmission that MS that is currently reporting the best radio channel carrier to interference ratio (CIR), subject to some fairness constraints. The idea in both of these access technologies is to optimally utilize the scarce radio resources, without completely starving the MSs that may be experiencing prolonged poor channel conditions [11, 12].

This fine-grained radio resource sharing at the MAC/link layer is a significant improvement on the bandwidth oscillation effects apparent in earlier technologies such as cdma2000 1xRTT, because it occurs at time scales that are much shorter than TCP's time scale, which is on the order of magnitude of an RTT. In theory, this gives TCP more time to adjust its internal state variables such as the moving average RTT estimate and the retransmission timeout (RTO) timer, and reduces the likelihood of spurious TCP retransmission events. Furthermore, both of these technologies reduce the system RTT through an increase in raw transmission rate as well as a decrease in the protocol signaling overheads.

However, despite the described improvements offered by packet-switched shared channel technologies such as HSDPA and 1xEVDO, continuing research into interactions with TCP has highlighted that problems arising from their inherent bandwidth variance are still present to some degree [13], and thus need to be addressed by the kinds of techniques discussed later in this chapter.

5.3.2 Wireless Local Area Networks

The challenges of carrying TCP over WLANs are different from those faced by the cellular radio access technologies described earlier. In WLANs, exemplified by the IEEE's 802.11b standard [14], latencies have tended to be orders of magnitude lower and transmission rates orders of magnitude higher in comparison with cellular technologies from the same era. This can be put down to the friendlier radio environment, including lower MS mobility, smaller distance between MS and access point (AP) than between MS and cellular BS, less interference due to typically smaller numbers of MSs attached to a WLAN AP than to a cellular BS, and a smaller number of other interfering APs. However, some issues still remain with the carriage of TCP over WLANs.

The first of these is the problem of unfairness in the TCP throughput of competing uplink flows [15], which comes about as a result of ACK packet congestion at the BS when the number of uplink flows exceeds a certain threshold. The IEEE 802.11 standard is based on a half-duplex wireless channel with equal transmission opportunities for a station and an AP. An excessive number of uplink flows leads to the situation where the AP cannot obtain all the downlink ACK transmission opportunities it needs, resulting in overflow of the AP buffer and a large number of ACK packet losses. TCP congestion control mechanisms are not invoked for ACK packet losses, except for the special case when all ACK packets in a single window are lost, invoking an RTO. Hence, it is the combination of this (1) ACK loss behavior of standard TCP and (2) the half-duplex wireless channel that yields the uplink unfairness. Namely, those "stochastically unlucky" uplink TCP flows that suffer an RTO due to an entire window's worth of ACK losses become unable to substantially increase their CWND from (1) for a long time, approximately equal to the duration of the ACK congestion event. Meanwhile, the "luckier" flows that sporadically lose some ACK packets instead of an entire window's worth continue to send with their normal CWND, and thus achieve an unfairly large throughput.

The second issue again pertains to TCP throughput unfairness, but this time concerns competing uplink and downlink flows [16]. The problem occurs regardless of the size of the AP buffer, and again finds its root cause in the combination of (1) and (2), as described earlier. With both uplink and downlink flows present, an AP that has reached congestion will begin discarding the downlink flows' data packets as well as the uplink flows' ACK packets. In such a situation, the congestion control mechanisms of standard TCP will dictate that the downlink flows act to reduce their throughput by lowering their CWND values. However, for the reasons described previously, the uplink flows will not be required to activate congestion control due to ACK losses, except for the infrequent whole-window-loss RTO events. Unlike their downlink counterparts, most uplink flows will continue to increase their CWND values as usual, resulting in significant throughput unfairness in favor of the uplink flows.

5.4 THE POSSIBLE SOLUTIONS: "TWEAKING" TCP, REPLACING TCP, OR INSERTING A PROXY

When confronted with the problems associated with carrying TCP over wireless access networks as described in the previous section, several courses of action are possible. These range from simple measures such as the use or disabling of specific TCP options, to the more moderate modification of TCP, or introduction of an intermediary node between sender and receiver, and finally to the extremes of either replacing TCP itself with a new, typically proprietary, transport protocol or "changing the underlying physical layer" by introducing new wireless access technologies that are friendlier to standard TCP. Each of these approaches is now described in turn.

5.4.1 USE OF THE TCP SELECTIVE ACKNOWLEDGMENTS (SACK) OPTION

The selective acknowledgment (SACK) option [16] available in the TCP specification modifies acknowledgments (ACKs) sent by the receiver back to the sender by including information on which segments have been received and which need to be retransmitted.

In this way, SACK, when combined with a selective repeat retransmission policy at the sender, can improve the otherwise poor performance of the original TCP Reno [17] variant in cases where multiple packets are lost in one window of data. Here the window of data refers to the outstanding TCP segments within a time period equal to one RTT.

Note that standard TCP (i.e., TCP NewReno) is also able to handle multiple packet losses occurring in a single window. However, SACK is more efficient than NewReno because the sender has a greater resolution of information, through the knowledge of exact packets that have been lost [18]. It is for this reason that SACK is commonly used in most modern TCP implementations.

The drawbacks of SACK include a limitation on the number of SACK blocks that one ACK can carry (due to the encoding of SACK information within TCP's option field), and its reactive nature in responding to packet losses [18]. The latter prevents SACK from being able to proactively avoid congestion, or infer the underlying cause of packet losses (e.g., is it a random radio event or a traffic-based congestion event?). It is for this reason that network researchers and engineers have looked beyond SACK for a more comprehensive solution to the problems of TCP over wireless.

5.4.2 Implementing IETF Recommendations: The PILC Working Group

The IETF PILC working group [19] was formed in order to partially mitigate some of the ill effects of wireless links that manifest themselves when TCP is used at the transport layer, without resorting to the more complex techniques discussed later in this chapter such as a significantly modified version of TCP or use of a proprietary transport layer protocol.

To this end, the PILC working group recommendations have focused on specific receiver- and sender-side configuration settings. A selection of some key recommendations is listed here, and the interested reader can find more details in [20]:

- Setting of an appropriate Sender and Receiver Window Size limit, based on the computed end-to-end Bandwidth Delay Product (BDP).
- Use of the Window Scale Option (RFC 1323), allowing Window Size to exceed 64 KB.
- Increased Initial Congestion Window (RFC 3390), given by CWND = min (4*MSS, max (2*MSS, 4,380 bytes)).
- Use of Limited Transmit to extend Fast Retransmit/Fast Recovery for TCP connections with small congestion windows (RFC 3042).
- Use of an IP MTU larger than the default size (576 bytes), allowing TCP to increase the congestion window (in bytes) faster.
- Use of Path Maximum Transfer Unit (MTU) Discovery, allowing a TCP sender to determine the maximum end-to-end transmission unit (without IP fragmentation) for a given routing path (RFC 1191 and RFC 1981).
- Use of the Selective Acknowledgment option (SACK) to achieve the benefits described earlier (RFC 2018).
- Use of Explicit Congestion Notification (ECN), in order to allow a TCP receiver to inform the sender of congestion in the network in a timely fashion (RFC 3168).

- Use of the Timestamps Option (RFC 1323) in order to avoid spurious timeouts by allowing the TCP sender to track changes in the RTT more closely.
- Disabling TCP/IP Header Compression (RFC 1144), because it does not perform well in the presence of packet losses expected on wireless links. Instead of transmitting entire TCP/IP headers, in RFC 1144 only the changes in the headers of consecutive segments are sent, so that a loss of a single TCP segment would cause a loss of synchronization between the transmitting and receiving TCP sequence numbers, and hence prevent TCP from taking advantage of the Fast Retransmit/Fast Recovery mechanism.

5.4.3 Modifying TCP

These approaches rely on the modification of the sender and/or the receiver TCP implementations. Such approaches are typically undesirable, as their deployment is hampered by the inertia of the large population of existing TCP versions in the global Internet, as well as by potential incompatibilities between new and old versions (e.g., fairness). Practical barriers aside, modified TCP approaches do offer neat and efficient solutions to the problem of carrying TCP over wireless networks, so we take some time to review a selection of such approaches next.

TCP Vegas 0 belongs to a class of TCP variants that is known as "delay-based" because of the reliance of their congestion control mechanism on RTT measurements. This is in contrast to the traditional TCP variants such as our standard TCP, NewReno, which are classified as "loss-based" because their congestion procedures are triggered by instances of packet loss. In comparison to standard TCP, Vegas introduces a different timeout mechanism, a change to congestion avoidance that attempts to control the connection's occupancy of network buffers, and a modified slow-start mechanism. By keeping a TCP connection's occupancy of network buffers close to some optimal level commensurate with the amount of end-to-end bandwidth available (using CWND adjustment), TCP Vegas aims to maintain a relatively constant RTT and then detect congestion at an earlier stage than traditional TCP variants by sensing unexpected increases in RTT.

In this way, Vegas attempts to eliminate the periodic self-induced segment losses that are so typical of standard NewReno TCP (i.e., CWND-probing "saw-tooth behavior"). The resulting TCP throughput behavior is more stable, and for illustration purposes this is shown in an idealized form comparing loss- and delay-based TCP in Figure 5.3. TCP FAST [22] is a more recent delay-based variant with similarities to Vegas, although it has been, thus far, mainly studied in the context of very high speed end-to-end optical links. In addition to their other benefits, it is noteworthy that the delay-based TCP variants have been validated as solutions to the significant throughput unfairness problems described earlier, when standard TCP is carried over WLANs. The authors of [15] show that in the case of delay-based TCP, with appropriate dimensioning of network buffers the harmful instances of ACK packet loss may be virtually eliminated due to the described regulation of CWND.

The complicating factor here, and reason why use of delay-based variants is not a "silver bullet" solution, is the well-researched problem of significant degradation of delay-based TCP performance when it coexists with the more aggressive

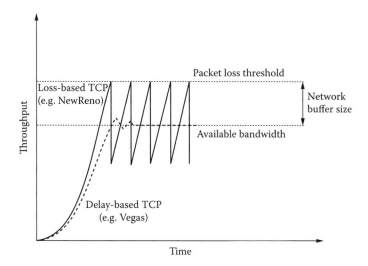

FIGURE 5.3 Comparison of throughput behavior between loss-based and delay-based variants of TCP.

loss-based TCP [23]. In fact, this is one of the key reasons why delay-based TCP has not been adopted for widespread use in the global Internet, and why more complex "hybrid" TCP variants have appeared over time combining some mixture of loss- and delay-based TCP characteristics.

There have also been numerous efforts in enhancing the original loss-based approach of TCP. One such proposal is TCP Forward Acknowledgment (FACK) [24], a loss-based TCP variant that separates congestion control algorithms from data recovery algorithms. It operates in conjunction with the TCP SACK option to keep an explicit measure of the total outstanding (unacknowledged) data in the connection. Because FACK more accurately controls the outstanding data in the network, it is less bursty and recovers from heavy loss more efficiently than standard TCP NewReno (with or without the TCP SACK option turned on).

Another enhanced loss-based implementation is TCP Westwood [25], which exploits two basic concepts: (1) end-to-end estimation of available bandwidth via sender-side analysis of the returning ACK stream, and (2) intelligent use of this estimate to dynamically set the congestion window. Upon reception of a triple duplicate acknowledgment, the congestion window and slow start threshold are set according to this dynamic bandwidth estimate, and in the more serious case of a RTO, CWND is set to one. Only sender-side modification of the TCP stack is required for implementation. Numerous studies have shown TCP Westwood to perform well in the presence of characteristics that are so typical of wireless environments, including packet loss caused by transmission or other errors ("leaky pipes"), large bandwidth-delay product ("large pipes"), and dynamically changing bandwidth ("dynamic pipes") [26].

Similar to TCP Westwood, TCP Jersey [27] employs a more refined bandwidth estimation algorithm as well as ECN support to make better decisions in controlling TCP sender behavior.

We conclude this section by describing TCP Eifel [28] a loss-based variant that does not change TCP's congestion control semantics, but instead focuses on improving the error recovery scheme. The Eifel algorithm is another example of a TCP implementation specifically designed for improving performance over wireless links, where connectivity may be intermittent and connection bandwidth may vary. The enhancement to TCP's error recovery scheme is shown to eliminate "retransmission ambiguity," thereby largely solving the problems caused by spurious timeouts and spurious fast retransmits, of the type described earlier. The retransmission ambiguity is the well-known problem of the TCP sender being unable to distinguish an ACK for the original transmission of a segment from the ACK for its retransmission. The implementation of the Eifel algorithm requires use of the TCP timestamp option and a small modification to the existing TCP sender, without any changes needed at the TCP receiver.

5.4.4 INTRODUCING AN INTERMEDIARY BETWEEN SENDER AND RECEIVER: SPLIT CONNECTION

Although TCP modification approaches of the kind just described in the above section do offer elegant solutions to many of the problems faced by TCP over wireless networks, the prevailing opinion in the literature and wider networking community has been that these benefits are offset by the practical cost of needing to make widespread changes to the enormous installed base of TCP stacks in the global Internet.

It is for this reason that more attention has been given to split-connection approaches that introduce an intermediary between sender and receiver, with the idea being to mitigate problems of TCP over wireless without needing to make major modifications to the protocol at the endpoints themselves.

This intermediary-based solution can be divided into two approaches: those that are transparent and those that are nontransparent. We now proceed to explore both of these in turn.

5.4.4.1 Nontransparent Approaches

Nontransparent split-connection approaches are less favorable than their transparent counterparts, because the intermediary establishes separate connections with both the sender and receiver in such a way that it requires both end systems to be aware of the intervening device, as well as, in some circumstances, changes to the endpoints' TCP stacks.

Application layer proxies for TCP-based applications such as Web surfing using the Hypertext Transfer Protocol (HTTP) are the simplest example of such an implementation requiring no TCP stack changes. These allow the proxy-to-server TCP connection to operate with a degree of independence from the client-to-proxy TCP connection, introducing benefits such as the ability of the proxy to efficiently prefetch HTTP content from the server, or the ability of the proxy to implement a variant of TCP customized to the access network underlying the client-to-proxy connection.

I-TCP [29] can be viewed as a typical example of more complex nontransparent split-connection approaches. The proposed indirect protocol model for mobile hosts in I-TCP suggests that interaction between a mobile TCP host (MH) and a TCP host on a fixed network (FH) should be split into two separate interactions: one between the MH and its mobile support router (MSR) over the wireless access technology, and another between the MSR and the FH over the fixed network. Using this approach, problems related to mobility and the unreliability of wireless links are handled entirely within the wireless domain. When an MH wishes to communicate with an FH, a request is sent to the current MSR (which is an intervening node between the MH and FH) to open a TCP connection with the FH on behalf of the MH. A separate connection is established using a mobility-aware variation of TCP that is tuned for wireless links, for communication between the MH and its MSR. The MSR spoofs an image of the MH to the FH so that the FH communicates with the MSR as though it were communicating with the MH directly.

TCP stacks on the fixed hosts are not modified, but in order to implement I-TCP on a remote host, MH applications must use special I-TCP calls instead of the regular socket system calls. This need for receiver-side modification to the TCP stack, along with the need for deployment of MSR nodes at a cell level (or even cell-cluster level), incurs too great a cost for this solution to be considered practically useful for a large-scale wireless data network.

M-TCP [30] is an interesting special case. Because it modifies the TCP stack on the MH to respond to notifications of wireless link connectivity, it cannot be considered a truly transparent solution. However, it is a special case that may be considered to be semitransparent because it introduces an intermediate node, the Supervisor Host (SH), that intercepts and manages the bidirectional TCP packets so as to effectively create separate split connections without endpoint awareness. The SH ensures that the TCP sender does not invoke congestion control mechanisms during wireless link loss, by sending an ACK containing a TCP receiver window size (AWND) update of zero bytes. This forces the sender into a state known as the "persist mode," where it will not suffer from retransmit timeouts that cause the retransmission timer to back off, or reduce its congestion window. When the receiver resumes normal operation, the SH transmits a nonzero AWND advertisement to the sender, which resumes transmitting at full speed. M-TCP relies on the SH, instead of the congestion window, to regulate transmission to the receiver.

5.4.4.2 Transparent Approaches

The transparent split-connection approach is considered advantageous over its non-transparent counterpart, because it requires no modifications to TCP hosts at either end of the connection, nor the awareness of the endpoints that their TCP connection has been split. It is for this reason that transparent PEPs have long been regarded in the literature as a useful and promising solution to the problem of TCP over wireless. It is also precisely why we devote the second part of this chapter to a detailed study of the design and measured network performance of a novel PEP we proposed in [31].

TCP Snoop [32] is another transparent solution, which takes a slightly different tack. Namely, it functions predominantly at the link layer as a transparent TCP-aware protocol designed to improve TCP's performance, particularly over hybrid networks of wired and single-hop wireless links. Snoop only requires modification to the base-station node of the wireless network through the deployment of a "snoop agent";

however, the TCP endpoints do not need to be aware of the agent, yielding true transparency. A buffer is maintained with all unacknowledged TCP segments transmitted across the wireless link, to facilitate more efficient retransmissions of lost segments. The intention is to suppress the MH's duplicate acknowledgments (corresponding to wireless link losses), in order to prevent unnecessary invocations of congestion control mechanisms at the TCP sender. While achieving transparency as per our earlier definition, in the context of large-scale cellular radio networks Snoop has a significant scalability disadvantage when compared to the PEP described later. That disadvantage is that it requires link-layer modifications at each BS, as opposed to a small number of PEP devices that deal with multiple "transparently intercepted" TCP connections.

5.4.5 Replacing TCP: Introducing Proprietary Protocols

TCP replacement is at the more extreme end of the spectrum of available approaches. It seeks to emulate TCP by replacing the actual protocol over the wireless portion of the end-to-end path, typically with some proprietary combination of the User Datagram Protocol (UDP) and transmission control implemented at higher layers. From a protocol stack point of view, this may be thought of as artificially pushing functions usually residing at the transport layer (TCP) higher up the stack. Such approaches eliminate the limitations of TCP and provide additional functionality such as data compression, at the cost of no longer being freely available in the public domain. These solutions are aimed at the telecommunications carrier and service provider market, having nontrivial purchase and implementation costs due to the proprietary nature of the software. Additionally, for optimal performance they tend to require the added operational and user inconvenience costs of maintaining a client installed on the user device.

VTP [33] is a proprietary protocol developed by Fourelle Systems and implemented by Venturi Wireless as part of a suite of tools for enhancing wireless TCP performance including protocol optimization, intelligent content compression, and high-speed caching. The protocol resides on top of UDP and completely replaces TCP over the wireless link portion of the network. Exact implementation details are unknown due to the proprietary nature of VTP, but it is claimed in [33] that it adds reliability to UDP while maintaining a standard UDP framework to ensure interoperability. The architecture requires deployment of a server in the network and a software client on the MH.

Developed by Flash Networks, Wireless Boosted Session Transport (WBST) [34] is a component of Flash Networks' NettGain Product. Similar to VTP, it is designed as a reliable transport protocol replacing TCP over the wireless link and requires the deployment of a NettGain server in the network. WBST applies bandwidth estimation techniques and rate-based control to increase the bandwidth utilization of the wireless link. The WBST sender continually maintains an estimate of transmission rate and monitors the ACK interarrival times. To accommodate changes observed in this interarrival time (which indicate a change in the available link bandwidth), the interpacket transmission delays are adjusted accordingly. The intention is to utilize all of the available bandwidth without causing congestion.

It should be noted that such TCP replacement approaches need to be restricted to only the wireless portions of a provider's access network, because they may cause unfairness issues when coexisting with standard TCP implementations.

5.4.6 Replacing the Wireless: Making New Technologies TCP Friendly

The final solution we examine may broadly be described as "if it's too hard or costly to change standard TCP to suit the wireless access, then change the wireless access to suit standard TCP." We have categorized this as a somewhat "extreme" solution because philosophically it violates the independence of protocol layers in the TCP/IP stack, which has long underpinned the guiding principles of Internet design, by calling for one protocol layer to be designed with knowledge of how it will interact with, and depend on, other(s).

However, when confronted by the significant downsides and complexity cost of most of the other solutions described so far, it becomes apparent that this may be the least unpalatable approach in the long term. Furthermore, it could be argued that even without a particular focus on TCP, the future of wireless is naturally moving along a technology evolution curve that is very beneficial to standard TCP. That is, exactly the kinds of changes necessary to improve the performance of TCP over wireless are already intrinsically included in future "beyond 3G" wireless access such as 3GPP Long Term Evolution (LTE) [35], WiMAX [36], and the fourth generation (4G) holy grail of IMT-Advanced [37].

For examples of this, we need look no further than the key areas of system latency and radio resource management time scale. In terms of the former, these new technologies are heading toward RTTs on the order of 10 ms [35, 37], which is the same order of magnitude as for typical wireline access through Asymmetrical Digital Subscriber Line (ADSL) or cable, and should ensure a more rapidly adaptive performance of standard TCP.

Second, the improvement (i.e., decrease) in the transmission time intervals for these new technologies means that the radio resource management time scale will be on the order of 1 ms [35], thus continuing to be significantly faster than TCP's time scale and alleviating serious bandwidth oscillation problems described earlier.

A word of caution is required here. None of the emerging technologies described in this section have yet reached the levels of mass-market penetration and maturity in large-scale live networks. As a result, sufficiently detailed measurement and analysis studies of TCP behavior over such networks of the type presented in [38] have not as yet appeared in a large enough number for us to confidently say that all potential "TCP-over-future-wireless" issues have been identified. Therefore, even with the described technological progress, it is conceivable that known TCP issues such as bandwidth variance and other as-yet undiscovered ones may arise. This would mean that some variants of the TCP enhancement techniques discussed in this chapter might continue to be relevant and necessary in the future.

5.5 CASE STUDY: A NOVEL TCP PERFORMANCE-ENHANCING PROXY

We devote the remainder of this chapter to examining the design and measured network performance of a proof-of-concept prototype for a novel PEP called the "TRL-PEP," first proposed in [31] and designed to make use of the most desirable functionality of existing algorithms in this field, as well as to add novel enhancements.

The main features of the TRL-PEP can be summarized as follows:

- It is a "clientless" solution with transparency to TCP connection endpoints, requiring no modification to, or intervention on the part of, sender or receiver.
- It combines into one solution different state-of-the-art ideas in the field described earlier: (1) "M-TCP"—use of TCP Persist mode and Zero Window Advertisements to pause the TCP sender while maintaining state variables (TCP sender control), (2) "TCP Snoop"—use of a buffer at an intermediate node, and local retransmissions over the wireless link to the receiver (TCP receiver control), and (3) "TCP FACK"—the use of the fact that presence of duplicate acknowledgments indicates that more data can be sent to the receiver (TCP receiver control).
- It proposes several novel algorithms including (1) efficient use of radio resources through aggressive link utilization and loss recovery mechanisms, (2) intelligent detection of loss coupled with SACK mechanisms where available, and (3) intelligent state prediction of TCP hosts.
- The version described here only operates on the downlink, optimizing the flow of data from the server to the end user device.

5.5.1 Architectural Considerations

The TRL-PEP would be located as close to the wireless TCP receiver as possible, while still being far enough from the endpoint that it may optimize traffic destined for a range of users. This is to ensure that all traffic destined for the wireless segment of the network passes through the TRL-PEP. The TRL-PEP works on the simple principle that it should attempt to fully utilize the wireless link to a receiver at all possible times. Network topology allows certain assumptions with regard to the application of this principle:

- Most TCP segment loss and delay variation experienced by the receiver is due to wireless channel errors rather than network congestion, which would only occur in the unlikely scenario of multiple users in the same radio cell simultaneously (1) obtaining excellent radio conditions and (2) trying to fill the wireless link with large amounts of traffic.
- Although TRL-PEP's main focus is mitigation of wireless channel errors, it still needs a basic congestion control capability to protect against the unlikely event of true congestion-based losses. This is achieved through a simple algorithm triggering a CWND reduction of X segments, whenever Y consecutive segments are lost (whether identified as lost by sequence number

inference or timeout). X and Y are dynamically configurable variables. As in standard TCP's congestion-avoidance phase, the CWND then grows back by one segment per RTT of loss-free operation.
- TCP segments are assumed to be delivered in order to the receivers, as there are no alternate network paths from the TRL-PEP to the wireless receivers.

5.5.2 Performance Enhancement Techniques Employed in the TRL-PEP

5.5.2.1 Aggressive Link Utilization

Once a connection is established between a sender and receiver through the TRL-PEP, data is requested by the TRL-PEP from the sender by actively spoofing ACKs to the sender, until the connection buffer at the TRL-PEP has been filled. The advertised window parameter, AWND, in the spoofed ACKs is reduced in proportion to the available buffer space, to the point that the TCP sender will stop sending data until the AWND is increased again. This is a standard TCP mechanism to control the flow of data from the sender. If the AWND is set to zero, the sender will enter the TCP standard Persist mode, which effectively pauses the sender while maintaining its state variables. TCP's conservative slow-start behavior is less appropriate in the typically high-bandwidth high-delay connection between a sender and receiver in today's broadband wireless networks. We observe a "thick long pipe" in cdma2000 and UMTS standards, compared to the "thick short pipe" analogy of wireline accesses such as ADSL or cable. Thus, the TRL-PEP transmits data to the receiver at the highest rate supportable by the wireless network. In the simulation studies reported here, this "feed rate" for the TRL-PEP to receiver link was separately pre-configured based on "typical" values of bandwidth and delay for each of the wireless access technologies under study. Although beyond the scope of the proof-of-concept prototype, a relatively simple extension could (1) facilitate a dynamic estimation of the optimal feed rate (e.g., via RTT measurements or ACK counting, as in TCP Vegas [21]or Westwood [25]) and (2) simultaneously support connections to receivers on different wireless access networks by separately estimating and managing the feed rate on a per-connection basis.

5.5.2.2 Additional Data Transmission on DUPACKs

When the TRL-PEP encounters a duplicate acknowledgment, it allows extra data to be transmitted. The rationale for this, similar to that used in the standard TCP NewReno's "Fast Recovery" phase, is that the reception of a duplicate acknowledgment infers that a packet has been received correctly by the receiver, so there is available capacity in the link for more data to be transmitted. This extra data pushed into the link, and the resultant duplicate acknowledgments, also act as a probing mechanism to determine if the link has been disconnected or overused.

5.5.2.3 Retransmission on Partial Acknowledgments

The TRL-PEP has the capacity to recover from multiple packet losses within the same window of transmitted data. This behavior is specific to the case where SACK information from the receiver is not available, and is governed by the following algorithm:

1. Upon a retransmission from a duplicate acknowledgment, a variable *dupack_target* is set to the highest sequence number sent to the receiver thus far (shown as A in Figure 5.4a). Note that there is no requirement that the packet numbered *dupack_target* has been successfully received.
2. When a new acknowledgment numbered less than *dupack_target* is received (known as a partial acknowledgment), the packet with the corresponding sequence number is immediately retransmitted (shown as "B" in Figure 5.4a).
3. Repeat Step 2 until a new acknowledgment numbered greater than *dupack_target* is received, at which point we return to normal TCP operation.

This algorithm is based on the premise that at the time that the partial acknowledgment B is received, the TRL-PEP has knowledge that a higher sequence numbered packet has been transmitted as indicated by the presence of a duplicate ACK. Thus, the received partial acknowledgment is an indication that the next-in-sequence packet was lost (assuming that packets and ACKs arrive in order).

5.5.2.4 Intelligent Use of Selective Acknowledgments

The TRL-PEP makes use of SACK data to retransmit only necessary packets to the receiver. To respond immediately to many concurrent retransmissions of data based on SACK information and quickly detect subsequent losses of these retransmissions, the TRL-PEP maintains two arrays of sequence numbers: *sack_seq[]* and *sack_target[]*. The *sack_seq[]* array records the sequence number of the segment that has been retransmitted in response to SACK, and *sack_target[]* records the corresponding acknowledgment number that ought to be received before the retransmitted data (use of these arrays is illustrated in an example below). Thus, it is possible to repeat the retransmission without waiting for an RTO. In this way, the proposed algorithm mitigates the loss of multiple transmitted and retransmitted segments within a single RTT period, unlike standard TCP. An example of this mechanism in operation is shown in Figure 5.4b, with a time-sequence graph of a TCP connection with the SACK option enabled. The vertical line segments marked S depict information from the receiver informing the sender (in our case, the TRL-PEP node in the middle of our end-to-end path) of exactly which packets have not been received correctly.

Consequently, in Figure 5.4b, the packets marked "A," "D", and "F" are all retransmitted as soon as the TRL-PEP receives enough information to determine that those packets have not been received. In this example, prior to packet "A" being retransmitted, *sack_target[0]* is set to the highest sequence number sent as shown by packet "B." Because it is the lowest outstanding segment, the sequence number of packet "A" is recorded in *sack_seq[0]* and then retransmitted. As shown in Figure 5.4, the same SACK information used in deciding to resend "A," is also used to indicate that the next sequence number to be resent is that of packet "D." This, then, is the lowest outstanding segment that has not already been retransmitted, and so is recorded in the next available position in *sack_seq[1]*. *sack_target[1]* is also set accordingly (i.e., to equal the sequence number of "B") and packet "D" is retransmitted.

SACK information continues to arrive with the duplicate acknowledgments and new data is sent accordingly, until the SACK information arrives at "C." The highest acknowledgment number indicated by the SACK information at "C" becomes

Enhancing TCP Performance in Hybrid Networks

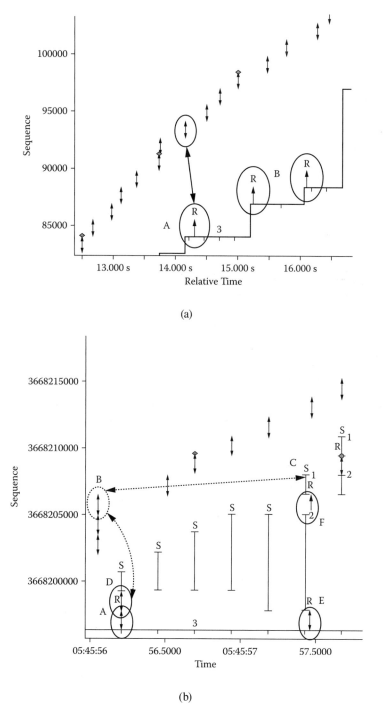

FIGURE 5.4 TRL-PEP algorithmic improvements: (a) Retransmission on partial acknowledgments and (b) intelligent use of selective ACKs.

greater than *sack_target[0]*, and the acknowledgment number is still less than *sack_seq[0]*. Assuming as before that packets (and acknowledgments) arrive in order, the TRL-PEP infers that the retransmission "A" of the segment has been lost and retransmits *sack_seq[0]* again, as shown by packet "E" (*sack_target[0]* is reset to the new value of highest sequence number sent). "C" contains two pieces of additional information: (1) packet "D" has been successfully received, so it is removed from the arrays; and (2) another packet is lost, as indicated by the SACK sequence number "gap" at "F." The corresponding sequence number is recorded as the new *sack_seq[1]*, *sack_target[1]* is set accordingly, and packet "F" is retransmitted.

5.5.2.5 TCP State Prediction

The TRL-PEP maintains the TCP sender and receiver states. This is achieved by observing the flags of packets received from the hosts and making use of knowledge of the standard TCP states and state transitions, as described in the original TCP specification, RFC 793 [39]. For example, at the beginning of a connection, should we receive a SYN flag, we know that the sender of that packet would be in the SYN_SENT state, and later, when we receive a SYN-ACK from the receiver, that the receiver would be in the SYN_RCVD state. This knowledge of the TCP end-host states is essential in maintaining integrity of the connection, as it allows us to manipulate the hosts for optimal data transfer and to gracefully establish and terminate connections.

5.5.3 SIMULATION RESULTS

A simulation of the TRL-PEP algorithms was developed in the *ns-2* discrete event network simulator. For the development of the TRL-PEP we installed *ns-2* (version 2.26) on a Linux Redhat 6.2 (Kernel 2.4.18) platform. To this standard installation we added custom-built simulations of GPRS (GSM) and 1xRTT (cdma2000) wireless links. *ns-2* contains a range of TCP implementations that we were able to enhance to include features such as dynamic window advertisement and TCP persist mode. We also enhanced the trace file format to include additional information such as advertised window, SACK, and other TCP options for use with a standard TCP analysis tool.

Tests were performed using simulated GSM/GPRS and cdma2000 1xRTT links for varying uniformly distributed packet error rates, packet jitter, and different TCP implementations in the network topology depicted in Figure 5.5. For baseline comparison purposes, our standard TCP implementation continued to be NewReno, although the SACK option was assumed to be enabled in order to make it consistent with the standard Linux BIC TCP implementation [40]. Figure 5.6 shows the typical performance improvement using simulated cdma2000 1xRTT, in the presence of the Bandwidth Oscillation effect particular to this technology, as described earlier. The TRL-PEP mitigates this behavior to clearly outperform other TCP implementations at different packet error rates. Typical performance improvement in GPRS (not shown here) was similar but not as pronounced.

In any comparison between the TRL-PEP and other schemes, it should be reinforced that the former requires no change to the sender- or receiver-side TCP

Enhancing TCP Performance in Hybrid Networks

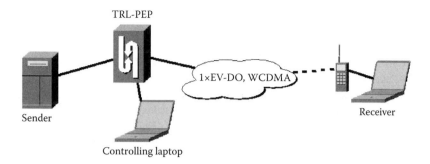

FIGURE 5.5 TRL-PEP test network topology.

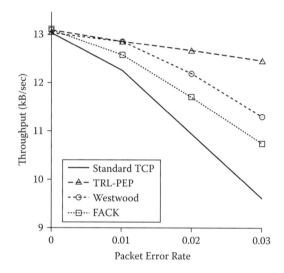

FIGURE 5.6 TCP throughput for a 1 MB FTP download over a simulated cdma2000 1xRTT link.

implementations, nor does it require the installation of any software clients on the user device. For a network provider, these represent significant advantages in favor of the TRL-PEP. The simulation was developed as a precursor to the "proof-of-concept" prototype. Simulation code was sufficiently modular to be directly used in implementation of the PC-based prototype, which was subsequently put into a test network for a field trial, the measurement results of which we detail in the next section.

5.5.4 Test Network Trial Results

The TRL-PEP prototype was implemented on a Knoppix 3.4 (Kernel 2.4.26) Linux operating system due to its freely available associated tools and libraries, which were required for development of the TRL-PEP. All code was written in C++ and compiled using GNU Compiler Collection (GCC) version 3.3.3. TRL-PEP simulation code was reused with little or no modification in the prototype implementation. The hardware used for the prototype was a modest desktop PC (Celeron 2.0 GHz processor) with a reasonably limited RAM (256 MB). The system had three 100Mbps Ethernet interfaces: a "wireless facing" interface, an "Internet" facing interface, and a "controller" interface. The latter was used for remote administration and monitoring, and had no functional purpose for the operation of the TRL-PEP.

The same topology from Figure 5.5 was used to test the TRL-PEP in controlled radio environment conditions over test networks based upon WCDMA (UMTS) and 1xEVDO (cdma2000). The radio channels studied (by way of channel simulator) were static, typical urban environment with 50km/h mobility (TU50) and a rural area environment with 100km/h mobility (RA100). CIRs were varied.

FTP throughput tests were conducted using the standard Windows FTP client with the receiver advertised window AWND optimally set to the given radio technology's bandwidth delay product. A Linux FTP server was used to study small (50 kB), medium (1 MB), and large (4 MB) file transfers. Throughput measurements were computed using the *tcptrace* analysis package, and time-sequence graphs examined using the Linux-based program *xplot*. The WAPT utility (v3.0) was used to automate Web page download tests, with the same Web page being used throughout all tests. This standard page consisted of a total size of approximately 300 kB allocated to text, a textured image background, two 10 kB images, two 20 kB images, two 40 kB images, and two 80 kB images. WAPT provided the functionality to repeatedly and independently download the test page and record each page transaction time. Each data point was constructed by averaging 40 runs. All files used in the tests were created to be uncompressible—including the images if lossy compression is not considered.

Tests were conducted comparing (1) the standard Linux BIC TCP implementation [40] (considered to be an improvement on the standard NewReno TCP because of the use of SACK options), (2) the TRL-PEP, and (3) a commercially available product relying on the TCP replacement approach (as discussed earlier, but without the content compression feature).

Figure 5.7 shows the 1 MB file size FTP mean throughput results obtained using WCDMA as the cellular radio access technology with a static fading model. The TRL-PEP provides up to a ~25% improvement in throughput over the SACK-enabled standard TCP and also slightly exceeds the performance of the TCP replacement product. The primary reason for the improvement is TRL-PEP's quick utilization of the dynamically expanding available bandwidth of the WCDMA link; conversely, standard TCP takes a long time to do so via slow start. The counterintuitive result of TCP replacement product throughput at the better CIR of 70dB being marginally worse than at 10dB CIR may be attributed to the variability of simulated radio channels. The observed discrepancy in our mean throughput estimator falls within the 95% confidence intervals computed for each data point (not shown). Recall that the data points represent the average over 40 repeated measurement runs under given radio channel settings.

Figure 5.8 shows the 4 MB file size FTP mean throughput results obtained using 1xEVDO in an RA-100 channel. In this case, the maximum throughput gain of TRL-PEP over the SACK-enabled standard TCP is ~20%, and it still marginally

Enhancing TCP Performance in Hybrid Networks

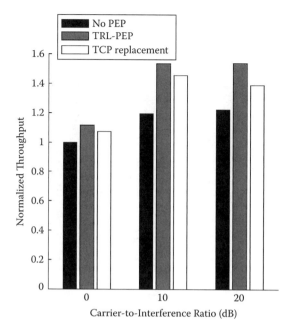

FIGURE 5.7 TCP throughput for WCDMA static scenario—1 MB FTP download.

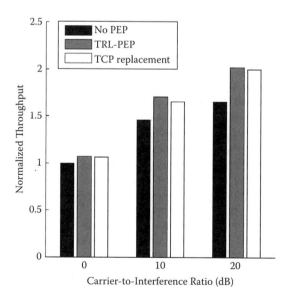

FIGURE 5.8 TCP throughput for 1xEVDO RA-100 scenario—4 MB FTP download.

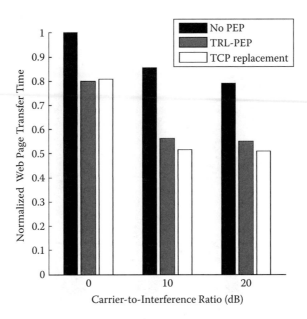

FIGURE 5.9 Web page transfer time for 1xEVDO static scenario—300 kB standard Web page download.

exceeds the performance of the TCP replacement product. As in the previous scenario, this occurs partly due to aggressive wireless link utilization, but also due to the TRL-PEP's advanced data recovery algorithms in the presence of packet losses. Figure 5.9 shows the results of the retrieval of our standard Web page in the static fading radio channel over 1xEVDO. The results indicate up to a 35% reduction in the overall time required by the TRL-PEP to retrieve the Web page over SACK-enabled standard TCP (as measured from the time the request is sent to the time the connection is completed).

In this instance, the performance of the TCP replacement product is either equal to or marginally better than that of the TRL-PEP, but as in Figure 5.7 and Figure 5.8, the two remain almost imperceptibly close. The improvement due to the aggressive utilization of the link by the TRL-PEP is even more significant here because standard HTTP dictates that multiple TCP connections are required to retrieve the Web page data, as opposed to the single TCP connection used in the FTP test. This leads to multiple "component TCP connections" experiencing slow-start phases, and thus presents a greater opportunity for the proxy to reduce the overall transaction time by ensuring that each such connection is able to bypass the wasteful slow-start phase.

ACKNOWLEDGMENT

The figures and some text in this chapter are based on M. Ivanovich, J. Li, and P. Bickerdike, "On TCP performance enhancing proxies in a wireless environment," *IEEE Communications Magazine,* Vol. 46, pp. 76–83, Sept. 2008.

REFERENCES

1. S. Floyd and T. Henderson, "The NewReno modification to TCP'S fast recovery algorithm," *RFC 2582*, IETF, 1999.
2. S. Ladha, et al., "On the prevalence and evaluation of recent TCP enhancements," in *Proc. of IEEE Globecom '04*, Vol. 3, pp. 1301–1307, 2004.
3. 3GPP2. (1999). Physical Layer standard for cdma2000 spread spectrum systems [Online]. Available: http://www.3gpp2.org/public_html/specs/C.S0002-0_v1.0.pdf
4. M. Yavuz and F. Khafizov, "TCP over wireless links with variable bandwidth," in *Proc. IEEE VTC '02*, Vol. 3, pp. 1322–1327, 2002.
5. M. Allman, V. Paxson, and W. Stevens, "TCP Congestion Control," *RFC 2581*, IETF, 1999.
6. G. Brasche and B. Walke, "Concepts, services and protocols of the new GSM Phase 2+ General Packet Radio Service," *IEEE Communications Magazine*, Vol. 35, pp. 94–104, Aug. 1997.
7. R. Chakravorty, J. Cartwright, and I. Pratt, "Practical experience with TCP over GPRS," in *Proc. IEEE GLOBECOMM '02*, Vol. 2, pp. 1678–1682, 2002.
8. H. Holma and A. Toskala, Eds. *WCDMA for UMTS*, 2nd ed. New York: Wiley, 2002.
9. P. Rysavy. (2004). Data capabilities: GPRS to HSDPA. 3G Americas White Paper [Online]. Available http://www.3gamericas.org/pdfs/rysavy_data_sept2004.pdf
10. M. Ericson, et al., "Transport channel switching for interactive TCP/IP traffic in WCDMA," in *Proc. IEEE VTC Spring-05*, Vol. 4, pp. 2240–2244, 2005.
11. 3GPP2. (2005). cdma2000 high rate packet data air interface [Online]. Available: http://www.3gpp2.org/Public_html/specs/C.S0024-A_v2.0_050727.pdf
12. 3GPP. (2002). High Speed Downlink Packet Access (HSDPA); Overall description; Stage 2," 3GPP TS 25.308 [Online]. Available: http://www.3gpp.org/ftp/specs/html-info/25308.htm
13. F. Ren, et al., "Improving TCP throughput over HSDPA networks," *IEEE Transactions on Wireless Communications*, Vol. 7 (6), pp. 1993–1998, 2008.
14. IEEE 802.11b, Part 11, "Wireless LAN Medium Access Control (MAC) and Physical Layer (PHY) specifications: Higher-Speed Physical Layer Extension in the 2.4 GHz Band," *IEEE*, Feb. 1999.
15. M. Hashimoto, et al., "Performance evaluation and improvement of hybrid TCP congestion control mechanisms in Wireless LAN Environment," in *Proc. ATNAC 2008*, pp. 367–372, 2008.
16. M. Mathis, et al., "TCP selective acknowledgment options," *RFC 2018*, IETF, 1996.
17. W. Stevens, "TCP slow start, congestion avoidance, fast retransmit and fast recovery algorithms," *RFC 2001*, IETF, 1997.
18. K. Xu, et al., "Improving TCP performance in integrated wireless communications networks," *Computer Networks*, Vol. 47, pp. 219–237, 2005.
19. IETF. (2003). Performance Implications of Link Characteristics (PILC) Working Group Charter [Online]. Available: http://www.ietf.org/html.charters/OLD/pilc-charter.html
20. H. Inamura and G. Montenegro, Eds. "TCP over Second (2.5G) and Third (3G) Generation Wireless Networks," *RFC 3481*, IETF, 2003.
21. L. S. Brakmo and L. L. Peterson, "TCP Vegas: End-to-end congestion avoidance on a global Internet," *IEEE Journal on Selected Areas in Communications*, Vol. 13, pp. 1465–1480, 1995.
22. C. Jin, S. Low, and X. Wei, "Method and apparatus for network congestion control," United States Patent & Trademark Office (2005–01–27).

23. T. Bonald, "Comparison of TCP Reno and TCP Vegas via fluid approximation," INRIA Research Report RR-3563, Nov. 1998.
24. M. Mathis and J. Mahdavi, "Forward acknowledgment: Refining TCP congestion control," *ACM SIGCOMM Computer Communication Review,* Vol. 26, (Oct), pp. 281–291, 1996.
25. M. Gerla, et al., "TCP Westwood: Congestion window control using bandwidth estimation," in *Proc. IEEE Globecom '01,* Vol. 3, pp. 1698–1702, 2001.
26. UCLA Computer Science Department. (2003). TCP Westwood home page [Online]. Available: http://www.cs.ucla.edu/NRL/hpi/tcpw/
27. K. Xu, Y. Tian, and N. Ansari, "TCP-Jersey for wireless IP communications," *IEEE Journal on Selected Areas in Communications,* Vol. 22, (May), pp. 747–756, 2004.
28. R. Ludwig and R. H. Katz, "The Eifel Algorithm: Making TCP robust against spurious retransmissions," *ACM Computer Communication Review,* Vol. 30, pp. 30–36, 2000.
29. A. Bakre and B. Badrinath, "I-TCP: Indirect TCP for mobile hosts," in *Proc. 15th International Conf. on Distributed Computing Systems,* p. 136, 1995.
30. K. Brown and S. Singh, "M-TCP: TCP for mobile cellular networks," *ACM Computer Communications Review,* Vol. 27, pp. 19–43, 1997.
31. M. Ivanovich, J. Li, and P. Bickerdike, "On TCP performance enhancing proxies in a wireless environment," *IEEE Communications Magazine,* Vol. 46, pp. 76–83, Sept. 2008.
32. S. Vangala and M. A. Labrador, "The TCP SACK-aware snoop protocol for TCP over wireless networks," in *Proc. Vehicular Technology Conf.,* pp. 2624–2628, 2003.
33. Fourelle Systems, Inc. (2000). The intelligent solution to optimizing Internet bandwidth. Technical White Paper [Online]. Available: http://www.techmarketingink.com/Fourelle_wp.PDF
34. O. Shaham, et al., "Rate control for advanced wireless networks," in *Proc. International Conf. on Third Generation Wireless and Beyond,* pp. 444–449, 2001.
35. 3GPP. (2008). High Evolved Universal Terrestrial Radio Access (E-UTRA) and Evolved Universal Terrestrial Radio Access Network (E-UTRAN); Overall description; Stage 2. 3GPP TS 36.300 [Online]. Available: http://www.3gpp.org/ftp/Specs/html-info/36300.htm
36. K. Etemad, "Overview of Mobile WiMax technology and evolution," *IEEE Communications Magazine,* Vol. 46, (Oct), pp. 31–36, 2008.
37. ITU-R, Recommendation ITU-R M.1645, "Framework and overall objectives of the future development of IMT-2000 and systems beyond IMT-2000," 2003.
38. E. Halepovic, et al., "TCP over WiMAX: A measurement study," presented at the MASCOTS '08 Conf., Baltimore, Maryland, Sept. 2008.
39. J. Postel, Ed., "Transmission Control Protocol," *RFC 793,* IETF, 1981.
40. L. Xu, K. Harfoush, and I. Rhee, "Binary increase congestion control for fast long-distance networks," in *Proc. INFOCOM 2004,* Vol. 4, pp. 2514–2524, 2004.

6 Femtocell—Home Base Station
Revolutionizing the Radio Access Network

Manish Singh

CONTENTS

6.1	Introduction	126
6.2	Femtocell Vision	126
	6.2.1 Why Femtocell?	126
	6.2.2 Addressing the 3C Challenge: Coverage, Capacity, and Churn	128
	6.2.3 The Fourth C: Cost	129
	6.2.4 Business Drivers for Femtocell	130
	6.2.5 Femtocell versus UMA	130
6.3	Network Architecture	132
	6.3.1 3G UMTS Network Architecture	132
	6.3.2 Network Architecture Evolution with Femtocells	133
	6.3.3 UMTS-Centric Network Architectures	133
	6.3.4 UMA-Based Architecture	134
	6.3.5 SIP/IMS-Based Architecture	135
6.4	The Signaling Plane in Femtocell	135
	6.4.1 Uu Interface Protocols	136
	6.4.2 Network Interface Protocols	136
	6.4.3 Remote Device Management and Control	137
	6.4.4 Security Aspects	138
6.5	Femtocell: Self-Organizing Network	138
	6.5.1 Licensed Spectrum	138
	6.5.2 Managing RF Interference	139
	6.5.3 Ping-Pong Effect	140
	6.5.4 Open versus Closed Access	140
	6.5.5 Emergency Calls	140
6.6	What's Next?	141
	6.6.1 LTE Femtocells	141
	6.6.2 Revolutionizing Network Rollout	142
Acknowledgments		143
References		143

6.1 INTRODUCTION

Femto comes from the Danish word *femten*, meaning 15. Femto is an SI prefix in the International System of Units denoting a factor of 10^{-15}, or one quadrillionth. In sharp contrast to a macrocell, a femtocell is a small wireless base station residing in a consumer's home. Femtocells are designed to provide excellent coverage in indoor environments and to work seamlessly with an existing wireless handset. Existing Internet Protocol (IP) broadband links (such as digital subscriber line [DSL] or cable) are leveraged to backhaul the mobile voice, video, short message service (SMS), and data traffic from the home and integrate with an existing wireless core network. Simply put, a femtocell is just another cell site that fully supports seamless handoff to and from the existing macro network and extends the wireless network into consumers' homes.

6.2 FEMTOCELL VISION

The Femtocell concept is not new; the idea of a home base station has been around since the early days of wireless networks. However, what was missing before is now a reality: voice over IP (VoIP), which enables the transfer of voice over an IP network. Another separate but equally important factor supporting large-scale femtocell deployment is broadband proliferation. Fifty-three percent of U.S. households have a broadband connection, and the percentage is even higher in other countries such as South Korea, where 89% of homes are served. Existing broadband connections form the backhaul to efficiently carry the mobile traffic from consumers' homes out to the wireless core network.

6.2.1 Why Femtocell?

Adoption of wireless data services is today's growth engine for wireless network operators. We have a "perfect data storm" in the making, fueled by a young Internet-savvy generation that wants immediate access to information anytime, anywhere. Mobile data users want instant access, from downloading the latest video on YouTube™ to updating their Facebook™ profiles with their latest pictures and videos. To meet this demand, wireless network operators have to find cost-effective ways to augment capacity in their networks.

There are two main techniques available to wireless network operators to add more capacity in their networks:

- Increase spectral efficiency: Measured in bits/sec/Hz. From General Packet Radio Service (GPRS) to Enhanced Rates for GSM Evolution (EDGE) to Wideband Code Division Multiple Access (WCDMA) and now Long-Term Evolution (LTE), radio frequency (RF) technology continues to improve spectral efficiency. LTE improves spectrum efficiency by four times over its predecessor, 3G WCDMA-based Universal Mobile Telecommunications System (UMTS).
- Cell splitting: Adding more cell sites and reducing subscriber density in a given cell site, thereby adding more capacity in the network.

Femtocell—Home Base Station

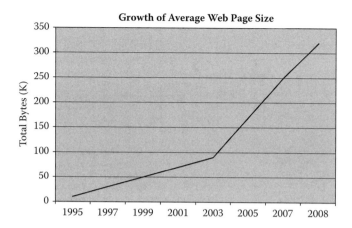

FIGURE 6.1 Average Web page size has tripled since 2003. (Domenech, J., A. Pont, J. Sahuquillo, and J. Gil, 2008. A user-focused evaluation of Web prefetching algorithms. *Computer Communications*. Used with permission.)

The reality is that the rate of growth of consumer data demand far outstrips the rate of spectral efficiency improvement. It took more than a decade for 3G WCDMA UMTS networks to be adopted, yet today mobile broadband traffic is growing by almost 400% every year. In addition, the average Web page size has tripled since 2003 [3], as shown in Figure 6.1, and the number of Web sites continues to grow exponentially.

To keep up with the growing demand, wireless network operators have to increase capacity through the construction of additional macro cell sites; that is, through cell splitting. This strategy is becoming much less attractive, however. Site acquisition costs are exorbitant and continue to rise as space on viable towers and buildings fills up, landlords exact high rent, and regulators impose onerous permitting requirements. Furthermore, public opposition to the building of large-scale base stations is increasingly common. Acquiring a site is only half the battle, though; sophisticated base station equipment must then be purchased, installed, insured, operated, and maintained. In sum, the net present value (NPV) of a cell site in the U.S. is estimated to be $500,000.

The 3G femtocell solution creates a new location for wireless base stations in consumers' homes or offices. A femtocell is a small customer premises equipment (CPE) device that plugs into a subscriber's existing broadband Internet connection and works with existing mobile handsets, as shown in Figure 6.2. Femtocells are low-power devices combining NodeB and Radio Network Controller (RNC) functionality and are self-configuring to minimize interference.

Femtocells provide a compelling solution to network operators for augmenting capacity in their networks in a cost-effective manner. In essence, a small space on a bookshelf or desktop in a consumer's home becomes a cell site and thus there are no site acquisition costs involved. The customer pays for the femtocell backhaul by using an existing broadband line (e.g., cable or DSL) that is typically charged at a flat rate, and hence there are no incremental costs to the femtocell subscriber. Electricity bills, which constitute a large operational expense item for macrocell sites, are paid for by the consumer as well.

- Offloads the most power hungry 3G users from the macrocell to femtocell
- Provides cell coverage indoors
- Low power and self-configuring
- Collapsed mobile network (RAN)
- Leverages customer broadband

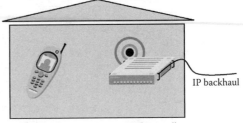

FIGURE 6.2 3G femtocell resides in a consumer's home and backhauls mobile traffic over IP broadband link.

In all, femtocells allow carriers to offload their existing macrocell networks because in-home wireless traffic is routed through broadband connections rather than through the operators' NodeB leased lines. Macrocellular network resources that were formerly charged with handling indoor users' mobile calls, text messages, and so on, are now freed up as that traffic is instead offloaded to femtocells and consumers' broadband connection.

6.2.2   Addressing the 3C Challenge: Coverage, Capacity, and Churn

The three Cs of wireless—coverage, churn, and capacity—are stifling 3G adoption. We discussed the capacity problem earlier; now we will turn our attention to the coverage issue in 3G networks.

3G wireless networks suffer from inadequate indoor signal penetration, leading to poor coverage in the indoor environment where consumers spend on average two thirds of their time. From an RF propagation perspective, buildings and walls have never been friends to a radio signal. Physical structures cause RF signal attenuation because of reflection, diffraction, scattering, and multipath signal fading, all resulting in poor radio signal reception in an indoor environment. In some cases, virtual dead spots can occur inside buildings even though there is great coverage outside; indeed, walls can attenuate RF signals by 10 to 20 dB depending on the type of construction. At the same time, studies have shown that an average consumer spends 50% to 60% of his or her time in an indoor environment and almost 70% of wireless calls are originated and terminated indoors.

The indoor coverage problem is amplified even more in 3G WCDMA UMTS networks when compared to their predecessor 2G Global Systems for Mobile Communications (GSM) networks. In most parts of the world, 3G UMTS networks are deployed in the 2.1 GHz spectrum range as opposed to 900 MHz or 1.8 GHz for GSM. As wireless networks are deployed in higher frequency bands (i.e., 2.1 GHz > 1800 MHz > 900 MHz), indoor signal penetration becomes worse because higher frequencies suffer higher levels of attenuation.

Poor coverage diminishes the quality of voice and video applications and slows down high-speed data services. Dropped calls and time-consuming downloads can lead to churn as high-value 3G customers, who expect a high quality of service, choose to pick up their landline phones or switch to other mobile carriers in search of uninterrupted voice calls, clear video images, and faster downloads. Churn is intricately linked to coverage and capacity and is a double-edged sword, costing carriers both in lost revenue—about $100 per month average revenue per user (ARPU)—and in new customer acquisition costs, about $400 per customer.

As small wireless home base stations, femtocells transmit at very low power yet create almost ideal indoor radio conditions. The very walls that are radio signals' foes actually become their friends as they attenuate RF signal propagation out of the home and thereby minimize radio interference with the existing macrocellular network or other nearby femtocells. Last but not least, one femtocell can support four to six simultaneous voice calls, which means that each member of a family of four can talk simultaneously on a femtocell.

6.2.3 THE FOURTH C: COST

Whether a consumer buys a femtocell or the operator subsidizes it, low cost is critical if femtocells are to meet the 3C challenge and achieve mass market adoption. As a consumer product, analysts project the target price point for a femtocell to be US$70 to $200, which means the bill of materials (BOM) cost needs to be at least half this amount. Cost is thus the fourth C for femtocells—and the technology is unlikely to flourish without a low per-unit cost.

A wireless base station has two key functions—modulation/demodulation (modem) and control. In traditional macro base stations depending on the capacity, the modem functions typically run on one or more Digital Signal Processing (DSP) processors. Similarly, control functions run on one or more application processors. To drive the BOM costs down for femtocells, integrated silicon supporting both modem and application functions in a single device are essential. Thanks to Moore's Law, a number of integrated silicon devices are now available that support both DSP and control functions in a single part.

There are two main classes of femtocell—residential and enterprise—and the main differentiating factor between the two is the number of active calls that the femtocell can support. A typical residential femtocell supports four to six simultaneous calls, and an enterprise femtocell can support a much higher number of users. The active call capacity directly impacts the total CPU cycles that are required to meet the performance and thus directly impacts the device's cost.

The cost of a femtocell can be further driven down if its functionality is integrated inside another CPE device such as a residential gateway, a set-top box (STB) or a DSL modem, as shown in Figure 6.3. CPE device convergence provides additional benefits to network operators as it is always easier to manage one CPE device as opposed to multiple devices. Thus, a femtocell access point (FAP) can be classified into two categories: standalone FAP and integrated FAP, where the femtocell function is embedded in another CPE device. Femtocell designers can drive costs down by sharing application processors, memory, network interfaces, and other elements

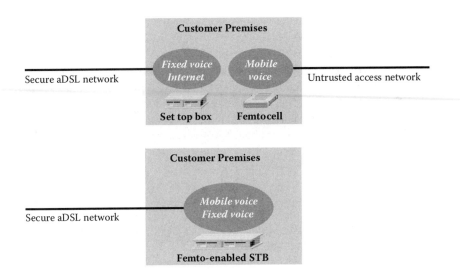

FIGURE 6.3 Standalone FAP versus integrated FAP where femtocell function is integrated inside a set-top box.

in an integrated FAP. In the future, fully integrated silicon devices are expected to support integrated FAPs where DSL modem functions, femtocell modem functions, and control plane processing are all integrated into a single silicon device.

6.2.4 Business Drivers for Femtocell

Growth in demand for mobile broadband data is fueling the demand for femtocells, which in turn provide tangible and intangible benefits to wireless network operators. Improving indoor coverage is expected to reduce churn and accelerate fixed mobile substitution (FMS), thereby increasing operators' average revenue per user (ARPU). In addition, wireless network operators can augment capacity to their existing 3G WCDMA UMTS networks by deploying femtocells. As described earlier, there are no site acquisition costs and 60% to 70% of mobile traffic can be offloaded from their macro networks, thereby creating additional and much-needed capacity.

6.2.5 Femtocell versus UMA

Fixed Mobile Convergence (FMC) provided the impetus for Unlicensed Mobile Access (UMA). UMA has been standardized by 3GPP under the Generic Access Networks (GAN) standard [2]. The basic idea behind UMA is to enable wireless handsets to work over a wireless local area network (WLAN) such as 802.11 (WiFi) in the unlicensed spectrum range. The premise is that UMA leverages the wide proliferation of WiFi hotspots through the embedding of WiFi chipsets in mobile handsets. Once the handset is in the vicinity of a WiFi hotspot, the UMA client attaches to the hotspot and, if access is granted, the client establishes a secure IP connection to the UMA Network Controller (UNC), after which all the mobile data and voice is routed via the WLAN hotspot to the UNC. The UMA client ensures that

TABLE 6.1
Femtocell vs. UMA Star Rating Comparison

Category	Femtocell	UMA
Handset availability	*****	**
Voice quality	*****	***
Data rates	***	*****
Handset's battery life	*****	***
Regulatory issues	*	*****
RF interference issues	***	*****
Over-the-top threat	*****	**

if the user moves in and out of a hotspot, calls are seamlessly handed to and from the macrocellular network. Similar to femtocells, UMA also offloads traffic from mobile cellular networks to WiFi hotspots and their associated backhaul networks.

Table 6.1 provides a star rating comparison of femtocell versus UMA; a single star denotes that a technology is at a significant disadvantage, whereas five stars denote a significant advantage.

As can be seen in Table 6.1, femtocells have certain advantages and disadvantages when compared with UMA. For one, all the necessary components for femtocell deployment are already present in many homes and offices today: a standard mobile phone and an IP broadband connection to backhaul traffic to the operator's network. The same generally holds true for UMA, but the main difference is that with femtocell a user need not upgrade to a new UMA-enabled handset to take advantage of better indoor coverage. Because femtocells operate in licensed spectrum, users' existing handsets work seamlessly in the femtocell environment. Furthermore, the cost of adding WiFi functionality to a UMA-enabled handset tends to drive up its BOM cost, which is a negative factor for widespread adoption. The disadvantage of femtocells, though, is that wireless operators have to deploy a new CPE device (the femtocell) for each indoor environment, whereas with UMA the many thousands of existing WiFi connections are easily leveraged.

There are also other differences between UMA and femtocells. UMA uses unlicensed spectrum, which makes it prone to interference and can result in deteriorated voice quality. Femtocells, on the other hand, use licensed spectrum and provide standard UMTS interfaces, and thus they leverage robust channel structures and deliver good quality voice with improved coverage. What's more, seamless handoff from the macrocellular network to femtocells and vice versa is already in place as the existing mobile core network is leveraged to deliver this function.

From the consumer's perspective, another important factor to compare is handset battery life. Because of the need for multiple chipsets, UMA handsets require more power when compared to 3G UMTS handsets and thus tend to drive battery life down. From the wireless network operator's perspective, a key risk factor that UMA poses is cannibalization of voice ARPU because UMA-enabled handsets potentially enable competitive over-the-top voice service delivery. From an RF interference point of view, UMA operates in unlicensed spectrum and therefore does not interfere with operators' existing macrocellular networks. On the other hand, because femtocells operate in the same UMTS licensed spectrum used in the macrocell network, dynamic Radio Resource Management algorithms are required to minimize interference with the macrocellular network.

6.3 NETWORK ARCHITECTURE

Historically, a regional wireless network build-out is done by strategically installing thousands or tens of thousands of base stations (NodeBs) across a city and along freeways and major roadways. These macrocells provide wireless coverage enabling seamless mobility in large-scale wireless networks. Network operators also strategically place microcells and picocells in dense urban population areas such as airports and multiple-dwelling units (MDUs) to address both coverage and capacity challenges. It must be noted that these cells, including picocells, are connected to RNCs over Asynchronous Transfer Mode (ATM) links including fractional E1/T1 or Ethernet in the case of IP-RAN, unlike femtocells which leverage consumers' IP broadband links for backhaul.

6.3.1 3G UMTS Network Architecture

Figure 6.4 shows a simplified 3G Public Land Mobile Network (PLMN) architecture [3]. A 3G PLMN is logically divided into a Radio Access Network (RAN) and a Core Network (CN). Key network elements inside the RAN are the RNC and NodeB. A NodeB is a base station and an RNC aggregates the traffic from NodeBs

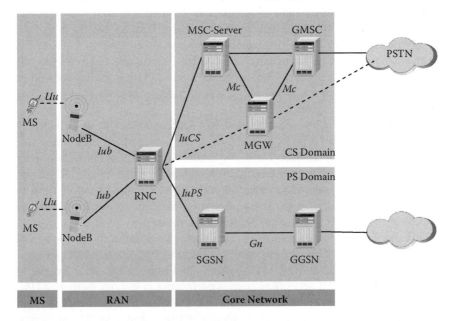

FIGURE 6.4 Simplified network architecture of 3G PLMN.

and delivers the aggregated traffic to the CN. The CN is logically divided into the circuit-switched (CS) domain and the packet-switched (PS) domain. Key network elements in the CS domain are the Mobile Switching Center (MSC) and Visitor Location Register (VLR). Key network elements in the PS domain are the Gateway GPRS Support Node (GGSN) and the Serving GPRS Support Node (SGSN).

The RNC delivers CS traffic, mostly voice and SMS, or texts, to the MSC over the Iu-CS interface. The RNC delivers PS traffic, mostly data, to the SGSN over the Iu-PS interface. The RNC also performs key Radio Resource Management (RRM) functions. The MSC anchors the calls in the CS domain and supports inter-RNC handoff. Similarly, the SGSN anchors the data calls in the PS domain and supports inter-RNC handoff to provide seamless mobility across RNCs. Another key element of the core is the Home Location Register (HLR)/Home Subscriber Server (HSS) that holds all the provisioning data including subscriber profiles and location information for routing incoming calls.

6.3.2 Network Architecture Evolution with Femtocells

With femtocells comes the challenge of integrating potentially millions of wireless home base stations with an existing mobile macrocell infrastructure. There are many different ways of achieving this integration. For simplicity and ease of understanding, we broadly classify these architecture options into the following three categories: UMTS-centric architectures, UMA-based architectures, and SIP/IMS-based architectures. Figure 6.5 shows the different network architecture options for integrating femtocells with a wireless operator's existing CN.

6.3.3 UMTS-Centric Network Architectures

In a nutshell, UMTS-centric architectures leverage the existing UMTS core and integrate femtocells directly into the core via a Femto Gateway. UMTS-centric architectures

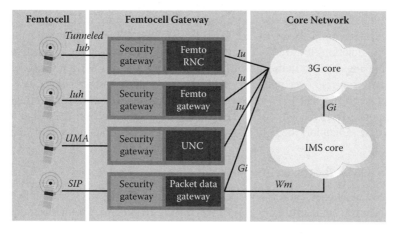

FIGURE 6.5 Different network architecture options for integrating femtocells with an existing core network.

essentially leverage the CN handoff functions as the mobile call is anchored at the MSC and/or SGSN and handoff is supported by these network elements. The drawback of these architectures is that they do not offload the CN as all the traffic from femtocells goes directly to the CN. At High Speed Downlink Packet Access (HSDPA) data rates, each femtocell could potentially pump 7.2 Mbps of traffic and as volume femtocell rollouts start, the aggregate mobile traffic would multiply, eventually requiring a major CN infrastructure upgrade. Regardless, this architecture is the easiest and simplest way of integrating femtocells with the existing mobile CN, and both Tunneled-Iub and Iuh-based architectures fall in this category of UMTS-centric architectures.

In the case of the Tunneled-Iub architecture, a femtocell is essentially a NodeB only and the RNC functions reside at the Femto RNC [4]. This architecture is similar to the macrocellular architecture and is most suited for enterprise femtocell deployments because the architecture does not scale well for high-volume residential femtocell deployments.

The Iuh network architecture was standardized by 3GPP at the end of 2008 [5]. Under this architecture, both RNC and NodeB functions are built into the femtocell, also referred to as a Home NodeB (HNB). Iuh is the interface between a femtocell and the Femto Gateway, which is also referred to as the HNB gateway (HNB-GW).

The Iuh interface essentially enables transfer of Radio Access Network Application Part (RANAP) signaling messages over Streaming Control Transport Protocol (SCTP) over IP, thereby eliminating some of the scaling restrictions inherent in traditional SS7 and SIGTRAN-based transport. For this purpose a new RANAP User Adaptation protocol layer has been standardized by 3GPP [6]. To support femtocell and UE registration between FAP and FGW, a new protocol Home NodeB Application Part (HNBAP) has been defined [7]. Femtocell provides a standard Uu signaling interface toward handsets, thereby enabling any existing 3G handset to work seamlessly with the femtocell. In this architecture most of the RNC functions like RAB Management, RRM, Mobility Management, and ciphering security functions are supported by the femtocell, so the femtocell is much more than just a NodeB.

The Femto Gateway (FGW) aggregates the traffic from tens of thousands of femtocells and delivers the aggregated traffic to the MSC over an Iu-CS interface and to the SGSN over an Iu-PS interface. FGW mirrors the CN functions toward femtocells and RNC functions toward the CN, thereby making the integration seamless and not requiring any changes in the CN infrastructure. Because traffic from femtocells can potentially traverse over public IP networks, for security reasons IPsec is used to encrypt the traffic. The FGW or a separate Security Gateway (SeGW) terminates the IPsec tunnels from the femtocells. The MSC and/or SGSN anchors the call in the CN and in case of handoff the core assists the handoff similar to an inter-RNC handoff.

6.3.4 UMA-Based Architecture

As briefly mentioned earlier, FMC led to the development of the UMA architecture, which later was standardized by 3GPP for GAN [2]. To integrate the mobile traffic via a generic IP network (like 802.11 WLAN) to the mobile core network, a Generic Access Network Controller (GANC) was defined. In UMA, GANC provides the Up interface toward mobile handsets and the standard Iu interface toward the mobile CN. In principle, UMA-based architectures extend the 3GPP GAN architecture to integrate femtocells in a similar way as UMA mobile handsets. The GANC, also commonly known as a UMA Network Controller (UNC), aggregates the traffic from femtocells and delivers it to the existing mobile core over a standard Iu interface. The UNC also provides support for security functions and terminates IPsec tunnels from femtocells.

It is worth noting that even in a UMA-based architecture, the femtocell still provides a standard Uu interface toward the handset, unlike the Up interface in the case of standard UMA. Furthermore, standard UMA implementation requires dual-mode handsets. To ensure that existing non-dual-mode handsets can work, femtocell supports the interworking function to interwork the Uu interface to the Up interface toward the UNC, thereby ensuring that with a UMA-based FGW architecture any existing 3G handset would work seamlessly with the femtocell.

Similar to UMTS-centric architectures, the UMA-based architecture leverages the mobile CN to anchor mobile calls and support handoffs. Traffic generated from femtocells traverses through the CN and thus the core is essentially not offloaded. The UMA-based architecture brings strategic advantage to network operators who already have commercialized UMA services, for these network operators can leverage their existing UNC assets to quickly and efficiently integrate femtocells into their CN.

6.3.5 SIP/IMS-Based Architecture

The SIP/IMS-based architecture integrates femtocells directly into the IMS core. Alternative architectures under this category include softswitch-based implementations where the femtocell is integrated to a softswitch via the SIP interface. Direct femtocell integration offers distinct advantages. First, the mobile CN is completely offloaded, so that as traffic scales it avoids a costly and disruptive CN infrastructure upgrade. Second, traffic latency challenges are mitigated because the number of hops a packet traverses is greatly reduced. Lastly, this architecture is most forward-looking in that it simultaneously enables solving the near-term 3G coverage problem while providing long-term options of delivering rich innovative services via an IMS core that leverages femtocells.

The biggest challenge that needs to be addressed with the IMS-based architecture is supporting seamless handoff between femtocell and macrocellular networks. As the mobile CN is completely bypassed, there is a need for a Voice-Call-Continuity (VCC)-like application in the IMS core to support handoff functions. The IMS-based architecture has the most appeal to carriers that have a strong IMS core and have both fixed and wireless assets.

6.4 THE SIGNALING PLANE IN FEMTOCELL

For large-scale successful deployment of femtocells, standards-based signaling interfaces are essential, and we broadly classify them into four key categories. The first one is the Uu signaling interface, as it is essential to ensure existing handsets work seamlessly with femtocells. Next we explore the signaling interfaces toward the network; these depend on the network architecture. A femtocell is a base station, but at the same time it is a CPE device and that brings with it a key requirement to

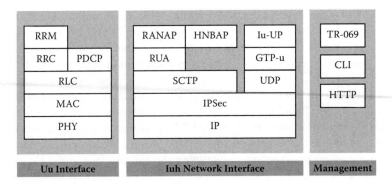

FIGURE 6.6 Iuh network architecture-based femtocell signaling interfaces.

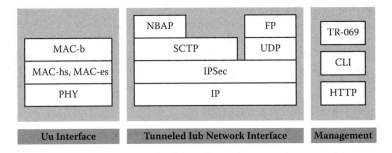

FIGURE 6.7 Tunneled Iub network architecture-based femtocell signaling interfaces.

have standard interfaces for configuration and remote management. Lastly, there are security aspects to consider.

6.4.1 Uu Interface Protocols

The interface between handsets and a femtocell is called Uu [8]. The protocol stacks on the Uu interface as shown in Figure 6.6 are comprised of RRM, Radio Resource Control (RRC), Radio Link Control (RLC), and Media Access Control (MAC). In addition, Packet Data Convergence Protocol (PDCP) performs header compression and decompression and is responsible for maintaining sequence numbers for data transfer.

Irrespective of the network architecture, the Uu interface protocol stack remains the same, and the only exception is the Tunneled Iub architecture. In the case of Tunneled Iub architecture, because the femtocell supports only NodeB functions, the signaling stack is comprised only of MAC and the higher layer functions RLC, RRC, and RRM are supported by the Femto RNC. Figure 6.7 shows the femtocell protocol stacks for Tunneled Iub architecture [4].

6.4.2 Network Interface Protocols

Figures 6.6 through 6.9 show the complete signaling stacks inside the femtocell depending on the network architecture. For the Iuh architecture, the signaling stack

Femtocell—Home Base Station

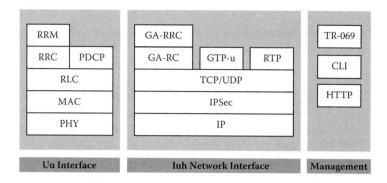

FIGURE 6.8 UMA network architecture-based femtocell signaling interfaces.

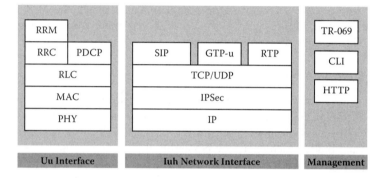

FIGURE 6.9 UMA network architecture-based femtocell signaling interfaces.

is comprised of RANAP and HNBAP. To be able to integrate millions of femtocells, the Iuh architecture defines a new adaptation layer RUA to enable transferring RANAP messages over SCTP. This enables scaling of the signaling interface to support large numbers of femtocells.

For SIP/IMS-based architectures, Figure 6.9 shows the signaling stacks. It is worth noting that for the SIP/IMS-based architecture, the handsets do not have to be SIP-enabled; existing 3G mobile stations can access the femtocell using the standard Uu interface because the femtocell has the interworking function to initiate corresponding SIP signaling messages toward the IMS core to set up multimedia sessions.

In the case of the Tunneled-Iub architecture, the femtocell supports Framing Protocol (FP) to transfer all the MAC channels over IP. The signaling interface is exactly the same as in a NodeB for IP-RAN.

6.4.3 Remote Device Management and Control

Unlike a NodeB, a femtocell is a CPE and will be deployed in the millions, which brings new requirements for the management plane. Remote device management and monitoring is essential for network operators to avoid truck rolls, otherwise the operational expenses to manage femtocell networks will far outweigh the benefits.

TR-069 is the Broadband Forum's technical specification for CPE Wide Area Network (WAN) Management Protocol [9]. It defines the protocol between a CPE and an Auto-Configuration Server (ACS) for remote device management. TR-069 is already used extensively in fixed broadband networks to manage CPE devices ranging from residential gateways to IPTV STBs.

The Femto Forum is liaising with the Broadband Forum to extend TR-069 for femtocells and develop a new data model for femtocells, which require complex configuration ranging from Internet Key Exchange (IKE) for IPsec to PLMN Identifier (PLMNID) and LAC for UMTS networks. One of the key requirements for successful large-scale deployments of femtocells is that they be "zero touch" devices in which a user can buy a FAP, plug it into his or her residential gateway, and have it start working. To achieve this plug-and-play goal, femtocells employ TR-069 to get all the configuration data from the ACS. Femtocells also use TR-069 to periodically upload statistics to the ACS and, in the case of an alarm, update the ACS. Network operators will use TR-069 for remote device management from configuration to monitoring to running diagnostics remotely.

6.4.4 Security Aspects

In wireless networks, security focuses primarily on securing over-the-air traffic. Femtocells bring a new dimension to security requirements as these devices operate in network operators' licensed spectrum yet traffic is backhauled over consumers' IP broadband links, which opens the possibility that mobile traffic will be carried over the public Internet. It is therefore essential to secure the IP data stream from a femtocell all the way to network operators' FGWs. To accomplish this, femtocells use IPsec for securing IP communication [10]; an IPsec tunnel is established between the FAP and the FGW before any signaling or data traffic is transferred between the two. IPsec tunnel establishment is done prior to downloading any configuration data from ACS to the femtocell as well.

6.5 FEMTOCELL: SELF-ORGANIZING NETWORK

Wireless network rollouts require careful RF planning to ensure that cell sites do not interfere with one another and that the limited spectrum is utilized to the maximum. Even after a network is operational, RF planners continue to monitor and optimize the radio network to achieve maximum efficiency. Adding a new cell site requires careful replanning and usually results in reconfiguring existing cell sites as well. Femtocells take the RF planning and cell site deployment complexity to a new level, however, and it is not humanly possible to deploy millions of femtocells inside consumers' homes using the same old RF planning techniques. As a result, femtocells require new capabilities and have to be self-organizing in the sense that they are completely aware of their local RF environment.

6.5.1 Licensed Spectrum

Femtocells operate in wireless network operators' licensed spectrum, which brings distinct advantages, as previously discussed, but at the same time brings some challenges as well. One of the challenges that network operators have to address pertains to regulation. In certain countries, a licensed technician is required to install a "wireless base station." If femtocells are classified as wireless base stations and not exempted, this type of regulation will defeat the femtocell value proposition because more "truck rolls" increase rather than decrease operational costs. In other countries, existing wireless licenses require network operators to maintain written records of all base station locations. The need to create and maintain records of every femtocell deployed would create an excessive burden on operators, again defeating the purpose.

Another important challenge that comes with licensed spectrum is the location of the frequency spectrum itself. Different wireless operators can own the same spectrum in different cities, and certainly in different countries. If a consumer packs a femtocell into his or her bag and travels to another location where the network operator does not own that same licensed spectrum, and then powers on the femtocell, the operator potentially could be considered in violation of spectrum usage. This critical requirement has led to global positioning system (GPS)-based location authentication being built into femtocells themselves. Based on the identified location, a network operator can lock down an unauthorized femtocell—essentially turning it into a brick. Once the femtocell is powered on, the ACS can request location information from the FAP prior to providing any configuration data. Once the ACS receives the location information from FAP, it can then determine whether to proceed with configuring the FAP or completely denying it authorization to power on and transmit.

6.5.2 Managing RF Interference

Because femtocells operate in the licensed spectrum they can interfere with the network operator's existing macrocellular infrastructure. For example, if a single frequency Code Division Multiple Access (CMDA) system is being operated where the macro and femtocell network utilize the same frequency band, then the power control algorithms of the femtocell can create interference with the macrocell, thereby degrading network capacity and quality of service. In MDUs such as condominiums and apartments, particularly, multiple femtocells can interfere not only with the macrocell network, but also with each other. Some carriers are planning to deploy femtocells on a different RF than what is used for their macrocellular network and thus avoid femto/macro interference altogether, but spectrum acquisition is costly. Self-organizing femtocell networks also have to be constantly aware of their ever-changing RF environment. In MDUs, a femtocell's neighboring RF environment could potentially change multiple times throughout the day as neighbors power on and power off their own femtocells.

Femtocells can also be equipped with sniffer capabilities to enable them to be RF aware. For instance, for a very short duration a femtocell can switch its operating mode and start functioning as a handset instead of as a NodeB. In this handset configuration, the femtocell "sniffs" its RF environment and populates a list of neighboring cells, their transmit frequencies or codes, and their power levels. Based on this sniffed information, the femtocell can then compute and select its own transmit code and power levels to ensure it minimizes interference with the neighboring cells. A femtocell can switch to handset mode only if it has no active calls and can do this periodically multiple times a day. Of course the duration in which a femtocell is in handset sniffer mode has to be very short to prevent triggering handsets that were camped on the femtocell to start cell reselection procedures [11].

6.5.3 Ping-Pong Effect

In addition to the challenges and policies described earlier, there still can be very tricky corner cases. As an example, if a femtocell is located in a house that is very close to a macrocell tower and has direct line of sight, it could result in some big dead spots being created in the network.

As users move in and out of femtocells, they are seamlessly handed off to the macro-cellular network. Consider the case where a user is moving up and down his driveway. In such a scenario, the call potentially could be continuously handed off between the femtocell and the macrocell, which is usually referred to as the ping-pong effect. The ping-pong effect directly increases the signaling load on the core network as the call is being handed in and out multiple times on very short notice. Intelligent femtocells mitigate this effect by either delaying the handoff signaling trigger to the CN or by slightly increasing the transmit power to ensure the call stays parked on the femtocell.

6.5.4 Open versus Closed Access

One policy decision associated with femtocells is the question, "Who really owns them–consumers or network operators?" One argument is that because the femtocell operates in a network operator's licensed spectrum, the network operator should own it. On the other hand, a femtocell uses the consumer's IP broadband link to backhaul the traffic from the home to the CN, so some argue that the consumer owns it. Note too that there is no way consumers would find it acceptable for any indiscriminate passerby's traffic to be suddenly carried over their private IP broadband link. This brings us to the key issue of open-versus-closed femtocells. An open femtocell is one where any user on the wireless operator's plan can use any femtocell in the operator's entire network. A closed femtocell is one in which only the user who has installed it in his or her home is permitted to use it.

For example, consider a corner case where a user buys a femtocell and installs it in his or her home, which happens to be close to a public bus stop. An open femtocell would then allow traffic from any passerby waiting at the bus stop. If the femtocell supports four active calls, it is very possible that at some point all four calls being carried by the femtocell are those of the people waiting at the bus stop, thereby leaving the user who actually bought and installed the femtocell in the home without available capacity for his or her own calls. Each network operator will have to define its particular policies on these matters as femtocells are deployed; no one universal policy will be in effect across all operators.

6.5.5 Emergency Calls

Because femtocells carry voice calls in licensed spectrum, they will be required to provide a 911-type emergency service. Alternative power sources or fallback to existing telephone infrastructure may be viable ways to achieve 911 availability. To meet E911 requirements, operators must be able to provide the location of the calling equipment to the Public Safety Answering Point (PSAP). Emergency 911 calls might put additional restrictions on open-versus-closed femtocells, too; even if a femtocell is closed, it might be required to carry emergency calls, a policy matter for regulators to address.

6.6 WHAT'S NEXT?

Femtocells are not limited to 3G WCDMA UMTS networks. For example, femtocell technology has opened up the possibility for cable Multiple System Operators (MSOs) to roll out their own wireless networks, and some operators are assessing the opportunity for WiMAX femtocells. In developing markets like China and India, operators are assessing 2G GSM femtocells, although it must be said that the business case for femtocells does not hold water if a strong broadband infrastructure does not already exist for backhauling traffic.

6.6.1 LTE Femtocells

3GPP standardization work for the next-generation "4G" LTE wireless networks is nearing completion and has already taken into account the Home eNodeB (HeNB)—LTE femtocell—requirements. LTE deploys Orthogonal Frequency Division Multiple Access (OFDMA) and multiple in, multiple out (MIMO) techniques to deliver 100 Mbps+ downlink peak data rates and 50 Mbps+ uplink peak data rates. In comparison, today's 3G High Speed Packet Access (HSPA) networks deliver 14.4 Mpbs downlink peak data rates, meaning that LTE promises an almost sevenfold increase in peak data rates.

One thing is for sure: There is, or most assuredly soon will be, enough demand for mobile bandwidth. There are enough applications and plenty of content that will quickly use up the 100 Mbps+ peak downlink data rates that LTE promises. However, consumers expect more for less and will not necessarily pay more for a sevenfold increase in data rates—and will definitely not pay seven times more for it. The good news is that LTE promises a fourfold increase in spectral efficiency over WCDMA, but for LTE femtocell economics to work, more network cost optimization will be needed in order to significantly lower the cost per bit.

Operators therefore have to address some basic questions as they develop their LTE network rollout strategies:

- How will they build more "capacity" in their networks to meet the demand?
- Where will they build more "capacity" in the network?
- How will they aggressively drive down the cost per bit?

The answers to these questions will require a deeper understanding of user behaviors and requirements than what currently exists. For example, as described previously in today's mobile networks, almost 60% of voice calls originate and terminate indoors. Will data usage patterns be any different?

To address some of these basic questions, mobile operators will bring to bear key learnings from predecessor 3G network rollout experiences. Let us recall some of those well-known lessons:

- Although 3G was promised for the year 2000, as reality turned out strong 3G adoption started only in late 2006—6 years after the original forecasts.
- 3G spectrum was expensive; mobile carriers paid billions of dollars for spectrum acquisition and thus the return on that investment was delayed by more than 5 years.
- That expensive 3G spectrum was not even "beachfront property." As it turned out, operators and consumers realized that poor indoor signal penetration in the 2.1 GHz band was a real problem.
- Citywide network builds required huge up-front CapEx investment, and until 3G adoption became mainstream, all that investment produced negligible revenues and returns.

Summing it up, to meet the growing demand, operators will have to find ways to augment wireless capacity in their networks in a very cost-effective and scalable manner that minimizes upfront LTE investment and risk.

6.6.2 Revolutionizing Network Rollout

Fortunately, mobile operators can leverage LTE femtocells as part of their LTE network rollout strategies—building their LTE networks one consumer at a time, one home at a time, and one femtocell at a time.

In particular, by using femtocells, operators can avoid huge upfront CapEx investments in building citywide and nationwide LTE networks. With LTE femtocells, operators can augment capacity where it is needed most—inside offices, homes, cafes, airports, and so on—while leveraging their existing 3G networks to provide citywide and nationwide coverage, albeit initially at lower data rates. It is important to note that LTE networks can overlay existing 3G networks and thus this rollout strategy changes the operator's business model completely, greatly reducing up-front investment and risk. Clearly there will be a tipping point when it will make economic sense for operators to build citywide macrocell LTE networks, but deploying LTE femtocells is a viable and innovative starting strategy.

Furthermore, to deliver true 100 Mbps+ downlink peak data rates via LTE femtocells, existing DSL or cable modems with 10 Mbps data rates will not suffice. However, by the time LTE networks are ready to be rolled out, Fiber to the Home/Curb/Building (FTTx) deployments will provide the baseline backhaul infrastructure. As a matter of fact, the early FTTx adopters are the ones most likely to be the early LTE femtocell adopters as well.

Lastly, LTE requires 20 MHz-wide channels and there are not many options available. In a large part of the world, LTE is likely to be deployed in the 2.6 GHz band, which means that the indoor coverage challenge will be even greater than it was in 3G WCDMA networks (which were mostly deployed in the 2.1 GHz band). Although these are still early days for LTE, the femtocell option provides a compelling alternative for how network operators can build out their networks—and they might just prove to be the most popular way to begin.

ACKNOWLEDGMENTS

I would like to thank my parents Kumud and Raj, my wife Swati, my sister Pooja, and my two lovely kids Anshul and Richa. They all are the source of my motivation in everything I do and provided me all the support for this work.

REFERENCES

1. J. Domenech, A. Pont, J. Sahuquillo, and J. Gil, "A user-focused evaluation of Web prefetching algorithms," *Computer Communications*, Vol. 30, 2213–2224, 2007.
2. D. Flinn and B. Betcher, "Re: latest top 1000 website data?", e-mail to author, Jan. 8, 2008. Available: http://www.websiteoptimization.com/speed/tweak/average-web-page/ 3GPP TS 43.318 v8.3.0 2008-08 and TS 44.318 v8.3.0.
3. 3GPP. Technical Specification Group Services and Systems Aspects, Network architecture TS 23.002, 2008-09.
4. 3GPP. Technical Specification Group Radio Access Network, UTRAN Iub interface: general aspects and principles TS 25.430, v8.0.0 2008-12.
5. 3GPP. Technical Specification Group Radio Access Network, UTRAN architecture for 3G Home NodeB Stage 2 TS 25.467, v8.0.0 2008-12.
6. 3GPP. Technical Specification Group Radio Access Network, UTRAN Iuh interface RANAP User Adaptation (RUA) signaling TS 25.468, v8.0.0 2008-12.
7. 3GPP. Technical Specification Group Radio Access Network, UTRAN Iuh interface Home NodeB Application Part (HNBAP) signaling TS 25.469, v8.0.0 2008-12.
8. 3GPP. Technical Specification Group Radio Access Network, Radio Interface Protocol Architecture TS 25.301, v8.3.0 2008-09.
9. Broadband Forum. Technical Report TR-069, CPE WAN Management Protocol 1.1. December 2007.
10. S. Kent and R. Atkinson, "Security Architecture for the Internet Protocol," *RTF 2401*, IETF. November 1998.
11. 3GPP. Technical Specification Group Radio Access Network, UE procedures in idle mode and procedures for cell reselection in connected mode TS 25.304, v8.3.0 2008-09.

7 Satellite Communications

Sooyoung Kim

CONTENTS

7.1 Introduction 145
7.2 Principles of Satellite Communication Systems 145
    7.2.1 Operational Principles and Characteristics 145
    7.2.2 Satellite Services and Frequency Bands 147
    7.2.3 Satellite Orbits 150
    7.2.4 Satellite Orbits and Spectrum Resources, and Standardization Issues 154
7.3 Current Status of Satellite Communication Technology 155
    7.3.1 Broadband Satellite Communications and Interactive Broadcasting 156
    7.3.2 Mobile Broadcasting to Handheld Terminals 158
7.4 Technologies for Future Satellite Communication Systems 159
    7.4.1 Hybrid and Integrated Networks 159
    7.4.2 Multicarrier Techniques 160
    7.4.3 Cooperative Diversity Techniques 161
    7.4.4 Layered Coding 164
References 165

7.1 INTRODUCTION

Satellite-based networks can contribute to the emergence and utility of next-generation networks (NGNs) in providing ubiquitous and universal broadband services to end users who require seamless generalized mobility. In this chapter, the basics of satellite communication systems are introduced, focusing on future satellite systems for NGNs. First, there is a brief introduction on the operational principles and characteristics of satellite systems, then satellite orbits and services in the context of international standardization issues are discussed. Next, an introduction to the current status of satellite communication technology is given. Finally, a few candidate technologies for future satellite communication systems are discussed.

7.2 PRINCIPLES OF SATELLITE COMMUNICATION SYSTEMS

7.2.1 OPERATIONAL PRINCIPLES AND CHARACTERISTICS

A satellite is an object revolving around a celestial body. For example, the moon is a satellite of the Earth. In our area of interest, a satellite is referred to as an artificial object revolving around the Earth with intended applications. The continuous revolution of satellites around the Earth is due to the equilibrium of two forces: the

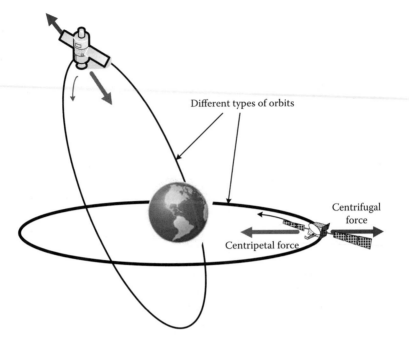

FIGURE 7.1 Operational principles of a satellite.

centripetal force on the satellite due to the gravitational force and the centrifugal force on the satellite due to the circular (more specifically elliptical) motion. The path, relative to the Earth, described by the center of mass of a satellite subjected primarily to natural forces (i.e., mainly gravity) is called an orbit. Various types of orbits are possible, as shown in a later section. Figure 7.1 shows the operational principles of a satellite.

The history of satellites for communication services began with a 1945 paper by Arthur C. Clarke, "Extra Terrestrial Relays" [1]. In this article, he proposed the idea of a geostationary orbit (GSO) satellite, which appears stationary with respect to the Earth's surface. This trick is possible even with the circular motion of the satellite, because of the fact that the Earth also revolves on its axis. Therefore, by controlling the orbital period of the satellite, or ensuring that the time required for a complete revolution around the Earth is equal to one revolution of the Earth on its axis, which is 24 hours, we can have a GSO satellite. The orbital period of the satellite is proportional to the distance from the Earth to the satellite, which is called the altitude (height) of the satellite. The altitude of the GSO satellite is about 36,000 km.

A satellite is generally located a great distance from the Earth; thus, high power is required to transmit or receive messages via a satellite, and a good line of sight (LOS) condition is also required. In order to secure good LOS conditions, a high elevation angle, ε, is preferred, which is defined as the angle between the center of the satellite beam and the horizontal plane of the Earth. The elevation angle is in the range of 0° to 90°. A 90° elevation angle indicates that the satellite is located directly overhead, and this is the best condition for communications. On the other hand, if the

Satellite Communications

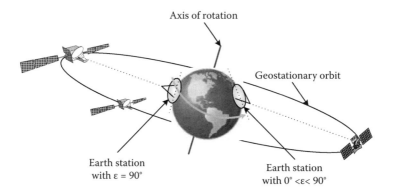

FIGURE 7.2 GSO satellite in the equatorial plane for various elevation angles.

elevation angle is low, it means the probability of an LOS condition is reduced due to various obstacles, such as mountains or buildings.

We established that a necessary condition for a GSO satellite is an orbital period of 24 hours, which results in an orbital height of 36,000 km. There is another necessary condition for a satellite that is completely stationary relative to the surface of the Earth. As shown in Figure 7.2, the rotation axis of the Earth is inclined about 23.5° with respect to the celestial plane; for this reason, the satellite orbit must be located in the equatorial plane if it is a GSO satellite. This means that all GSO satellites must be located in the same orbital plane, although the main target service areas can differ. Because of this, as the latitude increases, the elevation angle of the GSO satellite decreases, resulting in an undesirable communication condition. Although a GSO satellite may provide stationary and thus continuous satellite services, a GSO satellite is not a desirable option for users at high latitudes.

Figure 7.3 shows a satellite system and clarifies a few basic terms. The definitions of links differ slightly from those of terrestrial wireless systems. A satellite system consists of a space segment and a ground segment with various types of Earth stations. The space segment consists of a satellite in space and a station for telemetry, tracking, and control (TT&C). Although the TT&C station is physically located on the Earth rather than in space, its main purpose is to ensure the proper operating condition of the satellite, so it is categorized as a space component. There are various types of Earth stations, as shown in Figure 7.3, including gateways, fixed and mobile terminals for communications, and receiving terminals for broadcasting services. We refer to uplinks and downlinks between a satellite gateway and the satellite as feeder links, and uplinks and downlinks between a user terminal and the satellite as user links. On the other hand, we refer to links from the satellite gateway via the satellite to the user terminal as forward links, and links from the user terminal via the satellite to the gateway as return links.

7.2.2 Satellite Services and Frequency Bands

Satellite services can be classified by various methods. One of the most common methods is classification according to purpose; there are fixed-satellite services

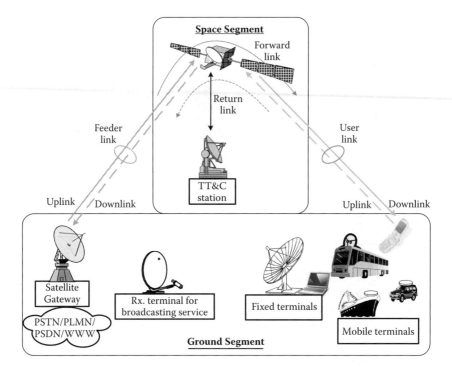

FIGURE 7.3 Satellite system and its associated components.

(FSS), mobile-satellite services (MSS), and broadcasting-satellite services (BSS). Satellite systems for FSS and MSS provide communication services to fixed and mobile terminals on the ground, respectively. Examples of MSS are voice, data, and multimedia services to handheld user terminals and high-speed multimedia services to terminals in moving vehicles. Satellite systems for BSS provide broadcasting services to fixed and mobile terminals.

Satellite services can also be classified according to the service coverage; there are international services, regional services, and domestic services. The coverage area of an international satellite service is the whole world; examples are the INTELSAT, INMARSAT, and the INTERSPUTNIK system [2–4]. The coverage area of a regional satellite service is usually a continent; examples are EUTELSAT, which mainly covers Europe, and ARABSAT, which mainly covers Arabian countries [5, 6]. Lastly, a domestic satellite service mainly covers a specific country, such as KOREASAT. It is noted that some regional and domestic satellite systems might also cover a global and regional area, respectively, for commercial purposes. It is also noted that many satellite systems also have a steerable antenna beam, which can cover a specific area for a telecommand signal during a given time period.

Because a satellite is located in space, the characteristics of satellite services are quite different from the services provided by other terrestrial wireless systems, although their main purposes are similar. In this context, we investigate the advantages and disadvantages of satellite services. The satellite altitude ranges from a few hundred kilometers to a few tens of thousands kilometers, resulting in a large coverage area on the ground, which can include a country or an entire continent. Three GSO satellites can cover the whole world, except for high latitudes, thus providing an excellent condition for global communications. Due to this large coverage area, a satellite system has an inherent broadcasting capability, and it can provide services with uniform quality, which are independent of the terrestrial infrastructure. In addition, a satellite is located beyond the troposphere, and service provision is unaffected by natural disasters such as floods or Earthquakes. Once a satellite is launched, the installation cost of the Earth station or user terminal is independent of the distance, so a satellite system is a good solution to providing a communication service to geographically isolated areas. It enables provision of telecommunication services in regions not covered by a terrestrial network. Areas not adequately covered by the terrestrial component include physically isolated regions, gaps in the terrestrial network, and areas where terrestrial systems are permanently or temporarily disabled due to natural disasters.

On the other hand, these advantages of satellite systems over terrestrial systems are accompanied by a few disadvantages. Because a satellite is located a great distance from the Earth, it incurs a high propagation loss, and thus a good LOS condition is generally required for secure communications. A long propagation delay is also a basic problem for satellite communications. For example, the round-trip delay of a GSO satellite approaches 0.5 seconds. We mentioned that satellite systems have an advantage due to their large coverage area. However, this can also be a disadvantage, because it can cause interference problems to adjacent areas, due to unwanted power emissions. The initial launch cost is high. Once a satellite is launched, simple or scheduled maintenance by a TT&C station such as attitude control is required, and we cannot upgrade or repair it.

Frequency bands for satellite systems are generally classified and designated by letters, as shown in Table 7.1. Satellite systems using L and S bands in the frequency ranges from 1 to 4 GHz incur a low propagation loss and can employ small size Earth terminals, so these systems are suitable for providing MSS and BSS to handheld terminals. Frequency bands allocated to satellite digital multimedia broadcasting (S-DMB) services and satellite components of the IMT-2000 system are included in these bands. C bands were used for most early commercial satellite systems to provide FSS and direct-to-home (DTH) broadcasting services.

Satellite systems using high-frequency bands above 10 GHz incur signal fading due to moisture in the transmission path, which is a common occurrence when it is raining. This is referred to as rain attenuation, which becomes increasingly serious as the operating frequency increases. Therefore, satellite systems using frequency bands above the Ku band require suitable compensation techniques for mitigating rain fades [7, 8], including classical methods such as power control and site diversity techniques, as well as adaptive transmission techniques. Even with serious rain attenuation and propagation loss in higher frequency bands, utilization of these bands is mandatory. This is not only because the demand for high-speed wireless multimedia service is increasing rapidly worldwide, but also because numerous currently operating GSO satellites already use frequency ranges below 18 GHz, and thus higher band alternatives should be investigated. Another future service requirement

TABLE 7.1
Frequency Bands for Satellite Communications

Band	Frequency Ranges (GHz)	Characteristics	Major Services
L	1–2	• Low propagation loss • Small size terminals	MSS, BSS (suitable for handheld terminals)
S	2–4	• Low propagation loss • Small size terminals	MSS, BSS (suitable for handheld terminals)
C	4–8	• Moderate available bandwidth • Many commercial satellites • Negligible rain attenuation • Large size Earth station (ES) antennas	FSS, BSS
X	8–12	• Moderate available bandwidth • Appreciable rain attenuation • Medium size ES antennas	MSS, FSS (mostly used for military purposes)
Ku	12–18	• Moderate available bandwidth • Serious rain attenuation • Small size ES antennas	FSS, BSS (DBS, DTH)
K	18–27	• Wide available bandwidth • Very serious rain attenuation • Very small size ES antennas	FSS, MSS (used for group terminals)
Ka	27–40	• Wide available bandwidth • Very serious rain attenuation • Very small size ES antennas	FSS, MSS (used for group terminals)
V	40–75	• Extremely serious rain attenuation • Very wide available bandwidth	Currently used for experimental and military purposes

is mobility. In order to provide high-speed multimedia, Internet, and high-definition television (HDTV) services to moving vehicles, we need a mobile broadband satellite technology with active array antenna techniques, which are investigated in more detail in a later section.

7.2.3 Satellite Orbits

As we mentioned, the motion of natural and artificial satellites around celestial bodies is due to the equilibrium of the centripetal and centrifugal forces on satellites with circular motion. The theoretical background on the circular motions of satellites was first introduced by Johannes Kepler (1571–1630), a German mathematician, astronomer, and astrologer. Kepler showed via experiments that there are three important laws governing the movement of celestial bodies. Subsequently, Isaac Newton (1643–1727) clarified these three laws via his famous theory of gravitation. He showed that the motions of objects around the Earth and other celestial bodies are governed by a single set of natural laws, and this demonstrated the consistency between Kepler's laws of planetary motion and Newton's theory of gravitation.

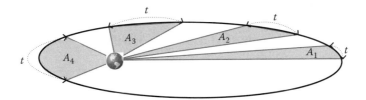

FIGURE 7.4 Kepler's second law, the law of equal areas.

Here, in order to understand the basic operational principles of a satellite and its orbit, we briefly investigate Kepler's three laws, without dealing with the theories in depth. More details on Kepler's laws and their relationship with Newton's laws can be found in Chapter 2 of [9]. By the first law, the orbit of a satellite around the Earth is elliptical, where the center of the Earth lies at one of the foci of the ellipse. A circular orbit such as GSO is a special case of an elliptical orbit, where the two foci are merged at a single point in the center of the circle. In an elliptical orbit, the farthest point of the orbit from the Earth is called the apogee, and the nearest point from the Earth is called the perigee.

Kepler's second law is also known as the law of equal areas. It states that the area, A, swept by all lines joining a satellite and the Earth is equal, during a given interval of time, so the rate, dA/dt, at which the satellite sweeps is constant. In Figure 7.4, all areas A_1 to A_4 swept by the satellite during a given time interval t are equal. According to the second law, the satellite moves faster near the Earth, so the area swept in a given time is equal to that swept at larger distances, where the satellite moves more slowly. This law is used to design a highly elliptical orbit (HEO) to cover high latitudes barely covered by GSO, due to the extremely low or negative elevation angle. If an HEO is designed such that the apogee is located directly overhead the target service area, then according to the second law, the satellite moves very slowly near the apogee and thus the service area has a comparatively good communication condition over the long term when there is a high elevation angle. A practical example of this type of orbit is the Molniya orbit used by Russia for communication services. The apogee and perigee of the Molniya orbit are about 40,000 km and 400 km, respectively.

Kepler's third law is also known as the law of periods. This shows the relationship between the time period T and size of the satellite orbit a. We express the relationship

$$T = 2\pi\sqrt{\frac{a^3}{GM}}, \qquad (7.1)$$

where a is the semimajor axis of the elliptical orbit, G is the gravitational constant, which is 6.67×10^{-11} m^3/kg sec^2, and M is the mass of the Earth. For a circular orbit, a represents the radius, so it is the height or altitude of a satellite from the center of the Earth. Table 7.2 shows the orbital periods and velocities of a satellite according to the heights for various circular orbits, which can be derived from Equation (7.1). According to Equation (7.1), the altitude of a satellite is proportional to the orbital

TABLE 7.2
Orbital Periods and Velocities of a Satellite for Various Circular Orbits

Altitude (km), a	Velocity of the Satellite (km/s)	Orbital Period, T
500	7,612	1 h 35 m 37 sec
1,000	7,350	1 h 45 m 08 sec
2,000	6,987	2 h 07 m 12 sec
5,000	5,918	3 h 21 m 19 sec
10,000	4,934	5 h 47 m 40 sec
30,000	3,310	19 h 10 m 51 sec
35,786	3,075	23 h 56 m 04 sec
40,000	2,932	27 h 36 m 39 sec

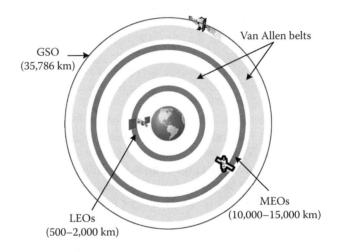

FIGURE 7.5 Classification of circular satellite orbits according to altitudes.

period, and thus it is inversely proportional to the speed of the satellite, as shown in Table 7.2. If the altitude of the satellite approaches 35,786 km, then the speed of the satellite is exactly equal to the rotation speed of the Earth on its axis, and the orbital period is 23 hours, 56 minutes, and 4 seconds, which corresponds to 1 celestial day. This is the altitude of a GSO satellite.

We usually classify satellite orbits according to their altitudes. There are low Earth orbits (LEOs), medium Earth orbits (MEOs), and GSOs.* Figure 7.5 shows this classification of orbits. Table 7.3 summarizes the characteristics of each. Note that there are gaps between LEOs and MEOs and between MEOs and GSOs where it is barely possible to have a satellite orbit. These gaps are the Van Allen radiation

---

* The abbreviation GSO sometimes denotes geosynchronous orbits, including all orbits synchronous to the rotation speed of the Earth, but not necessarily exactly equal to the rotation speed, and a geostationary orbit is represented by GEO, which denotes a geostationary Earth orbit. In this case, GEO is a special case of GSO.

TABLE 7.3
Satellite Orbits According to Altitudes and Characteristics

	LEO	MEO	GSO
Altitude (km)	500–2,000	10,000–15,000	35,786
No. of revolutions/day	11–15	2–4	1
No. of satellites required for global coverage	18–66	10	3
Example systems	Iridium, Globalstar	GPS	Most commercial satellites
Applications	Personal mobile communications, Earth observations, military applications	Personal mobile communications, global positioning service	Communication, broadcasting services

belts, which consist of ionized particles at a distance of 2,000 km–6,000 km and 15,000 km–30,000 km above the Earth's surface.

Compared to GSO, LEOs and MEOs are also known as non-GSOs, because they are not stationary with respect to a point on the Earth. GSO has the advantage that only three satellites are needed for complete coverage of the whole world, except for high latitudes. TT&C is comparatively easy, because the satellite is stationary to the Earth. On the other hand, very low elevation angles occur in areas with latitudes above 60°. Moreover, high transmitting power is required and there is a long propagation delay, and thus it is not suitable for mobile services to small handheld terminals. On the contrary, a non-GSO system can be designed to have high elevation angles for providing high-quality communication links. In addition, due to the relatively low delay, it may be suitable for providing high-quality voice services. On the other hand, because the satellites are not stationary and have smaller coverage, we need multiple satellites for continuous and global coverage. This also necessitates a complex handover mechanism between satellites.

We also classify satellite orbits according to the inclination angle with respect to the equatorial plane. GSO is in the equatorial plane, and satellite orbits that are not in the equatorial plane are denoted as inclined orbits. For example, the Globalstar system consists of eight orbits each with six satellites, resulting in 48 satellites in total, and each orbit is inclined 52° with respect to the equatorial plane.

A polar orbit is a special case of an inclined orbit. It is inclined 90° with respect to the equatorial plane, and the satellite passes over the north and south poles. As an example of near-polar orbit, the Iridium system consists of 11 near-polar orbits, with an inclination angle of 86.4°, each with six satellites, resulting in 66 satellites in total.

Lastly, as a special case of LEO, there is an orbit that lies in a plane maintaining a fixed angle with respect to the Earth–sun direction. This kind of orbit is called a sun-synchronous (helio-synchronous) orbit, and the satellite passes over a given place on the Earth at the same local time and maintains a consistent light condition. For this reason, it is usually used for remote sensing applications and meteorological and atmospheric studies.

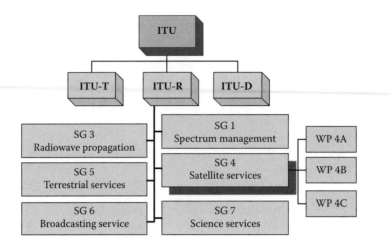

FIGURE 7.6 Study groups of ITU-R.

7.2.4 Satellite Orbits and Spectrum Resources, and Standardization Issues

Because satellite orbits and spectrums are limited natural resources, which are increasingly in demand from a large and growing number of services in various geographic areas and countries, global management of satellite orbits and spectrums is a vital issue for ensuring the rational, equitable, efficient, and economical use of these resources. The International Telecommunication Union (ITU) is an intergovernmental organization within the United Nations (UN), where governments and private sectors coordinate global telecommunication networks and services. One of the most important roles of the ITU is to ensure efficient management of satellite orbits and spectrum resources.

ITU has three sectors, radiocommunication (ITU-R), telecommunication (ITU-T), and development (ITU-D), as shown in Figure 7.6. Their activities cover all aspects of telecommunications, from setting standards that facilitate seamless global interworking of equipment and systems, to adopting operational procedures for the vast and growing array of wireless services, and designing programs to improve telecommunication infrastructure in the developing world [10].

ITU-R, which consists of several study groups (SGs) and working parties (WPs), has been developing recommendations for terrestrial and space-based wireless services, and each SG and WP has its own study objective. In 2008, ITU-R reorganized its SGs and WPs to reflect recent trends of convergence of communication and broadcasting services. The structure is depicted in Figure 7.6. Specifically, ITU-R SG 4 is responsible for developing recommendations for systems and networks of satellite services and has three WPs: WP 4A, for studying efficient orbit/spectrum utilization for FSS and BSS; WP 4B, for studying systems, air interfaces, performance and availability objectives for FSS, BSS, and MSS, including IP-based applications and satellite news gathering; and WP 4C, for studying efficient orbit/spectrum utilization for MSS and radio determination satellite services (RDSS).

Over a time scale of years, SG 4 of ITU-R has been developing standards and regulations to ensure efficient management of satellite orbits and spectrum resources.

TABLE 7.4
Classification of ITU-R Recommendation/Reports Series

Series Acronym (X)	Related Contents
BO	Satellite delivery
BR	Recording for production, archival and play-out; film for television
BS	Broadcasting service (sound)
BT	Broadcasting service (television)
F	Fixed service
M	Mobile, radio determination, amateur and related satellite services
P	Radio wave propagation
RA	Radio astronomy
RS	Remote sensing systems
S	Fixed-satellite service
SA	Space applications and meteorology
SF	Frequency sharing and coordination between fixed-satellite and fixed-service systems
SM	Spectrum management
SNG	Satellite news gathering
TF	Time signals and frequency standards emissions
V	Vocabulary and related subjects

Radio Regulations and Regional Agreements were implemented. These have been updated via World and Regional Radiocommununication conferences, respectively. Furthermore, a number of standardized recommendations and reports of SG 4 can provide vital information to satellite communication system designers, as they have been developed by experts from administrations, operators, industry, and other organizations from all over the world. Although implementation of recommendations or reports is not mandatory, the information they contain can be used to design and operate various satellite communications via efficient utilization of spectrum resources [11]. Recommendations and reports of ITU-R are generally named as recommendation (or report) ITU-R $X.Y\text{-}Z$, where the character X indicates a related part and is called a series, Y indicates a serial number, and Z indicates the number of revisions. Table 7.4 summarizes classification of the series of ITU-R recommendations and reports. For example, recommendation ITU-R S.1062-4 is titled "Allowable Error Performance for a Satellite Hypothetical Reference Digital Path Operating Below 15 GHz." Because it is categorized as the S series, it gives guidelines on bit-error rate (BER) design masks for fixed satellite services. It has been revised four times.

7.3 CURRENT STATUS OF SATELLITE COMMUNICATION TECHNOLOGY

The history of satellite communications suggests that for point-to-point services, satellites can only provide economical niche services beyond terrestrial coverage. On the other hand, for broadband and multicast services, which are increasing far more rapidly than unicast services, a satellite system is very effective and efficient. This is because a satellite system has an advantage over terrestrial services for delivery of identical content to users spread over a wide geographic area with uniform service quality and performance.

Recent service convergence trends mean that it is difficult to differentiate broadcasting and communication services in a classical manner. Thus, the recent technical specifications of satellite radio interfaces do not have specific roles for broadcasting or communications. Instead, they have been developed to provide both services at the same time. Another strong impetus for new technology is mobility, even with various technical difficulties. In this section, we review current satellite technologies in two main areas. The first is broadband satellite technology, which provides high-speed Internet connections and multimedia broadcasting and multicasting services (MBMS) to fixed and mobile terminals. The second is mobile broadcasting to handheld terminals.

7.3.1 BROADBAND SATELLITE COMMUNICATIONS AND INTERACTIVE BROADCASTING

Increasing demand for broadband services can be effectively handled by a satellite system using high-frequency bands such as the Ku and Ka bands. Due to the inherent characteristics of satellite communications such as wide coverage, broadcast mode of operation, and multicasting, a satellite system is one of the most effective means of providing high-speed Internet connections and multimedia long-distance transmission to fixed and mobile terminals. Indeed, to provide high-speed Internet and television (TV) services to maritime and air vehicles, a satellite system might be the only possible option. In this case, an active array antenna that is mounted on a moving vehicle is used to track a satellite and provide seamless connections.

In this kind of system, the volume of traffic in the forward link, which provides connections from the satellite gateway to the user terminals, is much greater than that in the return link, which provides connections from the user terminals to the satellite gateway. Recommendation ITU-R S.1709-1 proposes air interface characteristics, which can be used to implement broadband satellite networks [12]. In the recommendation, three air interface standards are summarized, as shown in Table 7.5. Many technological frameworks are based on the Digital Video Broadcasting (DVB) standards.

The DVB is a project by a market-led consortium in the TV industry, and the family of DVB standards has been standardized by the European Telecommunications Standard Institute (ETSI) [13]. The DVB via satellite (DVB-S) standard, which is specified as a forward link scheme for the first and second air interfaces in Table 7.5, describes an air interface for the satellite multiprogram TV service [14]. As a result of the rapid evolution of digital satellite communication technology since the introduction of DVB-S in 1994, the latest version of DVB-S, DVB-S2, has been published [15]. Remarkable improvements to DVB-S2 include a new channel coding scheme and higher order modulation, which are combined with an adaptive coding and modulation (ACM) mode, to compensate for rain attenuation in high-frequency bands. The new channel code in DVB-S2 is a concatenated code with low-density parity check (LDPC) codes and Bose–Chaudhuri–Hocquenghem (BCH) codes. The higher order modulation scheme is based on amplitude and phase shift keying (APSK)

TABLE 7.5
Air Interface Standards for Broadband Satellite Systems

	Standard Name		
Scheme	ETSI EN 301 790	TIA-1008-A	ETSI-RSM-A
Network topology	Star or mesh	Star	Star or mesh
Forward link scheme	DVB-S	DVB-S	High rate TDMA
Forward link data rate (Mbps)	1–45	1–45	100, 133.33, 400
Return link modulation	QPSK	O-QPSK	O-QPSK
Return link multiple access	Multifrequency (MF)-TDMA	MF-TDMA	FDMA-TDMA
Return link data rate	No restriction	64, 128, 256, 512, 1024, 2048 ksymbol/s	2, 16, 128, 512 kbits/s

up to order 32, which has been proven to be more efficient than quadrature amplitude modulation (QAM) for nonlinear distortion, due to the high-power amplifier.

The air interface in the first column of Table 7.5 is called a DVB- return channel by satellite (DVB-RCS), and in combination with DVB-S/S2 it provides two-way broadband satellite systems [16]. Because the current version of DVB-RCS does not consider mobile service environments, an advanced version for comprehensive support of mobile and nomadic terminals is currently under study. The second air interface in Table 7.5 is the Internet Protocol over Satellite (IPoS) standard that has been developed by the Telecommunications Industry Association (TIA) in the United States. The third air interface in Table 7.5 is the Broadband Satellite Multimedia (BSM) standard developed by ETSI. Important characteristics of the BSM architecture are that it is separated into satellite-dependent functions and satellite-independent functions. The purpose of this separation is to provide the capacity to incorporate future market developments, as well as the flexibility to include different market segment-based solutions in the higher layers. Among the several types of BSM air interface families, Regenerative Satellite Mesh (RSM)-A is based on a satellite with onboard processing (OBP) such as SPACEWAY by Huges, which supports a fully meshed topology. With RSM-A, data can be transmitted between any pair of satellite terminals in a single hop [17].

As a few examples of these systems, INMARSAT first started a Broadband Global Area Network (BGAN) service based on 3GPP waveforms. The BGAN provides circuit-switched and packet data services via a standard Ericsson 3G Release-4 Core Network (CN) coupled to a specialized Satellite Radio Access Network (S-RAN). The first generation of BGAN mobile terminals included a range of portable devices supporting packet data rates between 200 and 500 kbps and delivered by geostationary satellites to spot beams, which are similar in concept to terrestrial mobile network cells [18]. The Electronics and Telecommunication Research Institute (ETRI) in Korea developed a mobile broadband interactive satellite access technology (MoBISAT) system based on the DVB-S/DVB-RCS standard for satellite video multicasting and Internet via wireless local area networks (WLANs) [19]. The system can provide broadband services to passengers and crew of land, maritime, and air vehicles via installation of group user terminals with a two-way active antenna. For accessing Internet inside vehicles, WLANs can be provided, enabling users to use their laptop PC or PDA.

7.3.2 MOBILE BROADCASTING TO HANDHELD TERMINALS

Due to the inherent broadcasting capability of satellites, satellite TV services to fixed receivers are one of the most successful use cases of satellite communications. In the 1990s, the Worldspace system started a satellite audio broadcasting service to fixed receivers. However, services to mobile terminals could not be implemented quickly, due to technical difficulties such as the requirement for high service availability despite prevailing blockage conditions. As a result of various works on enhancing link availability over a decade, satellite audio broadcasting to mobile receivers in vehicles began in the early 2000s. Broadcasting of high-quality digital radio channels to mobile users was achieved by means of S-band frequencies around 2 GHz. This was called satellite digital audio radio service (S-DARS). Two companies in the U.S., XM Radio and Sirius, provided these services [20, 21]. The remarkable success achieved by XM Radio and Sirius was based on traditional countermeasures such as a high link margin, time, frequency and satellite diversity, and usage of terrestrial repeaters within urban areas.

For example, Sirius satellites transmit identical content from three spatially and temporally diverse signals to eliminate blockage or foliage attenuation, thus increasing link availability. The two active satellites transmit an identical signal at different frequencies with a 4-sec delay between them, which is inserted at the uplink Earth stations. In addition, a high elevation angle is achieved via adoption of a unique orbital configuration. The three satellites in the Sirius constellation are each in a highly inclined, elliptical geosynchronous orbit, which results in separation from the other two by 8 hours in ground track. With this configuration, the Sirius satellite constellation provides high elevation angles to users throughout its North American service area, with an average greater than 60°. Moreover, the satellite service is augmented in urban areas by terrestrial gap-fillers that rebroadcast identical content to that of each satellite [22]. The Federal Communication Commission (FCC) in the U.S. approved the merger of XM Radio and Sirius at the end of 2008.

More recently, in Korea and Japan, S-DMB service to handheld user terminals was successfully deployed via a GSO satellite [23]. The S-DMB system was based on code division multiplexing (CDM) technology specified as System E, which was described in Recommendation ITU-R BO.1130-4 [24]. High link availability was achieved via deployment of gap-fillers or repeaters in urban areas. Because no satellite diversity scheme is employed, more gap-fillers are required for this system than for the XM Radio and Sirius systems. A rake receiver in the user terminal can successfully resolve multipaths, including a direct path from the satellite and indirect paths from gap-fillers. Figure 7.7 shows the system configuration of the S-DMB system using the CDM technique.

In Europe the Unlimited Mobile TV concept was introduced [25], which is based on the ETSI standard of DVB—satellite services to handheld (SH) devices [26]. A handheld mobile terminal is supposed to receive broadcast signals from both the

Satellite Communications

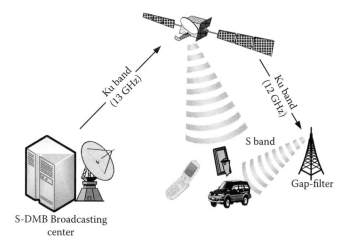

FIGURE 7.7 System configuration of S-DMB using CDM.

satellite and terrestrial repeaters, which are termed complementary ground component (CGC). The DVB-SH includes two transmission modes. The first is the orthogonal frequency division multiplexing (OFDM)/OFDM mode, which is based on the DVB-H standard [27]. The second is the time division multiplexing (TDM)/OFDM mode, which is partly based on the DVB-S2 standard. The former is more efficient in terms of the spectral aspect, whereas the latter is more efficient in terms of the power aspect. The OFDM-based DVB-SH standard is considered to be well adapted to the single frequency network (SFN).

7.4 TECHNOLOGIES FOR FUTURE SATELLITE COMMUNICATION SYSTEMS

Satellite broadcasting to mobile terminals is expected to become a killer application in the satellite industry. As discussed in the previous section, a couple of new techniques have already been proposed and deployed. In this context, we discuss a few candidate technologies and network architectures for future satellite systems, which are especially efficient for providing MBMS services.

7.4.1 Hybrid and Integrated Networks

ITU is currently developing a vision of NGN, with the clear objective of providing a means for the true integration of communication networks in their various forms. The purpose of this integration is effective utilization of the respective strengths of each network, within the context of their traditional roles and mandates. The hybrid and/or integrated satellite and terrestrial system enables an NGN by seamlessly interworking and cooperatively combining the most powerful aspects of each network.

Due to the increasing importance of these networks, ITU has studied various aspects of them, including clear definitions. Integrated systems are systems employing

MSS and terrestrial components, where the ground component is a part of the MSS system that is complementary to it and, in combination with the satellite component, provides an integrated service. In such systems the ground component is controlled by the satellite resource and network management system. It uses the same designated parts of the frequency band as the associated operational MSS system. The ground component is referred to as ancillary terrestrial component (ATC) in the United States and Canada and as CGC in Europe [28]. We previously investigated a couple of integrated systems, S-DARS by XM Radio and Sirius and the S-DMB system in Korea. The Unlimited Mobile TV, which is based on DVB-SH, can also be successfully realized on an integrated network architecture using CGCs.

An integrated system provides a combined (integrated) single network that uses both a traditional MSS link and terrestrial transmission paths to serve end-user handsets. With proper network planning and control of both the space and terrestrial components of the system, the operators can use the assigned spectrum extensively and efficiently to provide indoor and outdoor coverage in urban, suburban, rural, and remote areas, including direct service to small handsets. A typical integrated system comprises one or more multispot beam satellites and a nationwide or regional ensemble of terrestrial cell sites, where both the terrestrial and space components communicate with user terminals using a common set of MSS frequencies to provide ubiquitous coverage to end users with generalized mobility requirements.

The hybrid system utilizes a satellite component that is part of either an FSS or MSS network and the terrestrial component operates in a fixed, mobile, or nomadic mode. The satellite and terrestrial components do not necessarily operate in the same allocated frequency band. Examples of a hybrid system are a satellite digital multimedia broadcasting system with terrestrial repeaters and a WLAN or broadband wireless access systems with satellite access points. More advanced forms of these hybrid and/or integrated satellite and terrestrial systems are expected to be developed.

7.4.2 Multicarrier Techniques

Many terrestrial system standards for future implementation consider multicarrier-based modulation and multiple access techniques such as orthogonal frequency division multiplexing–frequency division multiple access (OFDM–FDMA; OFDMA). Considering that OFDM-type systems are largely used in terrestrial networks as a means to provide good spectral and energy efficiency over frequency selective channels, the idea of applying them to a satellite channel might seem to lack justification. OFDM transmission is one of the most effective means to overcome the intersymbol interference problem caused by multipath delayed signals. However, this is no longer the case for a conventional satellite system. Moreover, the high peak to average power ratio (PAPR) problem, which is the most serious problem for multicarrier-based transmission techniques, imposes a high burden on high-power amplifiers in satellite systems.

Even with the aforementioned obstacles to using multicarrier-type transmissions for satellite communication systems, there are a number of factors that make this technology attractive [29]. First, a satellite component of a future network could be regarded as coverage extension for service continuity of the terrestrial component with vertical handover. For cost-effective vertical handover, future satellite radio interfaces could be compatible with and have a high degree of commonality with the terrestrial interface. This might enable reuse of the terrestrial component technology to minimize the user terminal chipset and network equipment for low cost and fast development. Second, in broadband satellite communications, routing and capacity flexibility over the coverage area require the use of multiple carriers per beam with FDM. Third, in a scenario where an on-board switch is also used to route the TDM data from a given gateway to different beams, OFDM might simplify the architecture of the on-board switch. No complex carrier demultiplexing filters are needed to separate the carriers, because simple FFT engines can be used. Lastly, and more important, OFDM techniques could represent a means to increase the spectral efficiency by compacting carriers and eliminating guard-bands for applications requiring high-speed transmissions.

As we have noted, DVB-SH adopted OFDM transmission, which is the same signal format defined in DVB-H for terrestrial systems. The main reason for adopting OFDM stems from the fact that satellite and terrestrial transmitters form an SFN. This kind of configuration uses the available frequency resources efficiently, as the same frequencies are used for satellite and terrestrial transmitters. The same content is transmitted via the satellite and terrestrial link, as the transmitted waveforms are identical. Terrestrial transmitters can be fed by receiving the satellite signal, if the transmitted and received signal can be sufficiently isolated [30]. In addition to the high spectral efficiency of multicarrier transmission, we can increase the capability to allocate resources for adaptive usage, not only in terms of time but also in terms of frequency domains, as reported in previous research [31]. In order to reduce unfavorable distortions due to the high PAPR incurred in the multicarrier transmission technique, a number of approaches that were introduced for satellite applications can be applied [32–34].

7.4.3 COOPERATIVE DIVERSITY TECHNIQUES

Adaptive transmissions including power control and ACM are crucial techniques for all wireless systems. Their purpose is to control the transmitting resources such that the signal received has the required signal-to-noise ratio (SNR) with minimal energy consumption. The time-varying characteristics of wireless mobile channels necessitate adaptive radio interfaces to provide high quality and economic services. Because satellite bandwidth is a relatively scarce resource, adaptive usage of resources supported by various coding and modulation schemes is mandatory for an efficient and economical system. For example, the DVB-S2 standard adopted the ACM mode to mitigate rain fading. However, the performance gain achieved via these kinds of techniques can only be guaranteed when precise channel quality information (CQI) from the return link is available at the transmitter. The unidirectional nature of MBMS prohibits the use of control commands for power control and ACM. Therefore, in this situation, we should focus on downlink strategies to improve performance.

A transmit diversity technique is a means to provide performance gain when no CQI is available. Space time coding (STC) using several antennas is one of the most

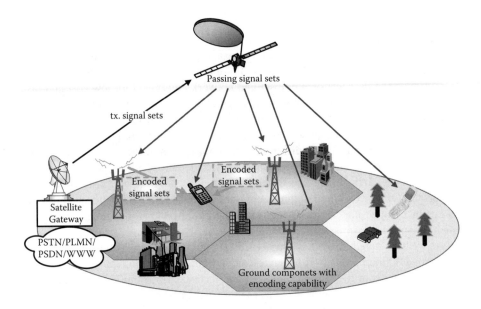

FIGURE 7.8 Integrated satellite and terrestrial network utilizing cooperative diversity.

effective means to achieve diversity gains. The gain can be maximized when it is reasonable to assume that the path components from other antennas are independent. By contrast with typical terrestrial wireless systems, we should not expect the same gain to be achieved in a satellite system, because the distance between the satellite and user terminal is much longer than that between the antennas, so every path seems similar. However, in integrated or hybrid systems, we can achieve diversity gains via the signal paths from the satellite and those from the ground components. At a user terminal in these networks, the signals from the satellite and ground components are independent. In this situation, the antennas are no longer collocated at the transmitter or receiver; rather, they are distributed at different ground components. In this scenario, we may apply an STC scheme to cooperatively achieve diversity gains with the satellite and ground components [35].

In the integrated network configuration in Figure 7.8, a multispot beam satellite in GSO and an ensemble of terrestrial cell sites with ground components are deployed. The satellite transmits data to the user terminals and all ground components. In order to achieve diversity gains via utilization of STC schemes, each of the ground components must transform the received signals into a given encoded signal format and retransmit them to the user terminal, as shown in Figure 7.8. With the system model in Figure 7.8, the ground components and satellite can cooperate to transmit STC signals, and the ground components can encode signals rather than serving as simple amplifiers. A user terminal can receive the STC-encoded signals. If the user terminal receives multiple signals from both repeaters and the satellite, then it can achieve STC gains using these signals.

As an example of utilizing STC schemes for cooperative diversity, consider the use of the Alamouti code, which was the first STC scheme proposed for two transmit antennas [36]. Assuming that the signal sequence $[s_i s_{i+1}]$ is transmitted during the two symbol period 2T, encoded signal formats of $[s_i s_{i+1}]$ and $[-s_{i+1}^* s_i^*]$ are used for the first and second antenna, respectively, where * represents a complex conjugate operation. In order to apply this method, the satellite is regarded as the first antenna and two different ground components are regarded as either the first or second antenna. Then, the satellite passes through a signal set consisting of $[s_i s_{i+1}]$ during the period 2T. Because the channel condition of the path from the satellite to a ground component differs so much from that of the path to the user terminal, without loss of generality, we can assume that the signal from the satellite to a repeater is almost error-free in the general case. In addition, the ground component has sufficient time to process the received signal, and thus the signal retransmitted from the ground component can arrive at the user terminal at almost the same time as the one transmitted directly from the satellite, as in the currently deployed S-DMB system.

After receiving the error-free signal $[s_i s_{i+1}]$ from the satellite, each ground component applies space-time encoding to $[s_i s_{i+1}]$. In other words, the first ground component transmits the signal set $[s_i s_{i+1}]$, whereas the second one transmits $[-s_{i+1}^* s_i^*]$, during the period 2T. By this means, a user terminal can achieve diversity gains via a combination of both signal paths from the satellite and ground components, wherever possible. A user terminal can receive various combinations of signal sets depending on the signal availability, mainly according to the location. If the user terminal receives two different signal sets, for example, $[s_i s_{i+1}]$ from the satellite or one of the ground components and $[-s_{i+1}^* s_i^*]$ from the other one of the ground components, then it can achieve maximal diversity gain via the ordinary decoding algorithm for the Alamouti scheme. That is, we can obtain an estimate of the transmitted signal, $\hat{s}_i$ and $\hat{s}_{i+1}$ as follows:

$$\hat{s}_i = h_1^* r_i + h_2 r_{i+1}^*$$
$$\hat{s}_{i+1} = h_1^* r_{i+1} - h_2 r_i^*, \qquad (7.2)$$

where, h_1 and h_2 are the complex multiplicative distortion (channel gain) for two different available paths at the STC decoder, respectively. r_i and r_{i+1} are the serially received signals at the terminal during the period 2T, which are expressed as

$$r_i = h_1 s_i - h_2 s_{i+1}^* + n_i$$
$$r_{i+1} = h_1 s_{i+1} + h_2 s_i^* + n_{i+1}, \qquad (7.3)$$

where n_i and n_{i+1} are complex random variables representing the receiver noise and interference during the period 2T. If any path is missing, then we can simply regard the channel gain of that path as zero.

In order to further increase the diversity gain, we can use an STC scheme for more than two transmit antennas. Generally as the number of transmit antennas is increased in the STC scheme, we expect an increase in the diversity gain. Similarly, we can achieve more diversity gain via adoption of efficient STC schemes [37], and cooperatively adopting channel coding schemes [38]. In order to reap this performance gain, the ground component must have additional encoding capability, rather than simply serving as a frequency converter and amplifier, as in the currently deployed system. The user terminal must have decoding capability. Due to this requirement, backward compatibility cannot be guaranteed if the user terminal cannot obtain a signal from the satellite. We also note that this performance gain must be conditioned on symbol time synchronization of two different signals. Achieving time synchronization is a very difficult problem for a system using multicarrier schemes such as OFDM.

7.4.4 LAYERED CODING

In this section, another adaptive technique for MBMS is introduced. Due to the unidirectional nature of MBMS, the receiver must operate adaptively on its own, without any control commands. The hierarchical modulation scheme is one of the adaptive schemes that can be applied to MBMS applications [39], but the purpose of this adaptability is to enable the service quality to be upgraded for a new terminal, while maintaining backward compatibility, rather than compensating for channel impairments. A layered coding scheme with concatenated error correction codes was proposed, where a receiver selected a suitable demodulation/decoding scheme for the channel condition, without any knowledge of the CQI from the return link [40].

The rationale behind the layered coding scheme can be explained as follows. In MBMS, channel coding is designed to address the worst-case fading condition, but this causes unnecessary processing complexity at the receiver for the majority of users. Alternatively, channel coding can address the average fading condition, but it cannot guarantee the quality of service of every user. Ideally, channel coding enables a user with a good channel to recover information with low complexity, while still enabling a user with a bad channel to achieve an acceptable BER at the cost of an increased complexity or additional decoding delay and power consumption. The layered coding scheme consists of several concatenated codes that can be separated and operate in different ways. The user terminal can choose to operate either with the symbols encoded using the high rate code or with the symbols encoded using the low rate code, according to the required BER and received SNR. The penalty associated with the latter choice is an increased decoding complexity and delay.

For example, serially concatenated turbo codes can be used as layered coding schemes [35, 40]. In these schemes, a receiver in the deep fading condition selects a decoding scheme for fully concatenated codes, enabling it to achieve a large coding gain, whereas a receiver in the mild channel condition selects a decoding scheme for a simple outer code only, thus reducing the decoder complexity. This layered coding technique enables each user terminal to select the optimum demodulation and decoding scheme, enabling the system to reduce the total transmission power. In other words, we can increase the cell coverage while using the same amount of power. Because the system broadcasts the signal with the strongest error correction coding scheme and the receiver selects a suitable decoding scheme without necessarily considering all redundancies involved, the system requires an increase in bandwidth and does not achieve coding gains. This is the major drawback of this layered coding scheme. Also, this method cannot guarantee backward compatibility.

REFERENCES

1. Arthur C. Clarke, "Extra Terrestrial Relays," *Wireless World,* pp. 305–308, Oct. 1945.
2. http://www.intelsat.com
3. http://www.inmarsat.org
4. http://www.intersputnik.com/
5. http://www.eutelsat.org
6. http://www.arabsat.com
7. S. K. Shin, K. Lim, K. Choi, and K. Kang, "Rain attenuation and Doppler shift compensation for satellite communications," *ETRI Journal,* Vol. 24, pp. 31–42, Feb. 2002.
8. Recommendation ITU-R S. 1061–1, Utilization of fade countermeasures strategies and techniques in the fixed-satellite service, 2007.
9. A. K. Maini and V. Agrawal, *Satellite Technology: Principles and Applications.* New York: Wiley, 2007.
10. http://www.itu.int
11. S. Kim and S. Kota, "ITU-R Standardization of Fixed Satellite Services (FSS)," in *Proc. International Workshop on Satellite and Space Communications,* 2007, pp. 247–252.
12. Recommendation ITU-R S.1709–1, Technical characteristics of air interfaces for global broadband satellite systems, 2007.
13. F. L. C. Ong, X. Liang, P. Pillail, P. M. L. Chan, G. Koltsidas, F. N. Pavlidou, E. Ferro, A. Gotta, H. Cruickshank, S. Iyengar, G. Fairhurst, and V. Mancuso, "Fusion of digital television, broadband Internet and mobile communications—Part I: Enabling technologies," *International Journal of Satellite Communications and Networking,* Vol. 25, pp. 363–407, July 2007.
14. ETSI EN 300 421 (V1.1.2). Digital video broadcasting (DVB); Framing structure, channel coding and modulation for the 11/12GHz satellite services, Aug. 1997.
15. ETSI EN 302 307 (V1.1.1). Digital video broadcasting (DVB); Second generation framing structure, channel coding and modulation systems for broadcasting, interactive services, new gathering and other broadband satellite applications, March 2005.
16. ETSI EN 301 790 (V1.4.1). Digital video broadcasting (DVB); Interaction channel for satellite distribution systems, April 2005.
17. ETSI TS 102–188 Technical Specification (TS) 102 188 (V1.1.2), Satellite. Earth Stations and Systems (SES); Regenerative Satellite. Mesh—A (RSM-A) air interface, July 2004.
18. P. Febvre, X. Bouthors, S. Maalouf, and D. Bath, "Efficient IP-multicast via Inmarsat BGAN, a 3GPP satellite network," *International Journal of Satellite Communications and Networking,* Vol. 25, pp. 459–480, Sept. 2007.
19. Y.-J. Song, P.-S. Kim, D.-G. Oh, S.-I. Jeon, and H.-J. Lee, "Development of mobile broadband interactive satellite access system for Ku/Ka band," *International Journal of Satellite Communications and Networking,* Vol. 24, pp. 101–117, March 2006.
20. J. Snyder and S. Patsiokas, "XM satellite radio-satellite technology meets a real market," presented at the 22nd AIAA International Communications Satellite Systems Conf. & Exhibit, Monterey, CA, May 2004.
21. R. D. Briskman and S. Sharma, "DARS satellite constellation performance," presented at the 20th AIAA International Communications Satellite Systems Conf. & Exhibit, Montreal, Canada, May 2002.
22. R. Akturan, "An overview of the Sirius satellite radio system," *International Journal of Satellite Communications and Networking,* Vol. 26, pp. 349–358, Sept. 2008.
23. S.-J. Lee, S. Lee, K.-W. Kim, and J.-S. Seo, "Personal and mobile satellite DMB services in Korea," *IEEE Transactions on Broadcasting,* Vol. 53, pp. 179–187, March 2007.

24. Recommendation ITU-R. BO.1130–4. Systems for digital satellite broadcasting to vehicular, portable and fixed receivers in the bands allocated to BSS (sound) in the frequency range 1400–2700 MHz, 2001.
25. N. Chuberre, O. Courseille, P. Laine, L. Roullet, T. Quignon, and M. Tatard, "Hybrid satellite and terrestrial infrastructure for mobile broadcast services delivery: An outlook to the 'Unlimited Mobile TV' system performance," *International Journal of Satellite Communications and Networking*, Vol. 26, pp. 405–426, Sept. 2008.
26. ETSI EN 302 583 v1.1.1, Digital video broadcasting (DVB); Framing structure, channel coding and modulation for satellite services to handheld devices (SH) below 3 GHz, March 2008.
27. ETSI EN 302 304 v1.1.1, Digital video broadcasting (DVB); Transmission system for handheld terminals (DVB-H), Nov. 2004.
28. Chariman, WP 4B, Report on the twenty fifth meeting of Working Party 4B—(Geneva, 24 September–1 October 2008), Nov. 2008.
29. A. Ginesi and F. Potevin, "OFDM digital transmission techniques for broadband satellites," presented at the 24th AIAA International Communications Satellite Systems Conf. (ICSSC), San Diego, CA, June 2006.
30. J. Krause and H. Stadali, "ETSI technical standards for satellite digital radio," *International Journal of Satellite Communications and Networking*, Vol. 26, pp. 463–474, Sept. 2008.
31. K. Lim, S. Kim, and H.-J. Lee, "Adaptive radio resource allocation for a mobile packet service in multibeam satellite systems," *ETRI Journal*, Vol. 27, pp. 43–52, Feb. 2005.
32. K. Lim, K. Kang, and S. Kim, "Adaptive MC-CDMA for IP-based broadband mobile satellite systems," in *Proc. IEEE 58th Vehicular Technology Conference*, Vol. 4, 2003, pp. 2731–2735.
33. K. Kang, S. Kim, D. Ahn, and H. J. Lee, "Efficient PAPR reduction scheme for satellite MC-CDMA systems," *IEE Proceedings-Communications*, Vol. 152, pp. 697–703, Oct. 2005.
34. S. Cioni, G. E. Corazza, M. Neri, and A. Vanelli-Coralli, "On the use of OFDM radio interface for satellite digital multimedia broadcasting systems," *International Journal of Satellite Communications and Networking*, Vol. 24, pp. 153–167, March 2006.
35. S. Kim, H. W. Kim, K. Kang, and D. S. Ahn, "Performance enhancement in future mobile satellite broadcasting services," *IEEE Communication Magazine*, Vol. 46, pp. 118–124, July 2008.
36. S. M. Alamouti, "A simple transmit diversity technique for wireless communications," *IEEE Journal on Selected Areas in Communications*, Vol. 16, pp. 1451–1458, Oct. 1998.
37. U. Park, S. Kim, K. Lim, and J. Li, "A novel QO-STBC scheme with linear decoding for three and four transmit antennas," *IEEE Communications Letters*, Vol. 12, pp. 868–870, Dec. 2008.
38. S. Kim, H. W. Kim, K. Kang, and D. S. Ahn, "A cooperative transmit diversity scheme for mobile satellite broadcasting systems," in *Proc. 4th Advanced Satellite Mobile Systems*, Aug. 2008, pp. 66–71.
39. H. Jiang and P. A. Wilford, "A hierarchical modulation for upgrading digital broadcast systems," *IEEE Transactions on Broadcasting*, Vol. 51, pp. 223–229, June 2005.
40. A. Levissianos, G. Metaxas, N. Dimitriou, and A. Polydoros, "Layered coding for satellite-plus-terrestrial multipath correlated fading channels," *International Journal on Satellite Communications and Networking*, Vol. 22, pp. 485–502, Sept. 2004.

8 Next-Generation Wired Access
Architectures and Technologies

Eugenio Iannone

CONTENTS

8.1	Introduction	167
8.2	Next-Generation Access Architectures	169
	8.2.1 WDM-PON Architecture	174
	8.2.2 GPON Architecture	177
	8.2.3 WDM-PON versus GPON Performance Comparison	178
	8.2.3.1 Capacity	178
	8.2.3.2 Optical Link Budget	179
	8.2.3.3 Security and Unbundling	180
	8.2.3.4 Flexibility	181
	8.2.3.5 Fiber Utilization	181
	8.2.4 Access Network Delayering	182
8.3	Key Technologies for Next-Generation Access	184
	8.3.1 GPON Optical Interfaces	185
	8.3.2 WDM-PON Optical Interfaces	188
8.4	A Case Study	189
References		192

8.1 INTRODUCTION

Different factors push the evolution of the access network from today's copper-based architecture to a fiber-based one. A primary driver is constituted by the diffusion of video services requiring a dedicated bandwidth of the order of several Mbit/sec, especially if video on demand is considered [1, 2].

Besides bandwidth increase, the penetration of ultrabroadband connections stresses the capacity of existing copper cables to support a high number of very fast xDSL connections (that is, high-speed connections on copper cable). In many cases, copper cables deployed before the diffusion of broadband access present nonnegligible pair-to-pair interference when xDSL is used. Often this phenomenon limits the number of connections possible in a given area, depending on the quality and length of the cables [3].

Finally, the need of the main carriers to reduce operational costs also drives usage toward a fiber-optic-based access network. As a matter of fact, optical transmission allows a completely passive access infrastructure to be implemented. Moreover, much longer spans can be realized via Passive Optical Networks (PON), opening the possibility of a delayering of the peripheral part of the network. In the delayered architectures, local exchanges are almost completely eliminated in populated areas (e.g., cities), initiating the passive access network directly in the metro core nodes [4, 5]. Eliminating a certain number of local exchanges further reduces operational expenses (opex) at the cost of a higher initial capital investment (capex).

Completely replacing the access infrastructure is perhaps the most capex-intensive operation that an incumbent carrier can face. Tier one carriers have several million terminations (for example, in a country like Italy, the network of Telecom Italia has about 20 million terminations) and their copper network arrives almost everywhere, both in cities and in the countryside. In this situation it is clear that the deployment cost is the main parameter characterizing a given access solution and a huge effort is directed toward cost reduction.

If civil works are needed to deploy new tubes where fiber cables have to be installed, this is by far the most expensive part of the deployment of the new access network. However, this situation is not very common, especially in the most populated areas, where the network deployment is likely to start. During the last 20 years, fiber deployment has been constant in the major cities of Europe, America, and Asia [6]. Moreover, even when fibers are not available, tubes are already in the ground, greatly facilitating fiber deployment. For these reasons, we do not consider civil works in our evaluation of the network cost; readers interested to this aspect can find a starting point in the bibliography [7, 8].

Beyond the capacity delivered to the end user and the cost, a variety of other parameters characterize an optical access network. Due to the huge investment required for deployment, next-generation access infrastructure has to be future proof. Thus, it has to accommodate changes in the service bundle offered to the end user and the related evolution of the network equipment. Nevertheless, waiting the full network implementation to experience a revenue increase is not a feasible strategy. The network has to be deployed gradually, in pace with the creation of new revenues.

In this situation, a possible strategy seems to individuate a long-term, future-proof architecture, and an evolutionary scenario that allows gradual capex expenditure in pace with new service introduction and a full exploitation of the copper network where this is still possible.

The migration of all telecommunication services toward a data-based paradigm opens the possibility to unify data transport on the same infrastructure, collecting traditional wireline residential and business users along with mobile users. Such convergence is possible by connecting the access points of the mobile network (the base stations) to the wireline infrastructure. This requires a further penetration of the fiber infrastructure, besides the capability of supporting a large variety of termination equipment.

From a bandwidth management point of view, the copper network ensures the physical separation of signals belonging to different users in the whole access segment: Each user has its own twisted pair. This is a good situation for several reasons. The privacy of the user signal is completely guaranteed because it is impossible to detect another user's signal from a user terminal. Moreover, the concept of unbundling is strictly related to this characteristic of the copper network, ensuring in this case perfect separation among signals belonging to different carriers. For these reasons it would be desirable to reproduce this situation in the next-generation access network as much as possible.

This brief discussion highlights the fact that, even if cost is always a key issue, cost and capacity are not the only important elements for a comparison among different network architectures. A wide variety of other elements must be considered, often depending on the particular case under analysis (consider, for example, the differences among regulatory rules in Europe, North America, and Asia).

8.2 NEXT-GENERATION ACCESS ARCHITECTURES

The copper network is a completely passive structure connecting every end user with the local exchange using a dedicated twisted pair. Replicating this structure using fiber optics would require direct connection of each end user with the local exchange using dedicated fiber optics (see Figure 8.1).

A similar architecture would have several advantages because it ensures maximum transmission capacity and physical user signal separation, and it allows unbundling to be replicated in the new infrastructure with the same strategy used in the copper network. In practice, however, this solution is considered unfeasible in many cases due to important differences between the copper pair and the fiber optic infrastructure.

With fiber deployment, the ideal strategy would be connecting each end user with the local exchange via the fiber infrastructure. However, in order to facilitate a gradual capital expenditure, a gradual fiber deployment strategy is often required. This means arriving with optical fiber in some intermediate point between the local exchange and the end user. At this intermediate point, active equipment converts the optical signal into an electrical one and the end user is reached with a traditional copper pair using a suitable xDSL format (generally VDSL2 [Very high Digital Subscriber Loop version 2]).

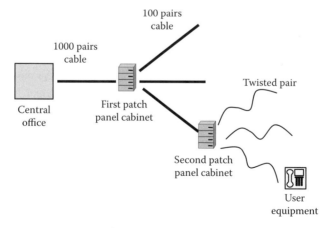

FIGURE 8.1 Example of a topology of the copper-based access network.

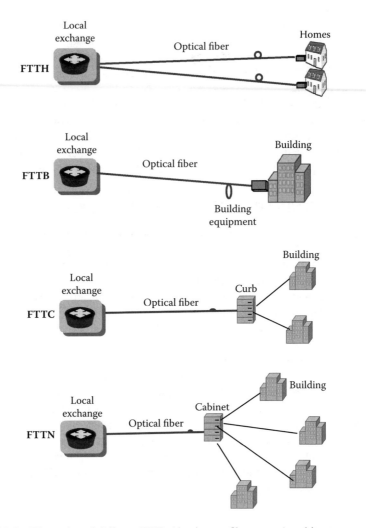

FIGURE 8.2 Illustration of different FTTx (that is, any fiber access) architectures.

Mixed optical-copper architectures are called different names depending on the distance between the intermediate point and the end user and the number of users connected to the single intermediate point equipment. In particular, the following nomenclature is generally used (see Figure 8.2), tailored mainly to urban areas, where the deployment of next-generation access will come first.

- *Fiber to the home* (FTTH) is an architecture in which there is no copper and the fiber arrives to the end user.
- *Fiber to the building* (FTTB) is an architecture in which the intermediate point is in the building basement or just outside a building and the number of end users linked to the same intermediate point is about 20 to 30. In this case, the intermediate node is generally within 50 m from the user.

- *Fiber to the curb* (FTTC) is an architecture in which the intermediate point serves a few buildings and the number of end users linked to the same intermediate point is about 100. In this case, the intermediate node is generally within 300 m from the user.
- *Fiber to the node* (FTTN) or *fiber to the cabinet* (FTTCab) is an architecture in which the intermediate point collects many buildings and the number of end users linked to the same intermediate point is on the order 400. In this case, the intermediate node distance from the user is generally beyond 300 m.

A single fiber is largely overdimensioned with respect to the needs of any group of nearby users: hundreds of multigigabit channels can be transmitted on a single fiber using Dense Wavelength Division Multiplexing (DWDM). However, DWDM techniques used in the backbone are far too expensive for the access network. To reduce costs, the single fiber capacity is also greatly reduced. Thus, adopting a network architecture that optimizes fiber use can be an important issue.

Moreover, in several practical cases the carrier either owns a limited number of fibers or it leases fibers from a third party (e.g., a utility company). In this situation it is important to optimize the number of fibers used in the infrastructure. The reduction of the number of fibers arriving at the central office also allows smaller optical patch panels to be deployed, resulting in lower optical connector failure rate and lower probability of routing error. Moreover, it is much easier to realize automatic patch panels if they are small, further reducing the local exchange maintenance cost. Finally, reducing the number of fibers arriving at the central office for a given capacity allows the switching machine located in the central office to interface with the network with a small number of high-bit-rate interfaces. This optimizes the operation of the local exchange switch machine minimizing its cost.

From the preceding discussion, it is clear that there are several reasons to push for reduction of fiber length in the access network. The solution to this problem is constituted by the so-called passive optical network (PON). In a PON, the fiber deployed in the field constitutes a logical tree, with the root in the local exchange

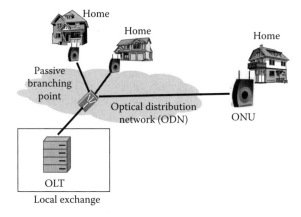

FIGURE 8.3 Passive optical network topology.

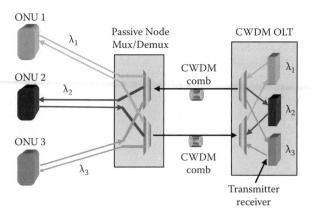

FIGURE 8.4 Functional scheme of a WDM-PON based on unidirectional transmission and CWDM wavelengths.

and leafs at the end users' locations (see Figure 8.3). At the branching points, an optical passive element is present, either a splitter in Time Division Multiplexing PON (TDM-PON) or a wavelength mux/demux in Wavelength Division Multiplexing PON (WDM-PON).

The optical device in the branching point has to be passive in the sense that it does not need power feeding it, thus in the case of WDM-PON a so-called a thermal mux/demux [9] is required. In the PON tree, every branch can be constituted either by a single fiber or by a fiber couple, depending on the adopted transmission technique. A single fiber is sufficient where bidirectional transmission is carried out; otherwise a fiber couple is used, one fiber for the upstream and the second fiber for the downstream. To effectively use a PON infrastructure, the optical signal traveling through the tree root has to be shared among the different end users, either via WDM or via TDM multiplexing.

In the case of WDM-PON, each end user is characterized by a different wavelength, as shown in Figure 8.4 [10, 11]. In general, the WDM-PON is realized by using unidirectional transmission, so that each PON branch is constituted by a fiber pair (bidirectional transmission is theoretically possible, but practically, reflections in the fiber infrastructures and the use of wide-band filters make it difficult to stabilize [12]).

In the case of TDM-PON, different users are assigned different time slots in a TDM frame and bidirectional transmission on the fiber infrastructure is realized using different wavelengths so that only a single fiber tree is needed, as shown in Figure 8.5. In Figure 8.5 a realistic Optical Line Termination (OLT) is shown where a set of PONs is managed by the same equipment. In the case of both TDM and WDM-PON, hundreds of PONs can be terminated on the same OLT, optimizing on one side real estate and power consumption and on the other the packet multiplexing operated by the OLT switch.

In the TDM-PON, another issue arises, due to the fact that different end users are at a different distance from the splitter at the tree branching point. In this condition, each end-user device (an Optical Network Unit [ONU]) has to transmit with a certain

Next-Generation Wired Access 173

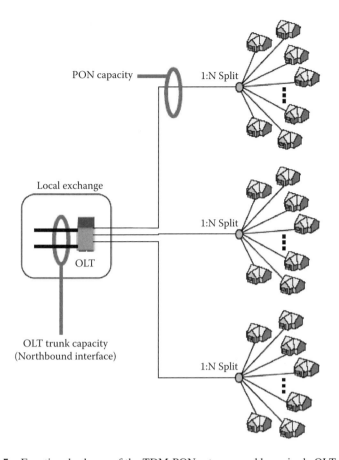

FIGURE 8.5 Functional scheme of the TDM-PON set managed by a single OLT.

delay with respect to the reference time frame to allow correct reconstruction of the signal at the branching splitter. This delay is continuously updated by a particular PON control procedure that measures the optical distance between the ONU and the branching point and drives the ONU asynchronous transmission.

A different implementation has been proposed for WDM-PON, but no standard has yet been produced. Extended standardization activity has been carried out on TDM-PON by a dedicated group of the International Telecommunications Union (ITU-T) called the Full Service Access Network (FSAN). Different standards follow the evolution of both optical technology and switching protocols, passing from APON and BPON, based on ATM, to GPON, capable of conveying multigigabit signals using either ATM or Ethernet protocols.

North American carriers such as Verizon, SBC (now AT&T), and others are looking to GPON to advance their FTTx rollout efforts. A different situation exists in Europe, where incumbent carriers generally have not yet committed to a clear strategy. However, a great number of FTTx field tests based on GPON are ongoing.

In the field of Ethernet-based PON, the Institute of Electrical and Electronics Engineers (IEEE) has generated an important standard (called GEPON) that has been largely adopted mainly in the Far East. NTT, the major carrier in Japan, and one of the largest telecommunications carriers in the world, began deploying GEPON-based FTTH access network equipment in 2003. In 2004, IEEE ratified the IEEE 802.3ah Ethernet in the First Mile specification that fully defined the GEPON technology deployed by NTT and later adopted by other carriers in Japan and Korea. A synthesis of the main TDM-PON standards is reported in Table 8.1, with some key optical parameters specified for each standard.

Hybrid schemes using both TDM and WDM can be also devised. These can be useful, for example, if a WDM-PON has been deployed in an FTTB configuration and the network has to evolve to FTTH. In this case, the building equipment can be substituted with a passive splitter and a TDM-PON can be deployed on the wavelength reaching every building. The splitting point of this TDM-PON is in the building basement, in this case.

8.2.1 WDM-PON Architecture

A simple architecture of a WDM-PON is shown in Figure 8.4. The OLT placed in the local exchange is constituted by an Ethernet switch managing a certain number of PONs and allowing grooming and quality of service (QoS) protocol management, and by a set of optical interfaces, one for each end customer. A different wavelength is associated with each end customer, so the OLT must have WDM-capable interfaces.

To avoid problems with reflections from the fiber infrastructure, unidirectional transmission is generally used. Thus, a fiber pair is needed on each branch of the PON infrastructure. A couple of athermal mux/demux is placed in the branching point and every signal is terminated at the final user site with an ONU tuned on the user wavelength.

The wavelength plan used in this architecture depends on the adopted WDM technique. If the Coarse WDM (CWDM) standard is used, generally no more than 16 wavelengths can be adopted. This is because common fibers have an absorption peak due to OH ions around 1390 nm, as shown in Figure 8.6, where the absorption profile of a standard fiber (ITU-T G.652) is superimposed on the CWDM frequency standard [13]. Assuming a sensitivity of –18 dBm of the receiving PIN, a transmitted signal power of 0 dBm, and a loss of 5 dB from the mux/demux, in order to achieve a reach of 10 km, a loss of less than 1.3 dB/km is needed. This requirement is not fulfilled by the CWDM channels around 1390 nm and 1410 nm, reducing the number of useful channels to 16.

Adoption of the CWDM standard has several advantages, including the possibility of using low-cost, robust transceivers, which are produced in very high volumes mainly for datacom applications, and the availability of low-cost athermal mux/demux. If a high wavelength number is to be used, DWDM is needed. In this case, it is easy to reach 32 or 64 wavelengths in the third fiber transmission windows at the expense of a higher interface cost.

One of the potential problems in deploying the WDM-PON architecture depicted in Figure 8.4 is the presence of colored ONU optical interfaces. As a matter of fact, every ONU receives and transmits a different wavelength. The management of 16 different ONUs during network deployment (in the case of CWDM PON) can increase operational costs, due to both spare parts management and in-field maintenance.

Next-Generation Wired Access 175

TABLE 8.1
TDM-PON Standards and Main Optical Characteristics

	Protocol	Data Rate	Reach	Split Ratio	Standard	Channel Insertion Loss
E-PON	Ethernet	1.25 Gbit/sec symmetrical	1000BASE-PX10: 10 km 1000BASE-PX20: 20 km	16 normal 32 permitted	IEEE 802.3ah	(Power Budget) PX-10U: 23dB PX-10D: 21dB PX-20U: 26dB PX-20D: 26dB
B-PON	ATM	622 or 155 Mbit/sec downstream @1490 nm 155 Mbit/sec upstream @1310 nm	Maximum 20 km	32 maximum	ITU-T 983.3	Class A: 20 dB Class B: 25 dB Class C: 30 dB
G-PON	ATM or Ethernet	2.488 or 1.44 Gbit/sec downstream 2.488 or 1.44 Gbit/sec downstream 2.488 or 1.44 Gbit/sec or 622 or 155 Mbit/sec upstream	Maximum physical reach 20 km	64 maximum 32 typical	ITU-T 984.2	Class A: 20 dB Class B: 25 dB Class B+: 28 dB Class C: 30 dB

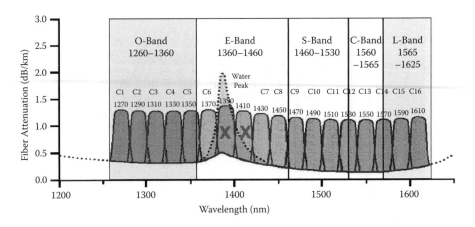

FIGURE 8.6 Standard CWDM wavelength comb and G.709 fiber attenuation profile. The channels that exhibit a large attenuation due to the water attenuation peak are also indicated.

To eliminate the need for colored interfaces, two different architectures can be individuated. The first is similar to the architecture of Figure 8.4, where tunable lasers are used both at the OLT and at the ONU. In this case, the same ONU can be used for all users, provided that a protocol is devised allowing it to tune on the correct wavelength. This simple solution is thwarted by the high cost of DWDM tunable lasers and by the difficulty of realizing CWDM tunable sources. A recent technology evolution, however, has opened up the possibility of using integrated laser arrays to implement low-cost tunable CWDM sources. We return to this point when discussing photonic integrated circuit (PIC) technology later in the chapter.

A second more sophisticated architecture of WDM-PON, allowing colorless ONUs to be used, is shown in Figure 8.7 [14]. The basic operating principle is based on automatic wavelength locking so that expensive components such as tunable lasers are not used at either the local exchange or the ONU. The working principle of this architecture is described here in the case of bidirectional transmission on a single fiber, but it can be easily applied when unidirectional transmission is used and the fiber infrastructure is composed of fiber pairs.

At the OLT, 2N lasers generate 2N wavelengths that are divided into upstream and downstream bandwidths, as shown in Figure 8.7. Downstream wavelengths are directly modulated and transmit the signal toward the end users. Upstream wavelengths are transmitted unmodulated up to the end user. Every ONU is equipped with an identical, colorless Fabry–Perot Laser Diode (FPLD).

When an unmodulated external optical beam is injected into the cavity of an FPLD, it will experience the internal gain from the semiconductor material. This process results in a reflected and amplified signal that contains the data modulation imparted by the pump current. Because the injected signal can be much larger than the internally generated spontaneous noise, the majority of the optical output power will be at the wavelength of the injected signal. The normal multiwavelength spectrum of the FPLD is transformed into a quasi single-wavelength spectrum similar to that of a DFB laser. This narrowband output signal can then be efficiently transmitted through a WDM

Next-Generation Wired Access

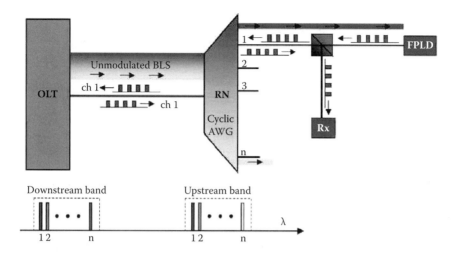

FIGURE 8.7 Block scheme of a WDM-PON based on unidirectional transmission and injection-locked FPLD. Base Lambda Signals (BLS) indicates the unmodulated wavelength comb produced at the OLT to inject the ONU FPLD (the figure is extracted from a white paper at the Novera site at http://www.noveraoptics.com).

communication channel. The external injected wavelength is called the "locking" or "seeding." Because the upstream signal is amplified by the FPLD gain, it is possible to design the system so that it is not disturbed by small reflections of the seed arriving in the downstream direction. Last, but not least, a single mux/demux has to be used in this system, processing the signal in the two directions. This is possible with the so-called cyclical Array Wave Guides (AWGs) [15] that are devised to contemporary multiplex one bandwidth in one direction and demultiplex the other in the other direction.

In this way, a colorless FPLD is forced to emit a colored signal by the incoming wavelength containing the wanted modulation and constituting the upstream signal coming from the ONU. It should be noted that, also in this WDM-PON architecture, every used wavelength is emitted by a colored source. However, all the sources are collected in the OLT, and the ONU is colorless.

8.2.2 GPON Architecture

The GPON standard, as like all other standards for TDM-PONs, is based on unidirectional transmission, conveying downstream and upstream signals on different wavelengths on the same fiber. In this case, the effect of reflection is limited by the great wavelength separation between the two signals. In addition to the GPON version with two wavelengths, there is a GPON standard conveying two downstream wavelengths. One of them (1490 nm) is used to convey the GPON downstream signal, and the second wavelength (1550 nm) is used for the distribution of analog CATV. This solution has been adopted in North America, for example, by Verizon. As a matter of fact, where CATV is practically the TV standard, this solution allows customers to maintain their home TV set. The frequency plan of the GPON standard is shown in Figure 8.8.

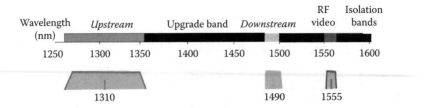

FIGURE 8.8 GPON standard wavelength allocation.

From a link budget point of view, the GPON standard defines different link budget classes with parameters that are reported in Table 8.1. The most recent documents regarding FSAN launched a discussion on the introduction of a new GPON class (called C+) with a link budget higher than 30 dB, but no decision has yet been made.

Splitter loss depends mainly on the number of output ports on the splitter and adds about the same loss whether traveling in the downstream or upstream direction. Each splitter configuration is assigned a particular maximum split ratio loss, including connectors, defined by the ITU G.671 standard and Telcordia GR-1209.

When using a 16 or 32 splitting ratio, the standard completely defines all the needed splitter characteristics. Because this is not true for 1×64 splitters, network designers must use a single 1×2 splitter interfacing two 1×32 splitters to make up the 1×64 configuration.

Although this is allowable with today's packaging, using Class B optics only leaves 5.35 dB of propagation loss. Therefore, even with the best fiber access infrastructure, where the spectral attenuation can be assumed to be 0.31 dB/km, only a 17.25 km PON network is achievable without including any of the connectors within the local exchange.

8.2.3 WDM-PON versus GPON Performance Comparison

At this point it is interesting to compare WDM-PON and GPON. In this section, we will limit our analysis to an architectural and performance comparison, delaying an economic comparison until the end of the chapter, when the impact of different technology factors is clearer. However, even without an explicit calculation, the main point driving the cost structure of WDM-PON and GPON should be clear: The GPON shares a single optical interface among all users, whereas the WDM-PON needs a different laser for each wavelength. The magnitude of the impact of this architecture difference on the cost depends on the source technology.

In order to correctly carry out the comparison, we have to concentrate our attention on a single WDM-PON architecture; here we will consider the simple architecture of Figure 8.4, where colored ONUs and unidirectional transmission are used.

8.2.3.1 Capacity

The capacity per user of a WDM-PON is easily evaluated: A single wavelength is dedicated to each end user. In general, a Gigabit Ethernet (GbE) signal is transmitted on each wavelength, assigning a capacity of 1.25 Gbit/s to each end user. It is worth noting that the WDM-PON has no particular advantage if part of the signal is constituted by pure broadcast (e.g., conventional IP-TV): The broadcast signal has to be replicated by the OLT on every wavelength and independently sent to each user.

The evaluation of the GPON capacity per user is not so simple, in that it depends critically on the bundle of services provided to the users. As a first step, let us assume that all the users have the same service profile composed by a 500 Mbit/s broadcast signal and a set of unicast services. Due to the intrinsic broadcast capability of the GPON fiber infrastructure, the broadcast signal can be delivered in a dedicated part of the downstream TDM frame that will be received by all end users. The remaining part of the downstream frame (e.g., 2 Gbit/s in our example) has to be divided among the users. Assuming a splitting ratio of 32 a bare physical bandwidth of 62.5 Mbit/s is assigned to each user. However, in order to evaluate the effective bandwidth per user, we have to take into account three elements:

- Statistical multiplexing at the end user switch.
- Statistical multiplexing at the OLT switch.
- Dynamic management of the bandwidth among different users performed by the OLT. This is a mechanism that allows the OLT to change assignment of the time slots of the TDM frame among different users on the ground of the instantaneous effective bandwidth usage [16].

It is difficult to precisely evaluate the impact of all these factors, but a rough evaluation can be carried out using classical packet traffic theory. In the case of a realistic bundle of services statistical multiplexing at the OLT allows a gain factor on the order of four and dynamic management of the bandwidth a gain factor on the order of two. The effective downstream bandwidth per end user can be obtained by multiplying the bandwidth guaranteed on the physical carrier by all the gain factors. The results in our example are on the order of 500 Mbit/s. It could be surprising that such a large bandwidth results from only 2.5 Gbit/s physical downstream signal, but we have to remember that the activity of the main data source is very low. For example, when we surf the Internet, we download Web pages and then we read them. Even if the download is quite heavy, we take a lot of time reading the page and during that time no signal is transmitted on the line. The result of these calculations could change if the downstream signal were mainly constituted by unicast video streaming (e.g., video on demand), but this does not seem to be a realistic situation for the time being.

8.2.3.2 Optical Link Budget

The transmission scheme of WDM-PON is quite simple: Attenuation is given by the loss of the mux/demux and by fiber propagation (taking into account connectors, patch panels, and other signal-losing elements that can be present in the access infrastructure).

Focalizing on CWDM PON, standard CWDM optics can assure a transmitted power of 0 dBm, and the receiver sensitivity depends on the used detector. Using a PIN, the sensitivity at 1.25 Gbit/s (assuming that a GbE is transmitted) can be about −18 dBm. This number increases to about −28 dBm using an APD.

Assuming use of 16 wavelengths, the worst channel experiences an attenuation of about 0.9 dB/km. Adding other contributions and a system margin, we can assume a loss of 1.3 dB/km. Inserting these numbers in a very simple link budget evaluation we obtain 10 km using a PIN at the receiver and about 22 km with an APD.

In the case of GPON, the link budgets are reported in Table 8.1 for the different GPON classes, and they are of the same order of magnitude of the budgets evaluated in the case of the WDM-PON. However, it is very interesting to investigate the physical effects thwarting GPON transmission, because the situation is much more complex than in the WDM-PON case.

In the case of A and B classes, the GPON span is essentially limited by the power budget. In optical access networks, reducing the ONU cost is a key issue. For this reason, in A and B class, GPON FPLDs are generally used at the ONU, whereas a PIN photodiode is present at the OLT. FPLDs are multimode lasers, so that mode partition noise due to the mode statistical fluctuation of the optical source arises [17]. This effect, if not suitably reduced by a careful design of the lasers, can greatly limit the GPON span. For these reasons, the key characteristics of FPLD to be used in GPON are specified in the standard to be able to match the required performances.

In the case of B+ and C classes, the situation is different. As a matter of fact, modal dispersion arises in the downstream link [18], constituting a first span limitation. There are two possible ways to handle this effect. The most diffused solution is to substitute FPLD with DFB lasers, accepting a cost increase. Specific DFB lasers are designed for use in GPON, with design requirements that are relaxed with respect to similar devices used in DWDM Long Haul (LH) systems in order to reduce the cost. As a matter of fact, perfect single-mode operation is not strictly needed to eliminate modal dispersion. As an alternative to DFBs, the adoption of electrical dispersion compensators (EDC) in the OLT has been proposed and demonstrated [19].

It is interesting to observe that standard forward error correction is practically ineffective in correcting the effects of modal dispersion. As a matter of fact, when the fluctuation in the modal structure of the source causes errors, they are concentrated in long bursts. It is possible to design a code conceived to correct long bursts of errors [20], but these codes are generally not used in optical communications.

As far as the downstream is concerned, in order to achieve the required power budget with the standardized classes of emitted power, the ONU detector has to be an APD. This also impacts on the cost of B+ and C GPON classes.

Another important phenomenon appears when the analog video overlay is present. In this case, due to the relevant power of this new channel, Brillouin effect arises [21] so that a careful design of the optics is needed to reach the specified performances.

8.2.3.3 Security and Unbundling

WDM-PON assigns a dedicated wavelength to each user. This is not exactly the same as the dedicated physical carrier that is assigned to the user in the copper network, but the situation is quite similar. Due to the presence of mux/demux in the branching point, every ONU receives only its own signal, so that a user cannot gain access to signals directed to other users. Moreover, a malfunction of a single ONU cannot influence the signals of other users, so a good degree of security is guaranteed.

As far as unbundling is concerned, different wavelengths could be assigned to different carriers if a suitable mechanism is implemented in the OLT. This is not exactly physical separation, but only the physical layer is common among different carriers and data security is guaranteed by the wavelength separation.

In the case of GPON, each user receives the signal directed to all users so that it is possible to gain access to the signal directed to another user simply working on the ONU. Moreover, if the ranging protocol of a single ONU does not work, the wrongly synchronized signal interferes with the signals from other ONUs in the branching point, damaging other users.

As far as unbundling is concerned, it is practically impossible using GPON, unless virtual unbundling is considered. In this case an amount of bandwidth in the shared downstream channel is assigned to a competing carrier under a defined service level agreement (SLA).

8.2.3.4 Flexibility

One of the main targets of next-generation access is to consolidate on the same access network all users' signals, from both fixed and mobile terminals. Even if a migration of mobile base stations (BS) to Ethernet backhauling seems to be very near, in today's situation, the majority of BSs have a specific mobile backhauling protocol [22]. Moreover, even when new BSs will use Ethernet, legacy equipment also has to be integrated into the new network. It also must be taken into account that many big business customers in Europe and North America use ATM data networks, whose backhauling also must be integrated into the access network.

If WDM-PON is deployed, the optical transmission infrastructure is completely protocol transparent, depending only on users' needs. If GPON is considered, this is not completely true, because all the incoming signals have to be consolidated on an Ethernet stream. This can be done by mapping the different protocols on GbE, but this adds more complexity to the network signal processing.

8.2.3.5 Fiber Utilization

Because bidirectional transmission is used in the GPON case whereas in our example of WDM-PON unidirectional transmission is adopted, the fiber infrastructure is clearly better exploited by the GPON. As shown later, unidirectional transmission can be used in WDM-PON, but it has a cost. As a matter of fact, in order to achieve a sufficient branching ratio, DWDM is needed, for example, 32 channels with a channel spacing of 100 GHz [23]. A possible design can individuate two different bandwidths to be used upstream and downstream, separated by a gap of about 800 GHz to prevent destructive interference from reflections. In this way, a branching ratio of 16 can be achieved. However, 100 GHz channel spacing requires cooled DFB lasers to be used in both the ONU and the OLT. This fact, besides the greater cost of the mux/demux, clearly influences the cost of the system. To cope with this problem, the use of a WDM comb derived from the filtering of a single broadband noise source [14] has been proposed, but it is not clear yet if a real cost advantage is achieved.

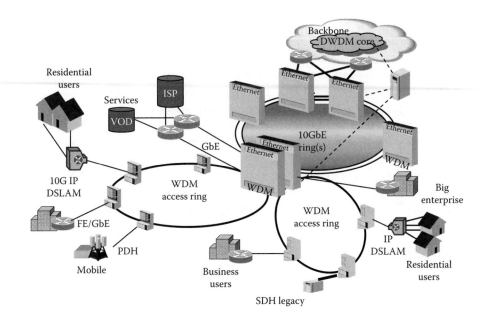

FIGURE 8.9 Three layers metro and access network.

8.2.4 Access Network Delayering

In dealing with next-generation access architectures, it is also necessary to analyze briefly a new network architecture that has been proposed by a few carriers (mainly British Telecom and Telecom Italia) to drastically reduce the operation cost of the peripheral area of the network. The architecture of metro and access network adopted by almost all European and North American carriers is shown in Figure 8.9.

Three network layers can be individuated:

- Metro core network.
- Metro access network.
- Access network.

The metro core network connects the LH nodes with the access area. Generally this network layer is constituted by one DWDM ring (seldom two or three interconnected rings as in the Verizon case) with an overall capacity generally exceeding 640 Gbit/sec and a length of between 150 and 500 km.

In European networks metro core rings are migrating from SDH (the European Standard for synchronous TDM transmission) [24] to carrier-class Ethernet and packet switching is already operational in almost all major cities. In North America SONET [25] rings (that is, TDM rings based on SONET, the U.S. standard for synchronous TDM transmission) are experimenting with an evolution allowing them to map packet traffic in the TDM frame, and the transition to pure packet switching will probably be much slower.

From a traffic point of view, a strong difference exists between Europe and North America, because in Europe incumbent carriers always operated both long-distance

Next-Generation Wired Access

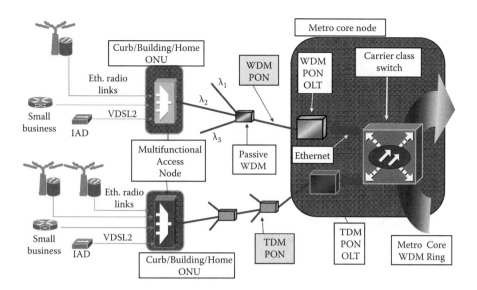

FIGURE 8.10 Access segment after the delayering.

and metro networks, whereas in North America, until recently, the two network segments were operated by different carriers. As a consequence, in Europe the metro core network of incumbent carriers performs pure transport, delegating all IP and higher level functionalities to the LH nodes. Thus, the metro core traffic is almost purely directed from peripheral nodes to the ring hub. In North America, where regional Bell operating companies (RBOCs) managed only the local network, the high-level functionalities are more distributed and a greater contribution of mesh traffic exists.

The nodes of the metro core rings are the hub for metro access rings, whose task is to groom the signal coming from the access and transport it up to metro core nodes. DWDM, CWDM, and TDM rings are diffused in the metro access area depending on carrier strategy and traffic needs. The local exchanges are located in the peripheral nodes of the metro access rings and they are the root of the access structure.

Starting with the observation that fiber-based access allows much longer spans to be realized (up to about 20 km in passive structures), a delayering of the network can be conceived with the target of eliminating local exchanges and the metro access ring and to place the roots of the access trees in the metro core peripheral networks. The resulting network architecture is depicted in Figure 8.10.

The elimination of a great number of local exchanges (for example, all the nodes located in cities) surely implies a large reduction of operational costs ahead of a relevant capex expenditure need for the network restructuring. In the current situation there is not a strong consensus among carriers on the effectiveness of the overall business plan of this operation, but in the hypothesis of wide penetration of the new video services, with a consequent increase in carriers' revenues, this could be a very appealing solution for the introduction of next-generation access.

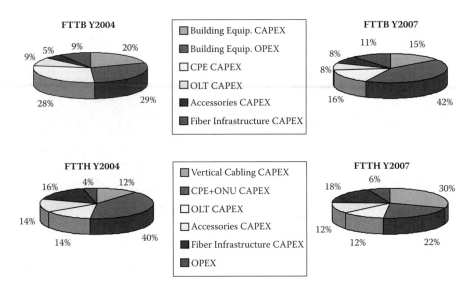

FIGURE 8.11 Cost structure of FTTB and FTTH installation. The evolution from year 2004 to year 2007 is evidenced and the typical environment of a medium European city is considered.

8.3 KEY TECHNOLOGIES FOR NEXT-GENERATION ACCESS

The problem that optical and electronics technology has to solve in the case of next-generation access is to allow Gigabit-class equipment to be realized and deployed in huge volumes at a cost comparable with today's Megabit-class equipment (e.g., xDSL DSLAMs). The first step to individuate the key technology challenges is to assess the cost structure of the new access network.

Figure 8.11 shows the cost structure of an FTTH and an FTTB network assuming deployment in 2004 and in 2007 [26, 27]. In both cases a GPON installation is assumed, exploiting B class GPON with a 32 splitting ratio. However, the results would not be so different if WDM-PON were considered, but for the greater impact of the OLT equipment that is almost comparable with the ONU.

To evaluate the infrastructure costs, a medium-sized European city is assumed as reference and the opex cost is evaluated over a period of 10 years. As assumed in this entire chapter, civil works are not included and the fibers are to be deployed in already existing tubes.

From Figure 8.11 it is clear that the greater part of the capex cost is due to the optical network termination: the building equipment in the FTTB case and the ONU in the FTTH case. On the other hand, the impact of the VDSL Customer Termination (DSLCPE) cost in the FTTB case has been quite reduced from 2004 to 2007. This last figure is the effect of a great cost reduction in xDSL chips over the last few years.

The cost of a WDM-PON ONU is almost completely determined by the optical interface, whereas in the case of the GPON ONU, two elements determine the cost of the equipment: the optical interface and the GPON chip, including the GPON physical frame processor, the ranging protocol processor, and the Ethernet switch, all in the same chip. From this brief discussion, it is evident that the optical interface technology is a key to the deployment of next-generation access networks, and the remainder of this section is devoted to these components.

8.3.1 GPON Optical Interfaces

GPON optical interfaces take the name of diplexers or triplexers when analog TV overlay is used.

A triplexer has one or more WDM filters dividing the three GPON wavelengths (1310, 1490, and 1550 nm) and one laser at 1310 nm, one digital receiver at 1490 nm, and one analog receiver at 1550 nm. In the case of a diplexer, a simpler filter is needed, without the 1490 nm output, and the analog TV receiver is not needed either.

It is also possible to integrate into the module a Transimpedance Amplifier (TIA) for the digital receivers to achieve the highest sensitivity. The front-end electronics for the analog receiver are usually located outside the optical assembly, due to its thermal dissipation and dimensions. The laser driver for burst-mode operation is also located on an external electronic board. The dimensions of the module can be very different relative to the adopted technology. In Figure 8.12 a triplexer block scheme is shown.

There are different approaches to realizing a GPON optical interface [28].

- Micro-optics: In this case diplexers or triplexers are made of discrete elements (TO packaged lasers and receivers) assembled together in a metallic package, coupled with a fiber with lenses and discrete thin film filters. Figure 8.13 represents the section of a triplexer realized by micro-optics techniques.
- Integrated optics: Optical elements (both active and passive) are integrated monolithically or assembled directly in die on a single substrate. Different approaches have been proposed with different balances between monolithic and hybrid integration [29–31]. Figure 8.14 represents an example of a triplexer realized by Planar Lightwave Circuit (PLC).

Several industries have pushed integrated optics as the way to drastically reduce the cost of GPON optical interfaces. This idea is based on the experience with

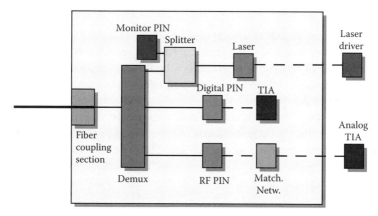

FIGURE 8.12 GPON triplexer functional scheme.

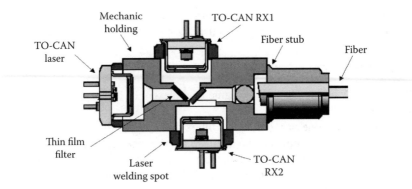

FIGURE 8.13 Cross-section of a GPON triplexer realized assembling individual miniaturized devices (micro-optics techniques).

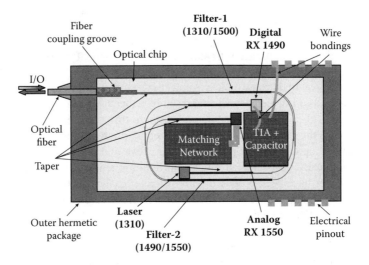

FIGURE 8.14 Top view of a GPON triplexer realized using monolithic and hybrid integration.

microelectronics, where the conjunction of high volumes and use of large wafers allowed the cost of electronic chips to be drastically reduced. However, technology evolution has shown that the parallel between optics and microelectronics cannot be perfect and commercial GPON uses almost all micro-optics components.

Beyond specific design elements, two main points differentiate optical components from electronic ones. Voltage and current sources can be integrated into electronic chips, but lasers cannot be monolithically integrated into PLC, with waveguides that are fabricated out of suitably doped glass. This forces us to adopt hybrid integration to accommodate the laser on PLC triplexers and diplexers. Moreover, even if interesting proposals exist at the research level to built photodiodes directly into a silica substrate [32], industrial processes still require hybrid integration for these elements. This means that the chip is no longer built only from a bare wafer using planar processes, but requires the integration of other chips [33–35]. This process increases the bill of material (BOM) weight on the cost of the component, whereas the BOM is inessential in the case of microelectronic chips.

Complete monolithic integration can be achieved by substituting silica-based PLC with InP-based integration. In this case, however, high-volume production (several million pieces a year) is much more difficult due to the combination of two elements. On one side, InP wafers are small (2 in. and 3 in.) and no experience exists in devising processes able to fabricate million pieces a year of complex integrated structures. On the other hand, yields of even simple InP-based components (like DFBs) are smaller with respect to traditional PLC. The combination of these elements generates a quite diffused consensus on the fact that today InP integration is not suitable for this application. Nevertheless, further evolutions in fabrication processes toward the use of larger wafers and the improvement of processes uniformity could change this situation.

Besides the difficulty in monolithic integration of active elements, even more important is the fact that whereas packaging is a trivial process for microelectronics, this is not true for optical elements, when fibers have to be coupled with integrated waveguides [33, 36, 37]. This provokes two consequences: The BOM increases due to the cost of the package elements and an important part of the manufacturing process (namely the fiber alignment and the final package) is not performed on the whole wafer, but it has to be realized component by component.

The effect of these two characteristics of optical component is twofold, creating a cost structure quite different from that of electronic chips. In the case of microelectronics, the cost of the final product is essentially constituted by the contribution of manufacturing equipment depreciation, with a smaller contribution from manufacturing operational expenses. In the case of integrated optical components, and more specifically optical GPON interfaces, the main cost element is the BOM, followed by the labor cost and the depreciation.

The difference between the two cost structures reflects the effect of volumes on the costs. In particular, the variation of the cost of a component is shown in Figure 8.15

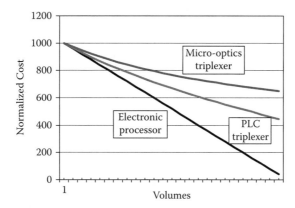

FIGURE 8.15 Cost of a triplexer versus the volumes for a microelectronics GPON processor. Both an integrated GPON triplexer (PLC technology) and a micro-optics GPON triplexer are considered.

versus the volumes for a microelectronics GPON processor, for an integrated GPON triplexer, and for a micro-optics GPON triplexer. The prices are normalized to better display the different trends. From Figure 8.15 it is evident how the cost of integrated triplexers scales with volumes a bit faster with respect to micro-optics components, but quite slower with respect to microelectronic chips, due to the effect of the importance of the BOM contribution and the presence of important manufacturing processes serially executed component by component.

8.3.2 WDM-PON Optical Interfaces

Differently from GPON, WDM-PON needs one optical interface per user, not only in the ONU but also in the OLT. For this reason, optical interface technology is a key in implementing a cost-effective WDM-PON.

A trivial way of realizing a CWDM PON is to use standard CWDM SFP transceivers. In this way, however, the cost of optical interfaces scales linearly with the number of users, being too high for a high number of users. Moreover, colored ONUs have to be used, increasing operational costs.

Both of these problems could be solved by a new class of components, called PICs [38–40]. The functional scheme of a transmitting PIC is represented in Figure 8.16; a bar integrating a certain number of lasers is coupled with a PLC mux/demux

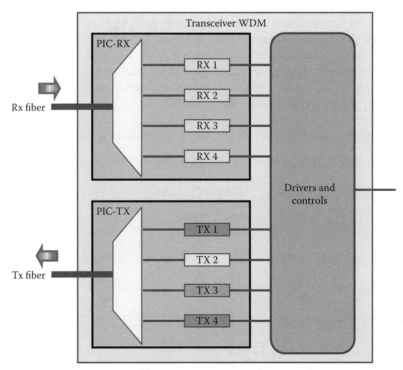

FIGURE 8.16 Functional scheme of a transmitting and a receiving PIC.

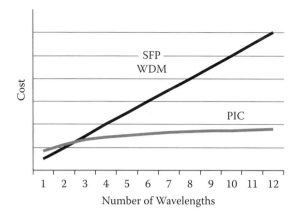

FIGURE 8.17 Comparison between the cost of a CWDM SFP and a PIC couple (TX and RX) with the same frequencies.

(generally an AWG) that launches the multiplexed channels into the output fiber. The scheme of a receiving PIC is identical, with a demux, a bar of photodiodes, and the WDM signal incoming from the input fiber. Both CWDM and DWDM PIC can be realized, and similar to CWDM SFP, it seems possible to implement CWDM PIC without thermal stabilization.

A key point of PIC manufacturing is the coupling process between the mux and the laser bar. Active coupling [33] ensures higher performances and production yield, but is much longer and seems more suitable for high-performance PIC (e.g., DWDM or 10 Gbit/s), whereas passive coupling [41] seems more suitable for low-cost CWDM PIC.

The cost behavior of a CWDM PIC is compared with the cost of a suitable number of SFP transceivers for different numbers of emitted wavelengths in Figure 8.17. To estimate the PIC cost, volumes on the order of 1 million and a process yield similar to that of an integrated triplexer are assumed. Besides reducing the cost of OLT colored sources, PIC technology can also be used to design colorless ONUs. In this case a PIC can be used like a tunable source, switching on only the lasers tuned on the desired wavelength.

Despite the evident advantages of PIC technology, volume production of these components has not yet started. For this reason all data based on the use of PICs have to be considered tentative and important changes can still happen.

8.4 A CASE STUDY

As a conclusion to this chapter, it is important to summarize all the information through an economical case study. In the case of GPON deployment, the case study is well consolidated and big changes are not expected, but in the case of WDM-PON the situation is quite different. As a matter of fact, the WDM-PON cost will critically depend on the cost of different types of PIC. Because PICs are currently under development, such a cost can be only an estimate, adding uncertainty to the economic comparison among different solutions.

In order to face this situation, we add to the comparison the sensitivity with respect to the PIC cost. In particular we analyze the case in which the PIC costs 10% more or less than our estimate. Comparing the reported data, it is possible to understand how much the PIC cost impacts on the final result.

In order to deduce data that will remain valid over time, if the prices of the involved systems change, we will always report relative prices, evaluated by dividing the considered price for a reference one. In a first approximation, prices of systems using similar technologies will change at the same pace in time, maintaining relative prices unchanged.

From an architectural point of view, we will concentrate our attention on a typical local exchange located in a city downtown and we will compare the cost per user of the next-generation access in different configurations. This allows us to appreciate both the difference among various architectures and the impact of the technology evolution.

The first step is to determine the type of FTTx architecture and the topology of the local exchange under consideration. In this case study we consider FTTB. The main parameters of FTTB in this case are as follows:

- 10,000 users per area.
- 20 users per building.
- Layer 2 switch with broadcasting and QoS in the CO.
- Maximum distance between CO and building of 15 km.
- G-PON splitters and WDM-PON mux/demux placed in outdoor enclosures.

The second step is to determine the characteristics of the traffic bundle. The customer bandwidth requirements are important, both in terms of peak and average bandwidth, but also in terms of packet statistics. The customers of a real network subscribe to a set of different traffic bundles. In this example we assume that each customer requires the same traffic bundle, with the following characteristics:

- Coexistence of video broadcast and video on demand in HDTV format with best effort Web browsing.
- Bandwidth profile
  - Maximum bandwidth on the physical carrier is 100 Mbit/sec.
  - Average bandwidth on the physical carrier is 20 Mbit/sec.

Three types of WDM-PONs are considered, depending on the OLT architecture. In the first case, standard CWDM SFP are used in both the ONU and the OLT. In the second case, a PIC with all 16 CWDM wavelengths is used, and standard CWDM SFP are used in the colored ONUs. In the third case, besides the use of PICs in the OLT, they are also used in the ONUs, where they work as tunable interfaces.

TABLE 8.2
Comparison between the Deployment Cost of GPON and WDM-PON in an FTTB Environment

	Equipment			Fiber	Total
	PIC estim.	Pic + 10%	Pic – 10%		
GPON	57%			43%	100%
WDM-PON w/SFP	239%			40%	279%
WDM-PON w/PIC 8×1 in the OLT	94%	98%	90%	39%	133%
WDM-PON w/PIC 16×1 in the OLT	70%	73%	67%	39%	109%
WDM-PON w/PIC 16×1 in OLT and ONU	101%	73%	67%	39%	140%

Note: All costs are normalized to the total GPON cost.

The last architectural point is to determine the capacity delivered to the user in the GPON and in the WDM-PON case. Using the results reported earlier and assuming the compression ratio of the building equipment equal to 4, the effective bandwidth per user is about 100 Mbit/s and 250 Mbit/s in the GPON and WDM-PON cases, respectively.

In this case study, costs for the carrier are considered (prices of the equipment, cables, installation, etc.). In the case of the GPON, the cost has been evaluated by averaging the figures presented by the winners in two GPON European tenders operated by incumbent carriers in the years 2006 to 2008. It is important to emphasize that these are small tenders, carried out to deploy extended field tests.

In the case of WDM-PON, cost estimation has been carried out at the development level working with equipment vendors that are either developing WDM-PON or evaluating its development. Three major equipment vendors contributed to this evaluation. The same procedure was carried out for the cost of the PIC.

The case study results are reported in Table 8.2, where the cost per user of the network is reported normalized to the GPON cost. In evaluating Table 8.2, it is important to remember that the cost of civil works is not included, and it is assumed that fibers are to be deployed in already existing empty tubes. As expected, GPON is the less expensive solution, at almost half the cost with respect to the WDM-PON implemented via standard CWDM SFP.

The PIC technology is expected to be quite effective in reducing this big gap, if development maintains its promise. In particular, in the better case, the solution using WDM-PON costs only 9% more than the solution using GPON.

To complete the discussion, an idea of what happens if other FTTx approaches are used can be derived from Figure 8.18. In this figure, the cost per user is derived for different numbers of customers per ONU and the cheaper WDM-PON configuration is considered. In particular, the FTTCab case is represented by a number of users between 200 and 500. From Figure 8.18 it is clear that the WDM-PON solution is quite less expensive in the FTTCab case due to the smaller capacity of GPON, which forces the use of multiple GPON in case of large cabs.

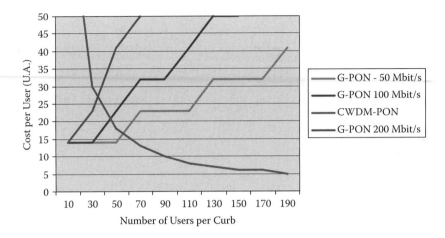

FIGURE 8.18 Cost comparison between GPON and WDM-PON for different number of users fed by the intermediate node and different effective user bandwidth. The sensitivity of the equipment cost versus the PIC cost is also reported when PICs are present.

REFERENCES

1. "IPTV systems, standards and architectures," *IEEE Communications Magazine*, Feb. 2008.
2. T. Sikora, "MPEG digital video-coding standards," *IEEE Signal Processing Magazine*, Vol. 14, pp. 82–100, Sept. 1997.
3. A. Leshem, "The capacity of next limited multichannel DSL," in *Proc. Sensor Array and Multichannel Signal Processing Workshop*, 2004, pp. 696–700.
4. S. Orlando. Next generation access network, le Architetture [in Italian]. Telecom Italia, Technical Workshop [Online]. Available: http://www.radiolabs.it/upl/File/Presentazione%20Orlando.pdf
5. S.-M. Lee, S.-G. Mun, M.-H. Kim, and C.-H. Lee, "Demonstration of a long-reach DWDM-PON for consolidation of metro and access networks," *Journal of Lightwave Technology*, Vol. 25, pp. 271–276, Jan. 2007.
6. "Annual fiber deployment," *Fiber Optics Weekly Update*, Sept. 17 2008. Available: http://findarticles.com/p/articles/mi_m0NVN/is_11_23/ai_99013219
7. H. A. Hmida, G. C. Cordner, A. Amer, and F. F. Shalan, "FTTH design and deployment guidelines for civil work, fiber distribution and numbering," OFC, March 2006.
8. A. Badoz, "Sustainable competition through fibre deployment," presented at Convergence Think Tank Seminar BERR–Department for Business Enterprise & Regulatory Reform, London, 2008.
9. J. Ingenhoff, "Athermal AWG devices for WDM-PON architectures," in *Proc. Lasers and Electro-Optics Society, 19th annual meeting of the IEEE*, 2006, pp. 26–27.
10. C.-H. Lee, W. V. Sorin, and B. Y. Kim, "Fiber to the home using a PON infrastructure," *Journal of Lightwave Technology*, Vol. 24, pp. 4568–4583, Dec. 2006.
11. E. Iannone. (2007, Nov.). Passive optical networks must evolve to survive," *Fiber Systems Europe* [Online]. Available: http://fibresystems.org/cws/article/magazine/32042
12. M. Suzuki, Y. Horiuchi, M. Hayashi, and T. Otani, "Optical network for high-quality broadband services," in *Proc. 9th Int. Conf. on Transparent Optical Networks*, Vol. 1, 2007, pp. 46–49.
13. ITU-T G.694.2, "Spectral grids for WDM applications: CWDM wavelength grid."

14. S.-J. Park, C.-H. Lee, K.-T. Jeong, H.-J. Park, J.-G. Ahn, and K.-H. Song, "Fiber-to-the-home services based on wavelength-division-multiplexing passive optical network," *Journal of Lightwave Technology,* Vol. 22, pp. 2582–2591, Nov. 2004.
15. H.-C. Lu and W.-S. Wang, "Cyclic arrayed waveguide grating devices with flat-top passband and uniform spectral response," *IEEE Photonics Technology Letters,* Vol. 20, pp. 3–5, Jan. 2008.
16. J. Jiang, M. R. Handley, and J. M. Senior, "Dynamic bandwidth assignment MAC protocol for differentiated services over GPON," *Electronics Letters*, Vol. 42, pp. 653–655, May 2006.
17. G. Kaiser, *FTTX Concepts and Applications.* Hoboken, NJ: Wiley Interscience, 2006.
18. F. Chang, "Understanding 1G EPON power budgets in IEEE formalism," presented at the IEEE 802.3 Plenary, San Francisco, CA, 2007.
19. H. Kim, F. Bien, J. de Ginestous, S. Chandramouli, C. Scholz, E. Gebara, and Laskar, "Electrical dispersion compensator for a Giga-bit passive optical network system with Fabry-Perot laser," in *Proc. J. Microwave Symposium,* 2007, pp. 207–210.
20. W. W. Peterson and E. J. Weldon, *Error-Correcting Codes*, 2nd ed. Cambridge, MA: MIT Press.
21. R. B. Ellis, F. Weiss, and O. M. Anton, "HFC and PON-FTTH networks using higher SBS threshold single mode optical fibre," *Electronics Letters,* Vol. 43, pp. 405–407, March 2007.
22. Recommendations TS 48.054, TS 48.004, TS 48.014 for the GSM and TS25.411 for UMTS.
23. Recommendation ITU-T G694.1.
24. ITU-T SDH Recommendations:
    ITU-T G.707: Network Node Interface for the Synchronous Digital Hierarchy (SDH)
    ITU-T G.781: Structure of Recommendations on Equipment for the Synchronous Digital Hierarchy (SDH)
    ITU-T G.782: Types and Characteristics of Synchronous Digital Hierarchy (SDH) Equipment
    ITU-T G.783: Characteristics of Synchronous Digital Hierarchy (SDH) Equipment Functional Blocks
    ITU-T G.803: Architecture of Transport Networks Based on the Synchronous Digital Hierarchy (SDH)
25. ANSI SONET Recommendations:
    ANSI T1.105: SONET—Basic Description Including Multiplex Structure, Rates and Formats
    ANSI T1.105.01: SONET—Automatic Protection Switching
    ANSI T1.105.02: SONET—Payload Mappings
    ANSI T1.105.03: SONET—Jitter at Network Interfaces
    ANSI T1.105.03a: SONET—Jitter at Network Interfaces—DS1 Supplement
    ANSI T1.105.03b: SONET—Jitter at Network Interfaces—DS3 Wander Supplement
    ANSI T1.105.04: SONET—Data Communication Channel Protocol and Architectures
    ANSI T1.105.05: SONET—Tandem Connection Maintenance
    ANSI T1.105.06: SONET—Physical Layer Specifications
    ANSI T1.105.07: SONET—Sub-STS-1 Interface Rates and Formats Specification
    ANSI T1.105.09: SONET—Network Element Timing and Synchronization
    ANSI T1.119: SONET—Operations, Administration, Maintenance, and Provisioning (OAM&P)—Communications
    ANSI T1.119.01: SONET: OAM&P Communications Protection Switching Fragment
26. C. Berthier and G. Wecker, "FTTH deployment cost, SOGETREL's experience," presented at the Digiworld Summit, Le Corum, Montpellier, France, 2006.

27. E. Iannone, "FTTH: Enabling technologies and deployments," presented at the ECOC Market Forum, Berlin, Germany, 2007.
28. W. P. Huang, X. Li, C.-Q. Xu, X. Hong, C. Xu, and W. Liang, "Optical transceivers for fiber to the premises applications: System requirements and enabling technologies," *Journal of Lightwave Technology*, Vol. 25, pp. 11–27, Jan. 2007.
29. Y.-L. Cheng, Y.-H. Lin, M.-C. Wang, F.-Y. Cheng, C.-S. Wu, and Y.-C. Yu, "Integrated a hybrid CATV/GPON transport system based on 1.31/1.49/1.55 /spl mu/m WDM transceiver module," in *Proc. Quantum Electronics and Laser Science Conf.*, Vol. 3, 2005, pp. 1678–1680.
30. L. Xu and H. K. Tsang, "Colorless WDM-PON optical network unit (ONU) based on integrated nonreciprocal optical phase modulator and optical loop mirror," *IEEE Photonics Technology Letters*, Vol. 20, pp. 863–865, May 2008.
31. A. Behfar, M. Green, A. Morrow, and C. Stagarescu, "Monolithically integrated diplexer chip for PON applications," in *Proc. Optical Fiber Communication Conf.*, Vol. 2, 2005, p. 3.
32. G. Wohl, C. Parry, E. Kasper, M. Jutzi, and M. Berroth, "SiGe pin-photodetectors integrated on silicon substrates for optical fiber links," in *Proc. IEEE International Solid-State Circuits Conf.*, Vol. 1, 2003, pp. 374–375.
33. A. R. Mickelson, N. R. Basavanhally, and Y.-C. Lee, *Optoelectronic Packaging*. Hoboken, NJ: Wiley Interscience, 1997.
34. G. E. Henein, D. J. Muehlner, J. Shmulovich, L. Gomez, M. A. Capuzzo, E. J. Laskowski, R. Yang, and J. V. Gates, "Hybrid integration for low-cost OE packaging and PLC transceiver," in *Proc. Lasers and Electro-Optics Society annual meeting*, Vol. 2, 1997, pp. 297–298.
35. S.-J. Park, K.-T. Jeong, S.-H. Park, and H.-K. Sung, "A novel method for fabrication of a PLC platform for hybrid integration of an optical module by passive alignment," *IEEE Photonics Technology Letters*, Vol. 14, pp. 486–488, 2002.
36. B. Morgan, J. McGee, and R. Ghodssi, "Automated two-axes optical fiber alignment using grayscale technology," *Journal of Microelectromechanical Systems*, Vol. 16, pp. 102–110, Feb. 2007.
37. H. S. Chuang, C. H. Chiu, M. Y. Cheng, and Y. C. Chuang, "Development of a vision-based optical fiber alignment platform based on the multirate technique," presented at IEEE International Conf. on Industrial Technology, 2008.
38. S. H. Oh, Y.-J. Park, S.-B. Kim, S. Park, H.-K. Sung, Y.-S. Baek, and K.-R. Oh, "Multiwavelength lasers for WDM-PON optical line terminal source by silica planar lightwave circuit hybrid integration," *IEEE Photonics Technology Letters*, Vol. 19, pp. 1622–1624, Oct. 2007.
39. D. J. H. Lambert, C. H. Joyner, J. Rossi, F. A. Kish, R. Nagarajan, S. Grubb, F. Van Leeuwen, M. Kato, J. L. Pleumeekers, A. Mathur, P. W. Evans, S. Murthy, S. K. Mathis, J. Baeck, M. J. Missey, A. G. Dentai, R. A. Salvatore, R. P. Schneider, M. Ziari, P. Mertz, M. Laliberte, J. S. Bostak, T. Butrie, V. G. Dominic, M. Kauffman, R. H. Miles, M. L. Mitchell, A. C. Nilsson, S. C. Pennypacker, R. Schlenker, J. Huan-Shang Tsai Webjom, M. Reffle, D. G. Mehuys, and D. F. Welch, "Large-scale photonic integrated circuits used for ultra long haul transmission," *Proc. 20th Annual Meeting of the Lasers and Electro-Optics Society*, 2007, pp. 778–77.
40. K. Masaki, et al., "InP integrated photonic circuits for digital optical networking," presented at the Joint Opto-Electronics and Communications Conf. and the Australian Conference on Optical Fibre Technology, 2008.
41. R. Hauffe, U. Siebel, K. Petermann, R. Moosburger, J.-R. Kropp, and F. Arndt, "Methods for passive fiber chip coupling of integrated optical devices," in *Proc. Electronic Components and Technology Conf.*, 2000, pp. 238–243.

9 Broadband over Power Line Communications
Home Networking, Broadband Access, and Smart Power Grids

Peter Sobotka, Robert Taylor, and Kris Iniewski

CONTENTS

9.1 Historical Perspective	196
9.2 BPL for the Developing World	198
9.3 BPL for Home Networking	200
9.4 BPL for Broadband Access	200
9.5 BPL for Wireless Backhaul	202
9.6 BPL for Smart Power Grids	202
9.7 BPL Technology	205
9.7.1 Power Grid Architecture	205
9.7.2 Overcoming Power Lines' Limitations	207
9.7.3 PHY and MAC for BPL Communications	210
9.7.4 BPL Network Devices	211
9.7.5 BPL Network Management	212
9.8 BPL Home Networking	214
9.8.1 Home Networking	214
9.8.2 Multidwelling Units	215
9.9 BPL Broadband Access	218
9.9.1 Passive Optical Network	218
9.9.2 PON-BPL Architecture	220
9.10 Smart Power Grids	225
9.10.1 The Need to Meter Power by Time of Use	227
9.10.2 Outage Management	228
9.10.3 Demand-Side Management Programs	229
9.11 BPL Networking Standards	232
9.11.1 IEEE	232
9.11.2 UPA	232
9.11.3 CEPCA	233

9.11.4 ETSI PLT 233
9.11.5 HomePlug Powerline Alliance.. 233
9.11.6 ITU G.hn.. 234
9.11.7 Research Programs 234
9.12 Conclusions.. 235
References... 236

9.1 HISTORICAL PERSPECTIVE

The ability to provide broadband Internet access over the same lines that deliver electricity is a compelling idea that opens up numerous new business opportunities for both private applications and communications industry [1–5]. The technology, called Broadband over Power Line (BPL) in North America or Power Line Communications (PLC) in Europe, is here, although many people are not yet aware of it. It is already fiercely competing with telephone companies' Digital Subscriber Line (DSL) technology and television cable Internet access in areas of preestablished connectivity, and is providing new connectivity where none currently exists. In areas with preexisting Internet connectivity, BPL is being used essentially as a "last mile" data-delivery technology in which data packets travel in a certain frequency range simultaneously with power supply currents. The method by which BPL customers receive data is similar to the technologies employed by telecommunications and cable providers, except that the last mile operates through community power lines. A basic principle behind BPL is shown in Figure 9.1. A BPL data communication signal of small amplitude but high frequency (typically in a 3–24 MHz range) is superimposed on the power signal operating at 50 to 60 Hz low frequency but high amplitude (from 220 V to 30 keV). All broadband services—video, data, and voice—can be transported through BPL circuits.

Power lines were originally devised to transmit electricity from power generators to a large number of consumers in the frequency range of 50 to 60 Hz. Data transmission was developed later. At first, data transmission over power lines was primarily done only to protect sections of the power distribution system in case of faults. In such an event, a fast exchange of information is necessary between power plants, substations, and distribution centers to minimize their detrimental effects. At the time, transmitting this information directly through the power lines was the fastest and most efficient method of relaying this information. The logic of relying on the power lines themselves was strengthened by the fact that power transmission towers are some of the most robust structures ever built. Thus, from a reliability perspective, any protection signaling scheme would be best served on such networks as well. Moreover, many remote locations were not hooked up to telephone networks. Thus, it was determined that signaling and exchanging information for power system protection purposes over existing power lines was the optimal solution.

Considering that data transmission over power lines has been around for quite some time, one might wonder why it is receiving such renewed attention recently, especially considering the data rate for protection purposes is at most a few kb/sec and is not comparable to the Mb/s data rate needed to support modern data-intensive applications. The answer is a combination of effects that took place during the last

Broadband over Power Line Communications

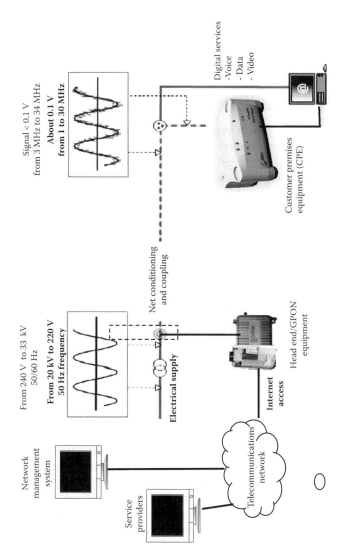

FIGURE 9.1 BPL technology concept.

rates are always significantly higher than phone line rates. Although electrification is an ongoing process worldwide, phone line expansion has stalled with the rise of cellular phone use. In other words, the cheap, easy way to reach the largest numbers in the developing world seems to be through power lines. Already, BPL research projects are being undertaken in the developing world, from Brazil, to South Africa, to Tanzania, helping establish BPL as one of the leading technologies in the spread of the Internet around the globe.

9.3 BPL FOR HOME NETWORKING

The benefits of BPL for the majority of people on Earth who do not yet have access to the Internet seem clear. But what benefits can BPL bring to those of us already well connected to Internet networks? The most immediate benefit of BPL for the average developed world consumer can be seen in its applications in networking within the home. Later in this chapter we also show that Internet access bandwidth and user costs can be significantly improved if BPL is used as an access technology.

BPL networking in the home is therefore serving two goals: providing a local home network with the advantages of the power line, and combining access and in-home network capabilities for service and system integration. There are several applications for a BPL network at home: shared Internet, printers, files, home instrumentation control devices, games, distributed video, and remote monitoring security. The key asset is, as always, "no new wires." Available products are in security, safety, and convenience service systems using BPL communications.

Home networking can be easily extended to multidwelling units (MDUs). BPL-based MDU gateways allow distribution of broadband signals over existing electrical wiring within hotels, apartment buildings, hospitals, or schools. The MDU gateway creates an Internet Protocol (IP) network throughout any type of building and enables the outside Internet signal to be connected to the electrical wiring or coaxial building wiring to provide a 200 Mbps connection for up to 96 end users (per gateway) in a building structure. The broadband building access solution is highly scalable by simply adding more MDU gateways to provide service to additional end users.

9.4 BPL FOR BROADBAND ACCESS

BPL is the use of the power line technology to provide broadband Internet access through ordinary power lines. A computer (or any other device) would need only to plug a BPL "modem" into any outlet in an equipped building to have high-speed Internet access. BPL can provide very high bandwidth access by working with other technologies based on optical fiber: passive optical networks (PON) and fiber to the node (FTTN).

The networking industry has pioneered high-speed, low-cost hybrid networks to deliver broadband connectivity and telephony and video services to residential customers by utilizing optical fiber and PON technology to deliver broadband to the node using optical fiber (FTTN) and BPL technology delivering connectivity to the home. This PON-BPL network enables the provision of triple-play services to the homes at a fraction of the cost of other technologies, including fiber to the home (FTTH) and DSL. PON-BPL networks allow carriers to effectively address new markets, countries, and cities where there is no owned infrastructure available for residential services. The use of BPL technology for the last 100 meters ensures fast execution of penetration strategy in selected areas.

The PON-BPL solution is based on a combination of two technologies. At the central office, a PON concentrator (OLT) is used to aggregate all fiber coming from the field toward the packet core. The distribution to the customer is done at three levels:

1. *Aggregation.* Fiber distribution infrastructure based on point-to-multipoint technology, PON (or point-to-point technology such as Ethernet) from the central office to selected poles at each area cell (determined by the low-voltage transformer and its serving area). A BPL gateway with optical termination capabilities (ONT) is terminating the fiber. Distances can be over 20 km.
2. *Access.* The BPL gateway is collocated with the low-voltage transformers and connected to 20 to 50 households, handling the traffic over the power line. Actual distance varies depending on the specific energy grid with typical distances of a few hundred meters.
3. *Home.* At the customer environment, the subscribers connect to the network using a home plug as described in the previous section.

Figure 9.3 shows the network serves large areas in a hierarchical manner, dividing them into small cells according to the structure of the energy grid.

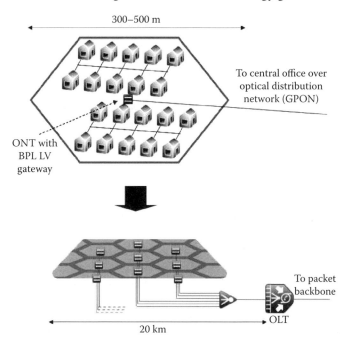

FIGURE 9.3 Data aggregation in PON-BPL networks.

By cooperating with power utilities, all right-of-way issues are addressed. While providing high-speed Internet access and digital entertainment services, the very same network can be used to create an Advanced Metering Infrastructure (AMI) for the power utility, paving the way to having a full intelligent grid, enabling the power utility with new capabilities never seen before. This technology has several benefits that include the following:

- *Cost effective:* Networks of this nature can be deployed not only quickly, but also at a fraction of the cost of FTTH.
- *Innovative:* FTTN solution for high-speed broadband services with a possible migration to FTTH in the future.
- *Simple:* Uniquely integrated BPL-PON solution that minimizes network complexity and the amount of network elements.
- *Full-service coverage:* Networks can be dimensioned for anything between basic broadband connectivity and uptake of triple-play services.
- *Fast time to revenue:* The delivery of broadband services simply requires customer premise equipment (CPE) in the home (self-install), providing carriers with fast time-to-market.

9.5 BPL FOR WIRELESS BACKHAUL

One other possible application of the BPL technology is to use it as the backhaul for wireless communications. This can be implemented by hanging Wi-Fi access points or cell-phone base stations on utility poles, thus allowing end users within a certain range to connect with equipment they already have. In the near future, BPL might also be used as a backhaul for WiMAX (802.16) networks.

Multiple Wi-Fi channels with simultaneous analog television in the 2.4 and 5.3 GHz unlicensed bands have been demonstrated operating over a single MV line. Furthermore, because it can operate anywhere in the wide spectrum, this technology can completely avoid the interference issues associated with utilizing shared spectrum while offering the greater flexibility for modulation and protocols found for any other type of microwave system.

9.6 BPL FOR SMART POWER GRIDS

As useful as BPL can be for providing and enhancing Internet access and networking for consumers, it can also provide great benefits to the power industry itself through the development of a "smart" electrical grid. The electrical grid is an interconnected system of lines and wires that links the electricity produced in power plants to consumers. Frequently referred to as a transmission and distribution (T&D) network, today's grid faces challenges in keeping pace with the modern digital economy of the information age, which requires higher load demands, uninterruptible power supplies, and similar high-quality, high-value services. Additionally, microprocessor-based technologies can alter the nature of the electrical load and result in electricity demand that is incompatible with a power system that was built to serve an "analog economy." This can lead to electric service reliability problems,

Broadband over Power Line Communications 203

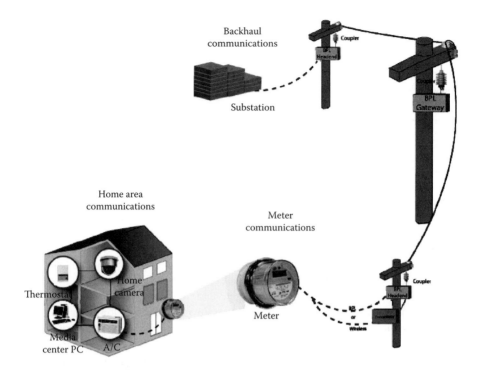

FIGURE 9.4 BPL for meter communications in smart power grids.

power quality disturbances, blackouts, and brownouts. However, rapid advances in communications and information technology now provide electric utilities with opportunities to invest in critical grid infrastructure upgrades that can serve the growing demand for high-quality, digital-grade electricity.

The "smart grid," then, is the application of communications and information technology (IT) to the electric power transmission and distribution networks themselves. A smart grid is more than smart metering, which entails replacing analog mechanical meters with digital meters. Smart meters are transformed when connected to a real-time, broadband, and Internet-enabled smart grid that extends from the power plants to each electrical outlet ("smart socket") or device attached to the grid. An example of a smart meter enabled by the BPL is shown in Figure 9.4.

The major driving forces to alter the current power grid can be divided into four general categories:

- It increases the reliability, efficiency, and safety of the power grid. It prevents outages, while lowering electricity bills and, consequently, CO_2 emissions.
- It enables the decentralization of power generation. Homes can be both energy clients and suppliers through providing consumers with interactive tools to manage their energy usage.
- It allows for the flexibility to power consumption at the client's side, and allows for supplier selection. It enables intentioned distribution between traditional and developing energy sources (solar, wind, biomass, etc.).

- It has the potential to create more new "clean" energy jobs related to renewable materials, plug-in electric vehicles, and other "green" technologies.

As global concerns about the price, availability, reliability, and environmental consequences of energy use increase, the electric power industry is undergoing a major transition. The existing transmission and distribution systems around the globe are decades old and include minimal monitoring and control over the grid. To address this aging infrastructure and create an electrical infrastructure that meets the growing and changing needs of consumers, utilities, and other major stakeholders, including regulators, governments, and customers, engineers are starting to explore the smart grid.

Central to the smart grid is the concept of BPL. This provides a stable communications platform on which to base the smart grid as it grants real-time access for utility applications. These applications range from usage information, proactive outage notification, and meter communications to load management and other distribution automation capabilities. BPL turns the existing Medium- and Low-Voltage (MVLV) infrastructure into a secure, high-speed data network capable of delivering any type of utility, or even consumer broadband data, to and from anywhere on the grid.

Fundamentally, the smart grid reduces operating expenses by lowering or deferring capital investment in the generation, transmission, and distribution of assets by smoothing out power consumption and reducing the accelerated wear on equipment used at or beyond capacity or at high temperatures. In addition, the smart grid detects emerging problems on the system before they affect service or result in costly systemwide failures. It automates healing of the grid by incorporating extensive measurements, rapid communications, centralized advanced diagnostics, and feedback controls that quickly return the system to a stable state after interruptions or disturbances. It also reduces outage durations by specifically identifying problems, which allows crews to be dispatched to the exact location with the exact equipment needed for repairs, and reduces peak demand by measuring, controlling, and shifting the consumption of power to lower peak usage and provide better, quantifiable information to customers. Finally, it protects the grid by incorporating protective systems to secure it against threats, lowers overall Operations and Maintenance (O&M) costs through more efficient work management, and responds to systemwide inputs and therefore has much more information about broader system problems.

By utilizing BPL as the communications backbone of the smart grid program, a utility does the following:

- Ensures future upgradability due to the open standards approach. Following this approach, any IP-enabled device can be integrated into the network, thereby supporting future requirements and new smart grid applications.
- Ensures network security and reliability by utilizing its own lines or assets and not being prone to unlicensed, and potentially tampered with, radio frequency (RF) bands.
- Ensures servicewide scalability because BPL is proven in dense urban areas in comparison to alternative technologies that might or might not function in urban downtown cores.

- Enables information provision on the utility lines because the utility's own lines are being used as a communications medium, providing valuable information such as outage information and noise profiles on the line.
- Reduces ownership costs by providing a single network that supports two-way communication across all smart grid applications and elements.

9.7 BPL TECHNOLOGY

BPL is a technology relying on sending and receiving RF signals over power lines to provide access to the Internet. Electrical power is transmitted over high-voltage transmission lines, distributed over medium voltage, and used inside buildings at lower voltages. PLC can be applied at each stage.

All PLC operate by superimposing a modulated carrier signal on the wiring system. Different types of PLC use different frequency bands, depending on the signal transmission characteristics of the power wiring used. Because the power wiring system was originally intended for transmission of electrical power, the power wire circuits have only a limited ability to carry higher frequencies. This is definitely a limiting factor when considering power line wiring.

Data rates over a PLC system vary widely. Low-frequency (about 100–200 kHz) carriers impressed on high-voltage transmission lines may carry one or two analog voice circuits, or telemetry and control circuits with an equivalent data rate of a few hundred bits per second over distances of several kilometers. Higher data rates generally imply shorter ranges, and modern BPL circuits can carry 200 Mb/s over distances up to 100 meters.

Power line networking devices were developed decades ago but remained on the sidelines due to problems with quality of service (QoS), low data rates, range limitations, interoperability, and high cost. All that has changed in the last few years. New advances in Application-Specific Integrated Circuit (ASIC) design, use of sophisticated signal-processing techniques, and recent standardization efforts have allowed engineers to overcome those problems [11–19]. Modern BPL equipment can provide data throughput of up to 200 Mb/s. As shown in Figure 9.5 at the distance of 500 feet (150 m) BPL link can provide a realistic throughput anywhere from 30 to 60 Mb/s, a significantly higher number than state-of-the-art wireless or xDSL connectivity.

9.7.1 POWER GRID ARCHITECTURE

High-voltage (HV) power lines or even extremely high-voltage (EHV) lines emanate from power plants and represent a wide meshed long-distance nationwide network. The term HV typically applies to voltages over 36 kV, and voltages over 300 kV are denoted as EHV. The EHV and HV levels establish a pure transmission network, as no customer premises are directly connected to the lines. The next lower level's task is bringing electrical power into cities, towns, and villages. Here MV is used, constituting a finer meshed network in comparison to HV. MV typically covers the range from 1 kV to 36 kV. Eventually, MV is transformed down to LV with levels below 1 kV for distribution to customer premises. The LV grid represents a very fine-meshed network, precisely adapted to the density of consumer loads.

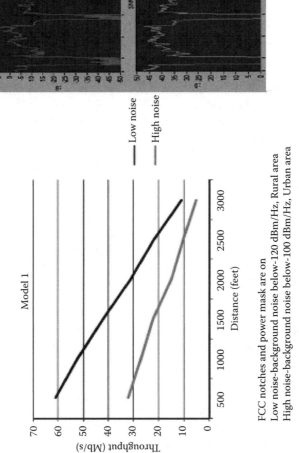

FIGURE 9.5 BPL signal characteristics.

From an MVLV transformer, power is transported over rather large service cables to the customer premises, where they end on a panel board. The latter serves as a central bus bar from which the indoor wiring to various appliances and wall sockets starts. Considering the total supply grid structure with respect to PLC, the transmission layer is of little interest today, as powerful wide-area links of optical fiber and copper-based technologies are omnipresent. Even most EHV and HV overhead lines are equipped with such fibers, spliced together with the ground wire on top of the masts. Thus, in all industrialized countries, high-speed backbone infrastructures with enormous capacities are likely available. However, the distribution of data streams to customers is still an unsolved problem, and technologies like PLC, wireless local area networks (WLANs), and digital subscriber lines (xDSL) are possible candidates to bridge the "last mile." Here the MV and especially the LV layers are interesting for PLC solutions. Due to different power distribution grid structures, it might be necessary to present an individual analysis of line characteristics with respect to PLC for selected regions.

BPL offers a unique possibility if there is no telephone infrastructure, and this feature is particularly important for less developed countries. The households being connected to the power supply grid may be provided with Internet or telephone service by means of PLC. Whereas in industrialized countries PLC is one alternative among others as a communications technology, it might be the only conceivable technology in less developed countries that can be realized with minimal investment. For historical reasons, the power supply grids of less developed countries originate from either the "European" or "U.S." power distribution system. The former European colonial rulers in many cases established their distribution systems in foreign countries. Countries extending their electrification or starting new electrification projects can choose between the U.S. and European models. Therefore, the structures of the power distribution grids also represent the fundamental structures of less developed countries. Nevertheless, it has to be taken into account that differences can result because the grid has a different degree of branching due to a different density of buildings.

9.7.2 Overcoming Power Lines' Limitations

Building power line technology is hard, as the power lines are not specifically designed for data transmission and provide a harsh environment for carrying high-speed data signals. Varying impedance, considerable noise, and high levels of frequency-dependent attenuation are the main issues. Fortunately, with progress in VLSI hardware and signal processing techniques, these issues have been solved and this technology today provides a reliable way of delivering high data throughput (up to 200 Mb/s). A frequency band of 2 to 34 MHz has been divided into several bands, as illustrated in Figure 9.6, in order to reduce noise and interference issues. Measured results using Corinex equipment indicate that effective data throughput of 40–50 Mb/s is achieved with proper noise filtering of the power lines.

For successful communication, the communication channel must be first modeled and analyzed accordingly. The channel between any two outlets in a home has the transfer function of an extremely complicated line network. Power line networks are usually made up of a variety of conductor types, joined almost at random, and

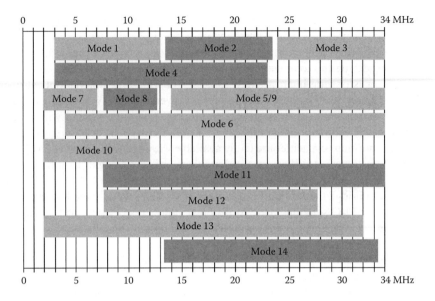

FIGURE 9.6 Frequency modes for BPL.

terminating into loads of varying impedance. Over such a transmission medium, the amplitude and phase response can vary widely with frequency. The signal might arrive at the receiver with very little loss over some frequencies, and it might be completely indistinguishable over other frequencies. Worse, the channel transfer function itself is time varying because plugging in or switching off of devices connected to the network would change the network topology. Hence, the channel can be described as random and time varying with a frequency-dependent signal-to-noise ratio (SNR) over the transmission bandwidth.

The location of the transmitter or the receiver (in this case the power outlet) could also have a serious effect on transmission error rates. For example, a receiver close to a noise source would have a poor SNR compared to one farther away from the noise source. The noise sources could be home devices plugged into the network.

Just like a wireless channel, signal propagation does not take place only between the transmitter and the receiver along a line-of-sight path. As a result, additional echoes must be considered. This echoing occurs because along a number of propagation paths, reflection of the signal often occurs due to the various impedance mismatches in the electric network. Each multipath would have a certain weight factor attributed to it to account for the reflection and transmission losses. All reflection and transmission parameters in a power line channel can be assumed to be less than one. The number of dominant multipaths to be considered is often not more than five or six because additional multipaths are usually too weak to be of any significance. This property results from the fact that the more transitions and reflections that occur along a path, the smaller its weighting factor would be. In addition, it has been observed from channel measurements that at higher frequencies the channel attenuation increases.

Noise in power lines is a significant problem for data transmission. Frequently, noise does not have properties similar to the easily analyzed white Gaussian noise of the receiver. Typical sources of noise are motors, fluorescent and halogen lamps, switching power supplies, and dimmer switches. Apart from these devices, ingress sources such as amateur radio transmission can render certain frequencies unfit for communication. The noise in power lines can be impulsive and frequency selective in nature. Due to high attenuation over the power lines, the noise is also location dependent. Recent studies have indicated that the noise in BPL systems can be typified into four categories:

1. *Color noise:* This type of noise has spectral density that decreases with increasing frequency. It is considered to be the sum of all low-power noise sources and can be time varying.
2. *Narrowband background noise:* This noise is mainly due to amplitude-modulated sinusoidal signals. This kind of interference is from broadcast stations in the medium- and short-wave bands, and is time dependent.
3. *Impulse noise that is synchronous with the generator's actual supply frequency:* This type of impulse noise repeats at multiples of the supply frequency of 60/50Hz. It has a short duration of about a few microseconds and a power spectral density that decreases with increasing frequency. The noise is caused from power supplies operating synchronously to the main's frequency.
4. *Impulse noise asynchronous with the main's frequency:* This is the most detrimental type of noise for data transmission. Its duration varies from a few microseconds to milliseconds and has a random arrival time. The noise spectral density of such impulse noise may be as much as 50 dB above the background noise spectrum. Hence, it is capable of wiping out blocks of data symbols during high data transmission at certain frequencies. It is caused by switching transients in the system network.

One of the key issues in data transmission through power lines is coupling through MV transformers. Both inductive and capacitive coupling can be used. Inductive coupling can be accomplished using magnetic materials like ferrite. It should be used at places with low impedance to provide maximum current. For high current ratings (over 100 A) at transformer substations, a capacitive coupling is the preferred technique. Space is typically limited and several feeders have to be served and the capacitive coupling also for signal coupling with the modem power supply.

For transformer substations with a small and compact distribution panel, a single-point coupling to the bus bar is an adequate solution. For older types of transformer substations with large distribution panels and many feeders, it is advantageous to apply a multiple-feeder coupling technique using a low loss power splitter/combiner. To serve a single-family home, a joint capacitive coupling is the preferred solution because of its simplicity. In MDUs, multipoint inductive coupling might be better.

Like every new data transmission technology, there have been concerns with interference with existing services. Many reports recommend several methods of interference prevention or elimination. Let us examine these briefly. The simplest technique to avoid interference would be to increase amplitude of the transmitted signal. However, this realistically is not an option, as increasing the amplitude increases noise levels for some other non-BPL services. In fact, BPL signal levels are far lower than the voltage noise already emanating from many power lines. Therefore, to minimize the emissions, BPL is frequently using wideband orthogonal frequency division multiplexing (OFDM) signaling to spread the noise energy across a wide band. At any given frequency, BPL is emitting a minimal amount of energy because OFDM divides any given signal into subcarriers, which are separated at the receiver without interference. OFDM technology enables operators to comply with the requirement that frequencies liable to interfere with existing customers be eliminated (notched out). The use of OFDM in BPL networks is not surprising, as this powerful technology has proven to be very useful against interference and impulse noise. Due to that advantage, OFDM been successfully applied in wireless and DSL communications.

9.7.3 PHY and MAC for BPL Communications

The communication in power lines can be divided into two main layers: the Physical layer (PHY) and the Media Access Control (MAC) layer. The Physical layer defines the modulation techniques to transmit data over the power lines and the MAC protocol specifies the resource-sharing strategy: the access of multiple users to the network transmission capacity based on a fixed resource-sharing protocol. Communicating at the PLC Physical layer demands robust modulation techniques like Frequency Shift Keying (FSK), Code Division Multiple Access (CDMA), and OFDM. For low-cost, low-data-rate applications, such as power line protection and telemetering, FSK is seen as a good solution. For data rates up to 1Mb/s, the CDMA technique could provide an effective solution. However, for high data applications beyond that, OFDM is the technology of choice for BPL. For MAC, there are generally two categories of access schemes:

- *Fixed access.* It assigns each user a predetermined or fixed channel capacity irrespective of whether the user needs to transmit data at that time. Such schemes are not suitable for traffic in bursts, such as data transmission that is provided by PLC.
- *Dynamic access.* These protocols can be classified into two separate categories: contention-based protocols where collisions occur, and arbitration protocols that are collision free. Contention protocols might not be able to guarantee a QoS, especially for time-critical applications, as collisions might occur and data might have to be retransmitted.

Transmissions using fixed access schemes assign to each user a predetermined or fixed channel capacity irrespective of whether the user needs to transmit data at that time. Such schemes are not suitable for sporadically heavy traffic such as the data transmission that is provided by PLC. Hence, dynamic access is provided for PLC.

Arbitration-based protocols are more capable of guaranteeing a certain QoS. However, contention-based protocols might actually provide higher data rates in applications that do not have stringent QoS requirements, like some data applications. This is because they require much less overhead compared to arbitration protocols (polling, reservation, token passing). Polling and Aloha are the two most studied protocols for medium access. Polling is a primary/secondary access method in which the primary station asks the secondary station if it has any data to send. Aloha is a random access protocol in which a user accesses a channel as soon as it has data to send. The transmitter waits for an acknowledgment from the receiver for a random period of time. It retransmits if it does not receive one. The main disadvantages of Aloha are the low throughput as the load increases and the lack of QoS.

Arbitration-based polling can handle heavy traffic and does provide QoS guarantees. However, polling can be inefficient under light or highly asymmetric traffic patterns or when polling lists need to be updated frequently as network terminals are added or removed. Similarly, token passing schemes are efficient under heavy symmetric loads. However, they can be expensive to implement and serious problems can arise with lost tokens on noisy unreliable channels such as power lines. Carrier Sense Multiple Access (CSMA) with overload detection has been considered for BPL. CSMA is a contention-based access method in which each station listens to the line before transmitting data. CSMA is efficient under light to medium traffic loads and for many low-duty-cycle bursty terminals. The primary advantage of CSMA is its low implementation cost because it is the dominant technique in today's wired data networks.

Collision detection (CSMA/CD) senses the channel for a collision after transmitting. When it senses a collision, it waits a random amount of time before retransmitting again. CSMA/CD used in Ethernet networks does enhance the performance of CSMA. However, on power lines the wide variation of the received signal and noise levels make collision detection difficult and unreliable. An alternative to collision detection that can be easily employed in the case of PLC is collision avoidance (CSMA/CA). As in the CSMA/CD method, each device listens to the signal level to determine when the channel is idle. Unlike CSMA/CD, it then waits for a random amount of time before trying to send a packet. Packet size is kept small due to the BPL's hostile channel characteristics. Although this means more overhead, overall data rate is improved because it means less retransmission. CSMA/CA is the chosen medium access protocol for the HomePlug standard that has been developed for in-home networking using power lines.

9.7.4 BPL Network Devices

BPL MV devices (MV access gateways) are used to inject and repeat the BPL signal onto the MV lines, via an MV coupler. The signal is repeated every 200 to 600 m, depending on noise and density. In some rural areas with low noise, the signal can propagate even farther. MV couplers take the output from BPL devices (typically coax) and capacitively couple onto overhead MV lines (or inductively couple for the case of underground cables). In the case of overhead couplers these should have ground disconnects for safety reasons. For the inductively coupled versions, the wire diameters must be known to maximize signal induction.

In smart grid networks, RF data collectors "collect" meter information from up to 1,000 electrical smart meters (with the associated RF communications module built in). In practical deployments, 1,000 m is rarely achieved, with 150 to 200 m being the typical maximum coverage in dense urban areas due to interference and obstructions.

The BPL equipment serves as a BPL head-end, either on an MV line or an LV line. It is connected to the Internet via an RJ-45 connector connected to a wide area network (WAN). The gateway injects a BPL signal onto an MV line with an MV coupler or lowers the voltage grid lines directly or with an LV coupler. The gateway operates on 110/240 V AC, 50/60 Hz and can inject the BPL signal through either its power cable or coaxial interface to a BPL coupler. MV BPL networks operate in the 2- to 34-MHz frequency range and are split into three bands for MV communications, typically referred to as Mode 1, Mode 2, and Mode 3. All three modes travel along the MV lines, alternating.

Limiting factors to be considered when deploying a BPL network are distance, noise, and attenuation. Figure 9.7 shows PHY bandwidth performance at various total line distances from the injection point to the meter. This test was done so that meaningful comparisons could be made between the performances of each frequency mode without penalty. It also provides a data set that shows the maximum capability of the technology.

The bandwidth available at the meter was not found to be a direct linear relationship of the distance from the injection point. However, the general trend was a proportional reduction in bandwidth for an increase in distance, due to simple transmission line losses. There are a number of contributing factors that lead to a variation in bandwidth results, including the following factors:

- Variations of background noise levels across the system.
- Different numbers of attenuation points, that is, the number of service wire junctions, four-way and T-junctions, and underground cable junctions.
- Variation in impedance relating to cabling types, and the number of circuits in a particular home.

Ideally, a BPL system should accomplish its communication objective while contributing minimum interference to its surroundings. Optimizing such a system would involve simultaneous consideration of many factors, including radiated interference, characteristics of background noise, how it is coupled into the BPL system, and the ability of coding and MAC strategies to deal with it. The goal of such an optimization would be to define the fundamental limits of BPL communication within two constraints: the existing radio noise background and the radiated interference limits imposed by governmental regulations. These ambitious goals have been largely solved by the BPL industry and mature products exist in a marketplace. Further research and development is currently conducted to build BPL networks that operate at 400 Mb/s and higher data rates.

9.7.5 BPL Network Management

Once physical connectivity is achieved, the next challenge for a broadband network is to allocate bandwidth per service, guarantee bandwidth for the users, and maintain the necessary QoS. At the fiber distribution part, the optical equipment, be it the PON OLT or the Ethernet switch, is capable of performing these tasks through well-known mechanisms such as IEEE 802.1p (strict priority), IP DiffServ, and state-of-the-art queuing mechanism. Virtual Local Area Network (VLAN) assignment will also guarantee proper isolation and separation. Different bandwidth packages can be assigned to different users and enforced downstream through these mechanisms.

Broadband over Power Line Communications

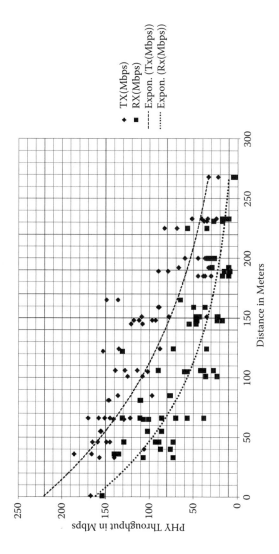

FIGURE 9.7 BPL bandwidth measurements.

In parallel, the BPL network can provide QoS features as well. The QoS features can be adjusted to guarantee the quality and compatibility of different types of traffic in parallel on the access networks. As an example, Corinex management software has been designed to ensure the quality of Voice over Internet Protocol (VoIP) flow with a minimum effect on the standard bidirectional data flow. The key parameters in the BPL equipment that is used to obtain an optimized performance of the QoS for access and in-building (one PLC hop) networks are the following:

- Minimum one-way latency step.
- Bandwidth limitation.
- CBR and VBR cells.
- Service classifier.
- Bandwidth limitation.

Each of these parameters can change the function of the network topology and the service requested by the Internet and VoIP provider. Together with the aggregation fiber network, this guarantees adequate performance of the provided broadband services, along with fairness and differentiated services.

Every active device in the BPL system has remote management capabilities based on Telnet and Simple Network Management Protocol (SNMP). The system requires Dynamic Host Configuration Protocol (DHCP) and Trivial File Transfer Protocol (TFTP) servers located at the Network Operations Center (NOC) to assign unique IP addresses and configuration files to each of the devices in the network. The acquisition of IP address and configuration file is automated within the device. Dynamic configuration of the units can be performed remotely using Telnet on a nonstandard port for security. Each head-end, repeater, CPE can be viewed through Telnet. Alternatively, these devices have SNMP built in, allowing remote monitoring of their performance. The Management Information Base (MIB) is provided with the equipment to integrate with existing Network Management System (NMS) systems such as HP OpenView and Jeizer. In addition, alarms can be configured in the equipment in fault-detection cases. Additional tools provided include SNR and Channel-Frequency-Response (CFR) monitoring of a link.

9.8 BPL HOME NETWORKING

With the explosive growth of the Web, we are increasingly moving toward a fully connected world. As broadband communication is brought to homes with accelerating speed and handheld devices get smarter, more popular, and better connected, the notion of being able to communicate with anything and anyone at anytime from anywhere is bound to become a reality. In this big picture, home networking is a natural next step in which both existing devices and future smart appliances are fully connected inside the house and accessible to the homeowners whenever needed.

9.8.1 Home Networking

Starting from the simple scenarios of sharing files, printers, and Internet connections, home networking is moving toward enabling multiplayer, multi-PC games, digital video and audio anywhere in the house, device automation, and remote diagnosis of home appliances. An example of BPL-enabled home networking architecture is shown in Figure 9.8. Different people have different ideas of what the killer applications for home networking might be. It is therefore important to provide an infrastructure for robust device connectivity to allow the construction of versatile applications on top of the infrastructure. Home networking introduces several new challenges in the area of dependability and robustness for home networks that are naturally heterogeneous and dynamic.

The BPL industry has responded to these challenges, providing robust technology that is cost effective to be deployed at every home or building. BPL technology can provide a permanent and stable video network backbone for streaming video, Internet Protocol TV (IPTV), and network gaming applications. The BPL devices distribute multimedia throughout the existing home electric network. Currently, this is one of the best technologies that provides reliable equipment for high-quality streaming video, audio, and data distribution throughout the whole home.

At present, the leading technology for home networks is wireless networking using the 802.11 standard. Power line networking, however, provides three major advantages over wireless. First, as in wireless networking, a power line network requires no new cables or special wiring. Any building that has normal electrical service is already wired for power line networking. There is no need to snake cables through walls or under floors, and no need for drilling or other structural work. Second, once the network signal is transmitted at any point into a wiring system, the network can be accessed from anywhere on the same electrical system. This means you can move from room to room or floor to floor in a building and still access the power line network from any ordinary wall socket. In fact, because electrical wires carry signals very well, a power line node actually can have greater range than a wireless access point. A power line node also avoids many of the problems of wireless signal loss caused by thick walls or other obstructions. Third, power line networking is more secure than wireless networking. Unlike wireless networking, a power line network's data signal remains confined to the physical wires. It does not broadcast into the surrounding spaces. This means that a potential electronic eavesdropper cannot drive by on the street and hack a power line network.

BPL networking at home is an application area serving two goals: providing a local home network with the advantages of the power line, and combining access and in-home network capabilities for service and system integration. The key asset of BPL technology is that it requires no new wires, but security and high bandwidth available through BPL are very valuable, too. Figure 9.9 shows a typical home connectivity setup using BPL modems. BPL Internet access can be quickly set up using commercially available equipment and existing wiring.

9.8.2 Multidwelling Units

The BPL gateway allows distribution of broadband signals over existing electrical wiring or coax cabling within hotels, apartment buildings, hospitals, schools, and other MDUs. The MDU gateway creates an IP network throughout any type of building and enables the outside Internet signal to be connected to the electrical wiring or coaxial

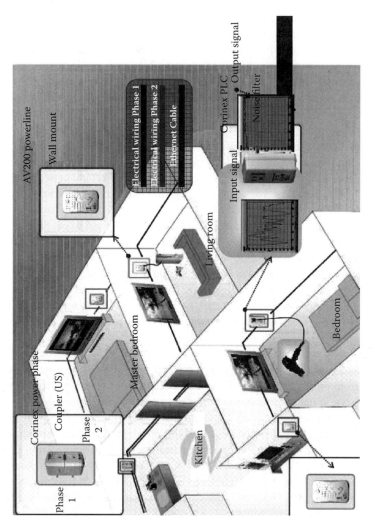

FIGURE 9.8 BPL-enabled home networking.

Broadband over Power Line Communications 217

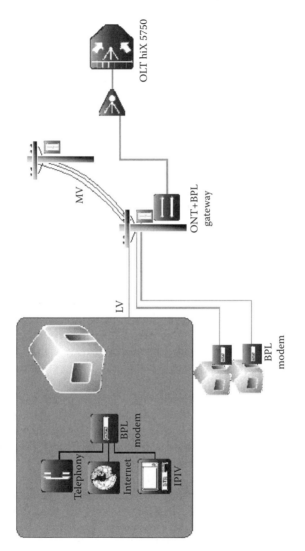

FIGURE 9.9 Home Internet access through a BPL modem.

building wiring to provide up to 200 Mb/s connection for up to 96 end users (per gateway) in a building structure. The broadband building access solution is highly scalable by simply adding more MDU gateways to provide service to additional end users. Figure 9.10 shows a networking architecture for an MDU that utilizes a BPL gateway.

In a power-line-based network, a head-end unit, preferably located close to the main electrical distribution board, controls the entire network. Access to the backhaul network (through the head-end) is provided via DSL, cable, or fiber installations. Power line modems available in all apartments that require networking will connect automatically to the head-end as soon as they are plugged into a power outlet as indicated in Figure 9.9. Each device, also known as a CPE, has an Ethernet interface for easy connectivity to any type of IP-addressable user equipment. Alternatively, a coaxial-based network is connected to the coaxial terminals within each apartment, and the MDU is connected to the main coaxial feed to the building.

9.9 BPL BROADBAND ACCESS

Throughout this book we have frequently referenced the importance of "last mile" technology. This last step in the network, where it actually enters homes and businesses, is where the majority of Internet bottlenecks occur that slow delivery rates, and is therefore the point in the network that can most benefit from an injection of high-speed communications.

9.9.1 Passive Optical Network

One of the leading technologies to employ for high-bandwidth Internet access is optical fiber, which, although expensive to install, is by far the most sophisticated technology, capable of handling the largest bandwidths. The benefits of FTTN technology are therefore not in debate from a technological point of view. What is critical, however, is exactly how the optical wiring is delivered within a given premises. Fiber can be directly connected to the home (FTTH), to the curb (FTTC), or to the building (FTTB). PON is typically the most effective way of delivering the split optical signal to individual users. PON technology utilizes optical splitters to enable a single optical fiber to serve multiple premises, with all users sharing available bandwidth.

PON technology is gaining worldwide acceptance, in particular in Asia and the third world, with massive service rollouts in the United States as well. It offers a significant cost advantage compared with FTTH. It has been demonstrated in applications of five homes per one transformer that the sustainable bandwidth of 15–27 Mb/s per home can be achieved. For the case of 100–150 homes, the data throughput drops to 1–4 Mb/s, which is still an impressive number compared to xDSL. A further advantage of PON-BPL architecture is that FTT-BPL can be upgraded to FTTH at a later time, should there be a demand for even higher bandwidth services.

A PON network requires no intermediate electronics or powering between the head-end and the subscriber's home and consequently is operationally simple. The main benefit of this type of architecture is that historically it has been significantly cheaper than other options, requiring lower power lasers and fewer fibers, reducing the capital (capex) and operational expense (opex) for service providers. The

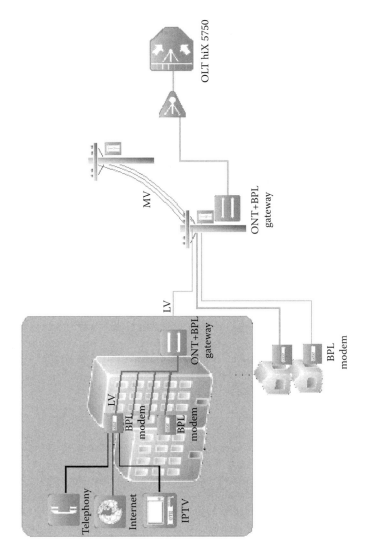

FIGURE 9.10 Multidwelling unit (MDU) Internet access using a BPL gateway.

drawbacks of a PON network are significant, though: Splitters reduce the distance possible between active electronics and make it so that more users are impacted by a single point of failure in the network. Also, as more users are added to the system, the bandwidth available for each given user decreases.

A typical PON architecture is shown in Figure 9.11. OLT represents an optical line termination unit. The access node is offering the last mile access via a PON interface. Traffic is aggregated using a passive optical splitter. Typical downstream bandwidth is 2.5 Gb/s and the upstream bandwidth can be as high as 1.25 Gb/s.

9.9.2 PON-BPL Architecture

In this section, we explore the combination of two technologies, PON and BPL, as shown in Figure 9.12. Currently, this technology has started to be deployed on all continents and shows great promise for offering universal Internet access to every home that lies in proximity to the optical fiber.

Until now, service providers have mostly been providing broadband to the home for standard Internet services requiring on the order of 1 to 4 Mbps. As higher bandwidth services, like IPTV, are being added, we are starting to see the demand for bandwidth increase as TV channels are utilizing 2–4 Mb/s each using MPEG 4 compression. The move to high-definition (HD) IPTV is on the horizon and will continue to increase the demand for bandwidth in the home even more. This means that service providers will need to launch cost-effective access initiatives to deliver higher bandwidth to the home while also requiring a strategy for distributing that bandwidth within the home environment.

Given the demand for bandwidth for the different applications to be deployed and the type of infrastructure and customers to be serviced, the question becomes this: What technology can deliver the best business case in a given area? Many access technologies can deliver broadband to the home or business using a combination optical fiber, wireless, coax, UTP5 cabling, and power line. The combination of PON and BPL technologies delivers good bandwidth in a cost-effective manner. In particular, FTTN and LV power line to the home or building are attractive.

The PON-BPL solution is based on a combination of two technologies. At the central office, a PON concentrator (OLT) is used to aggregate all fiber coming from the field toward the packet core. The distribution to the customer is done at three levels, as shown in Figure 9.13:

- *Aggregation*. Fiber distribution infrastructure based on point-to-multipoint technology, PON (or point-to-point technology such as Ethernet) from the central office to selected poles at each area cell (determined by the LV transformer and its serving area). A BPL gateway with optical termination capabilities (ONT) is terminating the fiber. Distances can be over 20 km.
- *Access*. The BPL gateway is collocated with the LV transformers and connected to 20 to 50 households, handling the traffic over the power line. Actual distance varies depending on the specific energy grid with typical distances of a few hundred meters.
- *Home*. In the customer environment, the subscribers connect to the network using a home plug as described in the previous section.

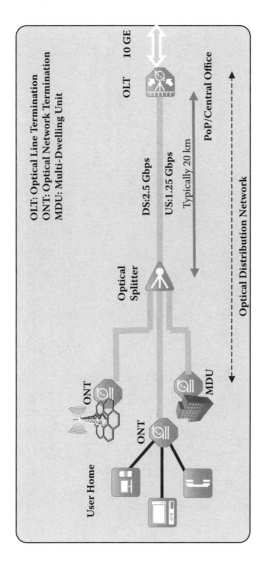

FIGURE 9.11 Passive optical networks (PON) concept.

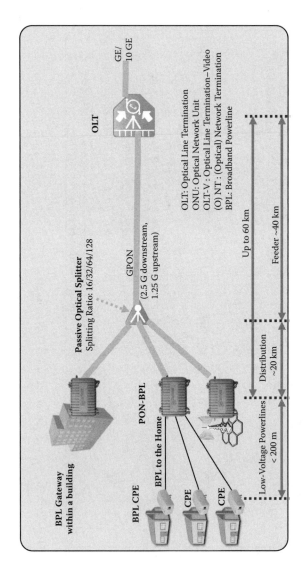

FIGURE 9.12 BPL–PON implementation.

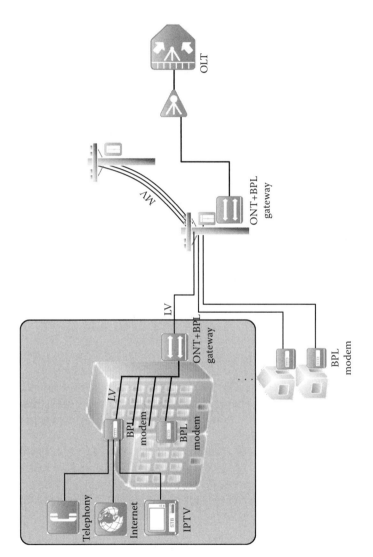

FIGURE 9.13 PON-BPL broadband access solution.

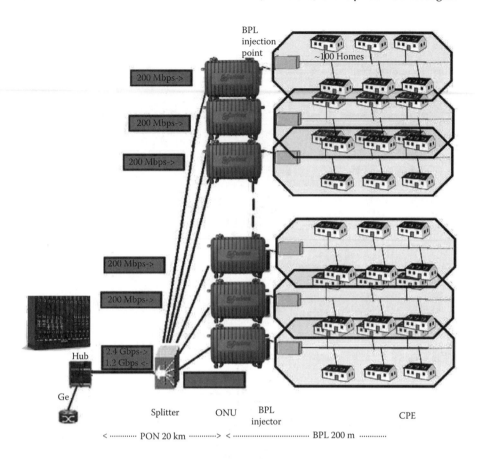

FIGURE 9.14 Example of PON-BPL implementation.

The combination of Gigabit PONs (GPONs) and BPL enables the telecommunication carriers and the power utilities to deploy an FTT-BPL (fiber to the BPL) broadband access network, using LV power lines for "last mile" access to the home, instead of leasing copper pairs.

The fiber distribution network provides downstream rates of 2.5 Gb/s, which can be shared among 16 or 32 BPL gateways (depending on the end user bandwidth requirements). BPL chipsets provide downstream rates of up to 200 Mb/s that can be shared among the end users. The combination of GPON plus BPL is a cost-effective way of meeting the bandwidth requirements for broadband access services such as Internet access, VoIP, and IPTV.

Examining the PON and BPL section a little more closely and starting from the OLT, we see we can have up to a 10 Gb/s input (Figure 9.14). The OLT has fiber strung from it out to a passive optical splitter that splits the signal up to 128 lines with aggregate bandwidth at 2.5 Gb/s downstream and 1.25 Gb/s upstream. From the split point we have fiber to the curb or pole. The ONT can be integrated into the PON-BPL gateway, which terminates the fiber and injects the signal onto the LV lines with maximum bandwidth at 200 Mbps (PHY) and practical bandwidths of up to 60 Mb/s.

Broadband over Power Line Communications 225

The maximum distance from OLT to ONT is 20 km, with the maximum distance on the LV lines running at 1 km. The main savings in this architecture versus FTTH is the reduction in the cost of laying fiber from the curb to the individual homes.

There are a few architectures on the BPL side that support different requirements. In basic broadband to customers in a highly dense MDU area, the LV lines are connected to up to 200 homes in MDUs. In this dense electrical topology, the provider is able to offer basic broadband service of up to 4 Mb/s per user, depending on customer penetration and utilization. The PON-BPL gateway in this architecture has a single 200 Mb/s silicon chip. Within the MDUs, the MDU gateway is used to deliver the broadband signal to consumers over the building's electrical or coaxial infrastructure.

In a lower density area the LV lines are connected to up to three dozen homes, which, with a penetration of 20%, translates to service to five to seven homes. In this topology, the provider is able to offer triple-play services of up to 30 Mb/s per user. Each home can use power line CPEs to receive the broadband signal and distribute throughout the home for connecting IPTV set-top boxes and computers.

In the preceding two examples, the provider is utilizing the LV power lines, so it needs to form a partnership with the electrical utility. The electrical utility would allow access to their lines in exchange for collecting meter data via the PON-BPL communications network. Carriers' broadband network and the utilities' meter reading network can coexist on the same infrastructure. VLAN technology is used to logically segment the networks, which allows the utility to manage its data while the carrier manages its network. In this scenario we can segment parts of the carrier or utility network even further to group types of customers based on any number of factors such as location or service type. The PON-BPL LV gateway looks as shown in Figure 9.15 and includes an integrated ONT (GPON card), Ethernet from ONT to BPL modules underneath, the power supply, Corning Fiber connector, and the LV line feed (which also powers the unit off 110/220-volt power on the pole).

9.10 SMART POWER GRIDS

The functioning of the smart grid depends on the transformation of the analog grid of yesterday into the digital grid of tomorrow. The smart grid must gather data collected from large numbers of intelligent sensors and processors installed on the power lines and equipment of the distribution grid. Examples include switches, circuit breakers, or transformers. The sensors will then provide information to central processing systems that both present the information to operators and use the information to send back control settings.

The foundation for this "closed loop control" is a reliable, two-way communication network, such as BPL. This allows for outage detection, automated fault isolation, and load management, which can improve the efficiency of power use and distribution. The smart grid strategy also includes the real-time metering of electric power (commercial and residential) and smart sensors on the network, such as smart thermostats, that reduce or shift use to off-peak periods to better manage a utility's fixed capacity to generate and transmit power.

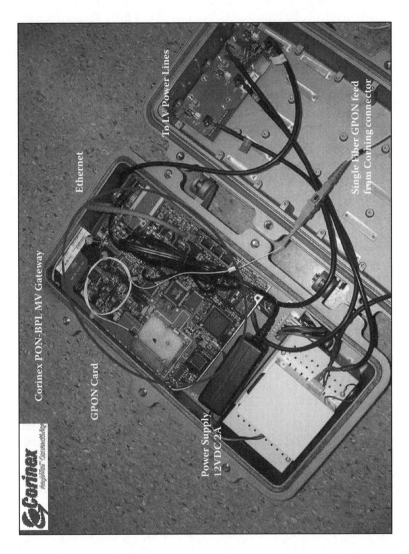

FIGURE 9.15 PON-BPL LV gateway.

A modernized smart grid becomes, in essence, a digital energy distribution system. "Wireline"-based smart grid networks (as opposed to those built on wireless technologies, which are less capable and less reliable) are unique in providing a full suite of applications and management tools, including AMI, outage management, distribution automation, demand response management, and security and surveillance.

The original power grid technology has its control systems embedded in the generating plants, transmission lines, and substations. In this structure information flows one way from the users and the loads they control back to the utilities. The utilities attempt to meet the demand and succeed or fail to varying degrees (when they are unsuccessful, brownouts, rolling blackouts, and uncontrolled blackouts can occur). The total amount of power loaded by the users can have a very wide probability distribution. This requires producers to maintain a number of spare generating plants in standby mode in case they are needed to respond to a rapid increase in consumer power usage. This one-way flow of information is expensive; the last 10% of generating capacity may be required as little as 1% of the time, and brownouts and outages can be costly to consumers.

Demand response (DR) is the next level of sophistication. This refers to mechanisms to manage the demand from customers in response to supply conditions, for example, having electricity customers reduce their consumption at critical times or in response to market prices. This can be as simple as timers to switch off electric water heaters during peak-demand periods, but such systems are unable to respond to contingencies. The full smart grid allows generators and loads to interact in real time, using modern information and communications technology. Managing demand to eliminate the peak fraction of demand eliminates the cost of generators, cuts wear, and extends the life of equipment. It also allows users to get more value from the system by putting their most important needs first.

9.10.1 THE NEED TO METER POWER BY TIME OF USE

Rising energy costs in recent years have stimulated energy marketers to move the cost of electric service from flat rates to charges based on the time of day and season of use. The change in pricing mechanisms also motivates consumers to move their electricity usage away from peak times, optimizing the utilities' infrastructure and capital demands. This change in pricing has driven the demand for energy metering to capture energy use on a real-time basis, both for billing purposes and for providing pricing information that will allow users to become active participants in the energy-savings process.

Utilities are responding by converting their mechanical meters to electronic "smart meters" that communicate energy use on a real-time basis across networks like BPL. The U.S. Energy Policy Act of 2005 and similar legislation worldwide recognized this need. They required smart meters for all federal installations and mandated that utilities and regulators produce plans for the transition.

A smart grid network will need to integrate the physical electronic meter in either a "wireless" or "wireline" fashion into the two-way communications network. Once the meter network is installed and tied into the grid system, real-time load monitoring and new aggregated power loads can be viewed and managed instantly across the grid: at the substation, feeder, neighborhood, and even household level. Tracking usage, managing peak loads, sending real-time market signals to the consumers, and verifying conservation programs are now all critical AMI-related benefits. More important, utilities need the AMI layer to drive other smart grid applications like DR. Demand management will systematically and operationally benefit every utility experiencing base load growth, and will increase generating costs.

Utility benefits from AMI programs can represent significant improvements in billing accuracy, and a notable reduction in energy theft. AMI programs also position utilities for Time of Use (TOU) and Critical Peak Pricing (CPP) programs that will drive future demand response programs being called for by regulators and consumers. AMI programs can also allow for the remote connection and disconnection of power in communities that experience high rates of residential turnover.

9.10.2 Outage Management

Outage management is really two operating functions that are created and enabled through the systemwide deployment of smart grid architectures: proactive outage detection and notification. With smart grid, almost every electric grid element becomes part of a private, secure digital communications network. As such, substation, feeders, cap banks, switches, transformers, and other elements can be viewed in their natural operating state. Baseline metrics are established and performance deterioration is identified and prioritized long before the point of failure. This can be used to determine a faulty insulator or a pending transformer failure.

A new ability to view power quality, operating temperatures, and DR program participation, as well as to aggregate loads, is available through Network Management Systems. Utilities begin proactive distribution and load management on enhanced scales and feeder efficiencies. Once the digital network and sensor points are established, immediate outage detection, or even proactive outage detection and notification, become the norm. The utility's NOC can now see both preoutage and outage alarms. Crew notification via e-mail, Short Message Service (SMS), phone, work ticket, and so on are all immediately triggered. In addition, through a unique database management process, utilities can also choose to notify affected customers around the failing network element. Utilities will even have the ability to automatically notify customers immediately after service restoration. When immediate response, proactive customer notification, and prioritized maintenance efforts are the norm, it will mark a tremendous customer service improvement, both perceived and actual.

Utility benefits around outage management can represent notable reductions in operating expense and improvements in performance. Other benefits are achieved in DR programs, once outage management is active. With all grid elements linked, monitored, and controlled, utilities have the view and processes to measure real-time power quality across each feeder and neighborhood. The controlled equipment consists of substation transformers, capacitors, breakers, reclosers, distribution transformers, and even aggregated meter loads.

Specific load control programs can be established that provide real-time capacitor bank and breaker-switch control, enhancing and balancing actual consumer demand with distribution load. A new level of power quality and load balancing combine to help utilities lower their line losses and improve their operating efficiency and costs. Separate management programs around transformer overloading; meter or retail theft; autoconnect and disconnect; and voltage, current, and phase monitoring and adjustments all combine to dramatically improve the customer experience and the utilities' operating efficiency. More specifically, follow-on programs around distributed generation, local renewable alternative energy, and even plug-in hybrid cars can all be integrated into the distribution grid and smart grid network. Command and control processes and dashboard management systems will already be in place to facilitate these new smart grid applications. These will allow the utility to deploy these new DG services using a multiyear phased gate program management approach. Utility benefits around DG include feeder line efficiency gain, dramatically reduced repair cycle times, and avoided capex for unneeded repair costs, through trouble isolation.

9.10.3 DEMAND-SIDE MANAGEMENT PROGRAMS

The demand for power at peak periods often exceeds base demand by over 50%. To handle these peaks, utilities overbuild their infrastructures and invest in expensive peak generation plants. Utilities that buy power must accurately project their demand or be faced with charges for purchased power that could be more than 30,000 times the cost of base power.

Demand-side management programs allow utilities to control peak demand by remotely trimming peak power consumption by turning up air conditioning set points, turning off noncritical appliances and equipment, and turning down lighting. Additionally, they can turn on power generation from alternative energy sources such as solar, battery backup, and so on to generate more energy during peak times without the need to purchase expensive peak power. Regulations in the United States and Europe have mandated these programs at both federal and state levels. The key to demand-side management programs is a communication backbone, such as BPL, through which power demand can be sensed quickly and commands can be sent to the remote-controlled appliances to decrease their consumption.

Utility savings can be achieved in several ways with an aggressive demand management program. Primarily the utility saves opex through the reduction of open market and wholesale peak power purchases. Secondarily, once demand management is an active program, nonconservation users usually pay a higher fee during peak times. So some margin increase for peak power occurs. Finally, the long-term strategic benefit of systemwide DR is gained through the massive capital savings tied to generation. Solid managed aggregated load reductions through the utility DR program can create a sustainable load reduction exceeding 1,000 MWs or more over a 5-year program build-out time frame. Interest and principal reductions around these generating plant offsets are substantial, and will eventually be required programs before future generation expansion gets approved in the utility's rate base.

Smart grids consist of three independent communications networks from the substation to, and inside, the home. Communications from the substation to the utility pole (or LV transformer) are called "backhaul communications," communications from the pole to the meter are called "meter communications," and communications within the home are called home area communications. Figure 9.16 shows each of the communications networks. Note that there might be overlap in some smart grid topologies.

Backhaul communications can be achieved through MV power lines, wireless (e.g., General Packet Radio Service [GPRS]), and narrow band power line. There are advantages and disadvantages to all three methods of communications, but essentially the only method that uses the utility's own infrastructure, which provides the utility with valuable data about the conditions of its own assets, supports a smart grid (as opposed to just AMI), works in dense urban areas, and has low operating costs (opex) is BPL.

Meter communications can be achieved through LV power lines, wireless (e.g., 800/900 MHz), and narrow band power line. There are advantages and disadvantages to all three methods of communications, but essentially the only method that uses the utility's own infrastructure, providing the utility with valuable data about the conditions of its own assets, delivers true real-time reads, supports the home area network component of the smart grid (as opposed to just AMI data), and supports broadband services is BPL.

As mentioned earlier, backhaul from the central data collector via MV lines (BPL) offers some unique advantages to the utility over any other type of backhaul, specifically in regards to noise profiling of the power lines, outage detection and management, and low opex costs. In Figure 9.16 the electrical meter is connected to the data collector via wireless technology to the wireless collector. The wireless collector feeds the BPL device, in this case a Corinex MV gateway (noise-resistant version or standard) via an Ethernet connection. Both the collector and MV gateway are pole mounted (or mounted inside a pad mount transformer).

The MV gateway connects to the MV line via an MV capacitive coupler that puts the signal on the MV lines. Every 200–600 m another MV gateway device is used to repeat the signal, all the way back to the MV substation where a fiber link connects back the NOC, where all of the management servers reside. This architecture allows for a broad coverage at a low cost, as not every transformer in the network needs to have a BPL device on it, perhaps one out of every three to six transformers, one at every data collector location, and then one device every few hundred meters.

In contrast to the MV backbone scenario where wireless was used for communications between the meter and the MV backbone, with this architecture BPL is used on the MV lines (as before) and on the LV lines directly to the home. In this architecture all we add is a BPL module inside any standard electrical meter, turning the meter into an IP-addressable device that can be fully managed on the network. In this architecture, we would use an MV access gateway at every MV/LV transformer to inject the BPL signal to every home. Because we are utilizing BPL to every home, the utility can potentially add broadband services to consumers (either directly or through an Internet service provider partner) and the meter can act as a signal booster (repeater) to ensure complete coverage in the home. On the LV side, we inject the BPL signal into all three phases of the LV lines to get complete coverage to all the homes. Additional bandwidth can be offered by utilizing multiple devices on the LV lines or taking fiber right to the MV/LV transformer, if desired.

Broadband over Power Line Communications 231

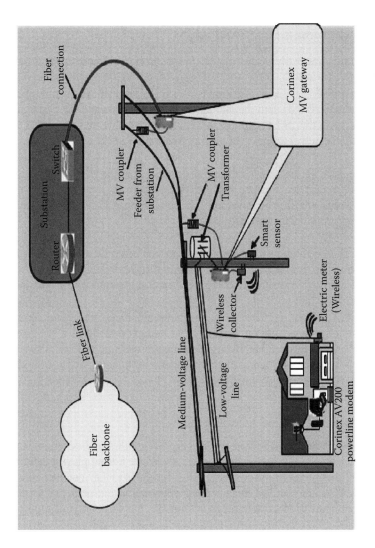

FIGURE 9.16 BPL implementation for meter communications.

When building a smart grid with BPL technology, the ability to sell off unused bandwidth or cost-share the financing of the network with a service provider is an opportunity for a utility. Utility services and carrier services can utilize the same infrastructure and can be separated via VLAN technology to ensure each organization sees only its information. The network can also be configured such that certain noncritical utility data can utilize nonpeak data traffic periods.

9.11 BPL NETWORKING STANDARDS

Several standards are evolving for power line networking. These include the standards presented by the Institute of Electronic and Electrical Engineers (IEEE), the Universal Power Line Association (UPA), the European Telecommunications Standards Institute (ETSI), and the HomePlug Powerline Alliance. It is still unclear which standard will finally be accepted as authoritative, so we explore all the various standards and research activities that might be relevant [20–27].

9.11.1 IEEE

The IEEE has been working on developing standards for PLC. The IEEE standards, still in working group form, are as follows:

- IEEE P1675—"Standard for Broadband over Power Line Hardware" is a working group working on hardware installation and safety issues.
- IEEE P1775—"Powerline Communication Equipment—Electromagnetic Compatibility (EMC) Requirements—Testing and Measurement Methods" is a working group focused on PLC equipment, electromagnetic compatibility requirements, and testing and measurement methods.
- IEEE P1901—"Draft Standard for Broadband over Power Line Networks: Medium Access Control and Physical Layer Specifications" is a draft standard for BPL networks defining MAC and physical layer specifications.

9.11.2 UPA

The UPA aligns industry leaders in the global PLC market and covers all markets and both access and in-home PLC technology to ensure a level playing field for the deployment of interoperable and coexisting PLC products to the benefit of consumers worldwide. A key aim is to promote among government and industry leaders the tremendous potential of PLC technologies to build a global communication society. The UPA aims to catalyze the growth of PLC technology by delivering UPA-certified products that comply with agreed specifications. The UPA focuses on time-to-market, guaranteeing high performance and maximizing the usage of the spectrum for both access and in-home audiovisual and data networking PLC applications to the benefit of all players in the PLC value chain. The UPA promotes products based on the UPA Digital Home Standard for home networking applications and on the OPERA standard for BPL power line access applications while contributing toward the upcoming IEEE P1901 standard.

9.11.3 CEPCA

The Consumer Electronics Power Line Communication Alliance (CEPCA) is a nonprofit corporation established to promote and continuously advance high-speed PLC technology to utilize and implement a new generation of consumer electronics products through the rapid, broad, and open industry adoption of CEPCA specifications. This goal will be furthered through the development of specifications enabling the coexistence between in-home PLC systems and between access PLC systems and in-home PLC systems. It will also be furthered by the promotion of high-speed PLC technologies to achieve worldwide adoption.

9.11.4 ETSI PLT

The ETSI produces globally applicable standards for information and communications technologies (ICT), including fixed, mobile, radio, converged, broadcast, and Internet technologies. ETSI's Power Line Telecommunications Technical Committee (TC PLT) aims to advance the necessary standards and specifications to cover the provision of voice and data services over the main power transmission and distribution network or in-building electricity wiring. The standards will be developed in sufficient detail to allow interoperability between equipment from different manufacturers and coexistence of multiple power line systems within the same environment. Harmonized standards will be developed to allow presumption of conformity with the relevant EU/EC Directives. The ETSI PLT followed a top-down approach to the standardization of power line telecommunications, concentrating initially on the higher levels of channel models and architectures.

9.11.5 HomePlug Powerline Alliance

The HomePlug Powerline Alliance is a trade group consisting of more than member companies. It was founded in March 2000 by leading technology companies to provide a forum for the creation of specifications for home power line networking products and services. The Alliance's mission is to enable and promote rapid availability, adoption, and implementation of cost-effective, interoperable, and standards-based home power line networks and products. The sponsors and members of the board of directors of the Alliance include Comcast, Earthlink, GE, Intel, Linksys, Motorola, Radio Shack, Samsung, Sharp, Sony, and Corinex, among others. The HomePlug Powerline Alliance has defined a number of standards:

- *HomePlug 1.0:* Specification for connecting devices via power lines in the home, with a theoretical speed of 14 Mbit/sec.
- *HomePlug 1.0 Turbo:* Faster, unofficial specification for connecting devices via power lines in the home, with a theoretical speed of 85 Mbit/sec.
- *HomePlug AV:* Designed for transmitting HDTV and VoIP around the home, with a theoretical PHY data rate of up to 189 Mbit/sec.
- *HomePlug Access BPL:* A working group to develop a specification for to-the-home connection.

- *HomePlug Command & Control (HPCC):* A low-speed, low-cost technology intended to complement the alliance's higher speed powerline communications technologies. The specification enables advanced, whole-house control of lighting, appliances, climate control, security, and other devices.

9.11.6 ITU G.HN

G.hn contributors are working to develop the next-generation standard for existing-wire home networking (a wired and complementary counterpart to a popular 802.11 wireless home networking standard). G.hn has targeted gigabit-per-second data rates and operation over all three types of home wires: coax, power line, and phone wires. G.hn proponents are working to make G.hn the future universal wired home networking standard worldwide, coexisting with and providing an evolution path from today's existing-wire home networking technologies including Multimedia over Coax Alliance, HomePNA 3.1 over coax and phone wires (already an ITU standard G.9954), HomePlug AV, UPA, and HD-PLC over power lines. A broad group of companies are part of the HomeGrid Forum launched in May 2008 to promote G.hn. ITU G.hn is at the time of writing close to finishing a specification for a single PHY that would support power line, coaxial, and home phone line networking. The promise of G.hn is an integrated circuit (IC) that can be used to network over any home wires. Some benefits of the final standard are expected to be lower equipment development costs and lower deployment costs for service providers.

9.11.7 Research Programs

BPL networking is the subject of numerous research programs that aim to build and standardize power line technology. The Open PLC European Research Alliance (OPERA) is a research and development project with funding from the European Commission. It aims to improve existing systems, develop PLC service, and standardize systems. Having achieved success in the first phase of its project, OPERA is now working on a second phase that will focus on PLC deployment in rural areas, MDUs (in combination with fiber-to-the-building, to offer high-end triple-play services), high-density areas (sharing a backbone connection between several MDUs to offer low-cost double-play services), and deployment to support utility-oriented services.

POWERNET was a research and development project with funding from the European Commission. It aimed to develop and validate plug-and-play Cognitive Broadband over Power Lines (CBPL) communications equipment that met regulatory requirements concerning electromagnetic radiation and can deliver high data rates while using low transmit power spectral density and working at a low SNR. The trials were conducted by deploying the new CBPL demonstration units in the field, and measurements were taken to demonstrate the improved performance of the system. The main target of dissemination was the IEEE P1901 standardization activities. The project results have been integrated into the retained standards discussions in the working group. Beyond OPERA and POWERNET, there are numerous smaller research programs worldwide aimed at achieving similar research objectives.

9.12 CONCLUSIONS

Through this review of BPL's applications, it is clear that BPL represents great financial opportunities for those taking advantage of, and implementing, this rising technology. One of the most obvious ways BPL can do this is by replacing wireless or cable-wired home networks with a power-line-based home network, thereby eliminating the need for excessive cable wiring through the house and avoiding the wireless interference that reduces the technology's safety and increases users' frustration. Unlike DSL or cable modems, BPL is not anchored to head-ends or central offices. It also enjoys certain advantages over wireless when it comes to underground environments and penetrating through walls and buildings. As a result, BPL can be deployed into areas where other technologies cannot reach, for either economic or technical reasons. The BPL technology is quickly maturing, and there are several specifications that have been developed through industry consortia and standards through the IEEE and other international bodies. The development of these standards is expected to promote interoperability between different technologies and economies of scale that will further accelerate BPL deployment.

In addition to delivering more efficient broadband access in areas that are already Internet wired, BPL can help meet some of the most dire needs of planet. In the area of global development, BPL can serve a central role in expanding Internet access to the majority of the world still living without it. In addition to this, through the implementation of "triple-play" technology, the same technology can provide remote and impoverished communities with both video and phone service. In other words, BPL can provide an enormous technological leap to poor and remote communities around the world.

Perhaps most important of all, in this time of environmental peril and skyrocketing energy costs, it is absolutely essential that we find ways to reduce our overall energy use and make more efficient the energy use that is still necessary. In order to accomplish this, we will need both a sophisticated monitoring system and an informed and empowered population of consumers. The real-time, two-way communications that will be available with a true smart grid will greatly enhance the efficiency of our power grid, at the same time informing and empowering the consumer in ways never before thought possible. In addition to this, the smart grid will enable consumers to be financially compensated for their efforts to save energy and to sell energy back to the grid through net-metering. By enabling distributed generation resources like residential solar panels, small wind farms, and plug-in hybrid electric vehicles, the smart grid has the potential to spark a revolution in the energy industry by allowing small players like individual residents and small businesses to sell power to their neighbors or back to the grid. The same will hold true for larger commercial businesses that have renewable or backup power systems that can provide power for a price during peak demand events, typically in the summer when air conditioning units place a strain on the grid. This "democratization of energy" has the potential to save energy, generate increased revenues, and reduce greenhouse gas emissions, developments that in this day and age are almost essential.

REFERENCES

1. K. Doster, *Powerline Communications.* Upper Saddle River, NJ: Prentice Hall, 2000.
2. H. Hrasnica, A. Haidine, and R. Lehnert, *Broadband Powerline Communications.* Hoboken, NJ: Wiley, 2004.
3. M. Gebhardt, F. Weinmann, and K. Dostert, "Physical and regulatory constraints for communications over the power supply grid," *IEEE Communications Magazine,* Vol. 41, pp. 84–90, May 2003.
4. W. Liu, H. Widmer, and P. Raffin, "Broadband PLC access systems and field deployment in European power line networks," *IEEE Communications Magazine,* Vol. 41, pp. 114–118, May 2003.
5. N. Pavlidou, A. J. Han Vinck, and J. Yazdani, "Power line communications: State of the art and future trends," *IEEE Communications Magazine,* pp. 34–40, April 2003.
6. A. Majumder and J. Caffery, "Power line communications," *IEEE Potentials,* Vol. 23, pp. 4–8, Oct.–Nov. 2004.
7. "Internet Usage Statistics: The Internet Big Picture, World Internet Users and Population Stats," Internet World Stats [Online]. Available: http://www.internetworldstats.com/stats.htm
8. International Telecommunication Union. (2007). Measuring village ICT in sub-Saharan Africa. ITU, Geneva. [Online]. Available: http://www.itu.int/ITU-D/ict/statistics/material/Africa_Village_ICT_2007.pdf
9. S. Ntuli, H. N. Muyingi, A. Terzoli, and G. S. V. Radha Krishna Rao, "Powerline networking as an alternative networking solution: A South African solution," presented at IEEE 2006.
10. J. Anatory, N. H. Mvungi, and M. M. Kissaka, "Trends in telecommunication services provision: Powerline network can provide alternative for access in developing countries," in *Proc. IEEE Africon 2004,* July 2004, pp. 601–606.
11. M. Götz, M. Rapp, and K. Dostert, "Power line channel characteristics and their effect on communication system design," *IEEE Communications Magazine,* pp. 78–86, April 2004.
12. F. Corripio and J. Arrabal, "Broadband modelling of indoor power-line channels," *IEEE Transactions on Customer Electronics,* Vol. 48, pp. 175–183, Feb. 2002.
13. J. Bausch, T. Kistner, M. Babic, and K. Dostert, "Characteristics of indoor power line channels in the frequency range 50–500kHz," in *Proc. 2006 IEEE ISPLC,* 2006, pp. 86–91.
14. P. Henry, "Interference characteristics of broadband power line communication systems using aerial medium voltage wires," *IEEE Communications Magazine,* pp. 92–98, April 2005.
15. D. Liu, E. Flint, B. Gaucher, and Y. Kwark, "Wide band AC powerline characterization," *IEEE Transactions on Consumer Electronics,* Vol. 45, pp.1087–1097, Nov. 1999.
16. M. Zimmermann and K. Dostert, "A multipath model for the powerline channel," *IEEE Transactions on Communications,* Vol. 50, pp. 553–559, April 2002.
17. M. Karduri, M. D. Cox, and N. J. Champagne, "Near-field coupling between broadband over power line (BPL) and high-frequency communication systems," *IEEE Transactions on Power Delivery,* Vol. 21, pp. 1885–1891, Oct. 2006.
18. C.-K. Park, J.-M. Kang, M.-J. Choi, J. Won-Ki Hong, Y. Lim, and M. Choi, "An integrated network management system for multi-vendor power line communication networks," presented at IEEE International Conf. on Information Networking, Okinawa, Japan, 2008.
19. Y.-J. Lin, H. A. Latchman, M. Lee, and S. Katar, "A power line communication network infrastructure for the smart home," *IEEE Wireless Communications Magazine,* pp. 104–111, Dec. 2002.

20. IEEE P1901, Draft Standard for Broadband Over Power Line Networks: Medium Access Control and Physical Layer Specifications [Online]. Available: http://grouper.ieee.org/groups/1901/
21. http://ist-opera.org/
22. http://ist-powernet.org/
23. http://upaplc.org/page_viewer.asp?section=About+Us&sid=8
24. http://www.cepca.org/about_us/
25. http://www.etsi.org/WebSite/AboutETSI/AboutEtsi.aspx
26. http://portal.etsi.org/plt/Summary.asp
27. http://portal.etsi.org/plt/ActivityReport2007.asp

10 Ethernet Optical Transport Networks

Raymond Xie

CONTENTS

10.1 Introduction 239
10.2 Ethernet Technology Overview 240
    10.2.1 Invention of Ethernet 240
    10.2.2 Ethernet Bandwidth Growth..241
    10.2.3 The Ethernet Frame 242
    10.2.4 Half-Duplex and Full-Duplex Ethernet 244
    10.2.5 Ethernet Repeaters and Hubs...245
    10.2.6 Ethernet Bridges and Switches 245
    10.2.7 Evolution of Ethernet Header: VLAN, PB, and PBB 246
        10.2.7.1 Virtual LANs (IEEE 802.1 Q)..247
        10.2.7.2 Provider Bridges (IEEE 802.1AD) 247
        10.2.7.3 Provider Backbone Bridges...249
10.3 Ethernet Optical Transport Technologies...250
    10.3.1 Ethernet over Fiber 250
    10.3.2 Ethernet over WDM 251
    10.3.3 Ethernet over Next-Generation SONET/SDH (EoS)........................251
    10.3.4 Ethernet over MPLS and MPLS-TP 252
    10.3.5 Ethernet Transport over PBB-TE...254
    10.3.6 Ethernet Optical Transport Standards Activities.............................254
    10.3.7 Ethernet Transport with Intelligent Optical Layer 255
10.4 Carrier Ethernet for Mobile Backhaul..257
    10.4.1 Carrier Ethernet Definition..257
    10.4.2 Carrier Ethernet for Mobile Backhaul: Motivations and Challenges....260
        10.4.2.1 Synchronization 261
        10.4.2.2 Delay and Jitter 261
        10.4.3.3 Reliability..263
References..263

10.1 INTRODUCTION

One of the fast growing technologies in the context of wireless edge and optical core is Ethernet, either from a local area network (LAN) application perspective or from a wireless backhaul solution point of view. In fact, one of the hottest markets in the telecommunication industry today is metro Carrier Ethernet transport, which aims to provide a single converged metro Ethernet network that supports business services, residential triple-play, wholesale bandwidth, and wireless backhaul in a ubiquitous, economic, and efficient way. As a result, even the latest advances in optical core technology are centered on how to optimize Ethernet transport capabilities. To understand the technology trend of wireless edge and optical core, it is very important to understand the history and evolution of Ethernet, and how it could impact the design and architecture of wireless and optical networks. This chapter provides an overview of Ethernet technology and outlines various optical transport approaches to carry Ethernet over wide area networks (WANs). It further introduces Carrier Ethernet, and finally discusses motivations and challenges to use Carrier Ethernet for wireless backhaul applications.

10.2 ETHERNET TECHNOLOGY OVERVIEW

10.2.1 INVENTION OF ETHERNET

Ethernet has been the dominant choice of technology for LANs. The twisted-pair Ethernet cable is well recognized by all professionals who need access to a network. Essentially all desktop and laptop computers come with a built-in Ethernet socket today. It has been a long way since Ethernet was originally invented (U.S. Patent 4063220) by Robert Metcalfe, David Boggs, and others from Xerox Palo Alto Research Center (Xerox PARC) in the early 1970s.

Ethernet was originally designed to enable computer communications over a shared broadcast medium such as coaxial cable. It utilizes a key communication mechanism called Carrier Sense Multiple Access with Collision Detection (CSMA/CD). A good analogy would be conversations among multiple people sitting in a dark room. In this case, the air served as the shared broadcast medium. When one person wanted to talk to another person in the same room, he needed to call out his name first and start to talk because he could not see the other person. During this period, all other people could hear his talking, but had to wait until he finished before they could initialize another conversation. Collision happened when two or more people started to talk at the same time. Each one of them stopped and waited for a short random period of time before he or she started to talk again. Because the waiting time was random, there was a good chance that one person started to talk while others were still in the waiting stage. This collision detection and backoff scheme would allow communication among these people in the dark room. In the Ethernet world, the coaxial cables attached to multiple computers serve as the shared broadcast medium, and the short random period is measured in microseconds, and exponentially increased backoff timers are used for computers to restart transmission. This simple idea of a cable broadcast system allowed a network of computers to communicate with each other at high speed in a very economical way. From this early concept, Ethernet has evolved into a more complex networking technology involving point-to-point links connected by Ethernet hubs and switches for improved reliability, efficiency and manageability.

The first Ethernet standard was released in 1983 by the Institute for Electrical and Electronics Engineers (IEEE) 802.3 Committee in a document called "CSMA/CD Access Method and Physical Layer Specifications." During the early years of Ethernet

Ethernet Optical Transport Networks

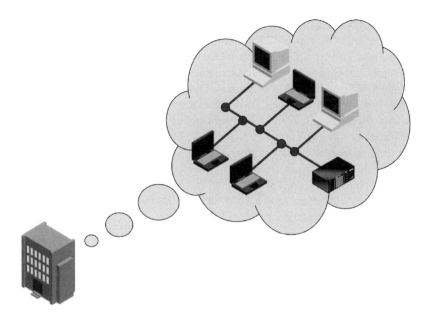

FIGURE 10.1 Shared medium Ethernet LAN.

in the 1970s and 1980s, the share broadcast medium evolved from thick coaxial cables (10Base5) to thin coaxial cables (10Base2) to unshielded twisted-pair cables (10BaseT) in 1990. Because of the low cost and easy installation of unshielded switched-pair cables, 10BaseT became very popular and eventually dominated the LAN technology in office buildings (Figure 10.1) since 1990. In 1993, IEEE released 802.3j standard (10BaseF), which enables a pair of optical fibers to be used as a transmission medium for Ethernet. This 10BaseF technology extends the operational distance to 2,000 m, which further advances Ethernet's popularity in office parks and university campuses.

10.2.2 Ethernet Bandwidth Growth

Early Ethernet standards are defined to support bandwidth of 10 Mbps. In 1995, the 100 Mbps 802.3u 100BaseT standard, often known as Fast Ethernet, was released from IEEE. It supports three transmission mediums: 100BaseTX, two pairs of CAT 5 twisted pair; 100BaseT4, four pairs of CAT 3 twisted pair; and 100BaseFX, a pair of multimode optical fibers. In the 1990s, 10/100BaseT Ethernet was quickly adopted as the most widely deployed LAN network and the explosion of Internet and business applications continues to drive the need for Ethernet bandwidth growth. The Gigabit Ethernet standard that supports bandwidth of 1,000 Mbps was released by IEEE with 802.3z and 802.3ab specifications in 1998 and 1999, respectively. It supports a comprehensive list of transport media ranging from single and multimode optical fibers to shielded and unshielded twisted pairs. Some high-end personal computers like Apple's Power MAC G4 and PowerBook G4 even feature a built-in 1000BaseT interface. The demand for Ethernet bandwidth continues, and the first 10Gigabit Ethernet standard was released in 2002 with the 802.3ae specification that supports different reaches of optical fiber interfaces. It was enhanced later with the 802.3ak specification in 2004 and 802.3an specification in 2006, which added InfiniBand-type cable and twisted-pair copper interfaces, respectively. 10Gigabit Ethernet ports are primarily deployed to aggregate Fast Ethernet and Gigabit Ethernet traffic and transport them over WANs, which typically consist of SONET/SDH (Synchronous Optical Networking/Synchronous Digital Hierarchy) and DWDM networks. There seems to be no end to the potentials of Ethernet bandwidth, which amazingly not only has caught up with the mostly deployed 10Gbps SONET/SDH or Optical Transport Network (OTN) transport pipes, but also might exceed the transport capability with the recent 40Gigabit and 100Gigabit Ethernet efforts. The 40GigE/100GigE standard has been driven by the IEEE P802.3ba task force, and a preliminary draft was released in December 2008. Some Ethernet pioneers are even talking about Terabit Ethernet, with 10 times the capacity of 100Gigabit Ethernet, to be realized and deployed in 2015 [1].

10.2.3 THE ETHERNET FRAME

The fundamental building block of Ethernet traffic is an Ethernet frame. The Ethernet frame is like a carriage carrying user data and other necessary control information to enable the data link layer. A basic IEEE 802.3 Ethernet Media Access Control (MAC) frame is shown in Figure 10.2. It consists of the following fields:

- *Preamble (7 bytes):* The preamble field consists of seven octets (56 bits) of alternating ones and zeros, 10101010. The pattern is designed to indicate the beginning of an Ethernet frame so that a receiving device can detect the presence of a frame and start to process it accordingly.
- *Start Frame Delimiter (SFD) field (1 byte):* The SFD field has a fixed value of 10101011 that marks the end of the preamble of an Ethernet frame. The SFD field is designed to break the preamble pattern of alternating ones and zeros and signal the start of the actual frame.
- *Destination MAC Address field (6 bytes):* A MAC address is a globally unique identifier assigned to network devices on a permanent basis. It is typically described in six groups of two hexadecimal digits, separated by hyphens (-) or colons (:), for example, 01:23:45:67:89:ab. The Destination MAC Address identifies the station that is to receive the transmitted frame. It can be a single station or a multicast address for multiple stations.
- *Source MAC Address field (6 bytes):* The Source MAC Address field uniquely identifies the transmitting station. Its format is the same as the Destination MAC Address field.
- *Ethernet Type/Length field (2 bytes):* The Ethernet Type or Length field indicates either the number of bytes of data contained in the MAC Client Data field if the value is less than or equal to 1,500 or the type of protocol carried in the Ethernet frame if the value is greater than or equal to 1,536.
- *MAC Client Data and Pad field (46~1,500 bytes):* The MAC Client Data field contains the user data to be transmitted from the source to the destination. The minimum size of this field is 46 bytes and the minimum Ethernet frame size is 64 bytes, measured from the Destination MAC Address field

Ethernet Optical Transport Networks

802.3 MAC Frame	Preamble	Start of Frame Delimiter	Destination MAC	Source MAC	Ethertype/ Length	Payload	CRC32
Pattern	10101010....	10101011	12:34:56:78:90:AB	12:34:56:78:90:AB	<1500 Length >1536 Type	Data plus Padding	32-bit CRC
Length	7 bytes	1 bytes	6 bytes	6 bytes	2 bytes	46~1500 bytes	4 bytes
Total Length	8 bytes				64 ...... 1518 bytes		

FIGURE 10.2 Basic IEEE 802.3 Ethernet MAC frame.

through the frame check sequence. If the field is less than 64 bytes, then the Pad field adds extra bytes to bring the frame to its required minimum length. The maximum size of this field is 1,500 octets. Jumbo frames that can carry up to 9,000 bytes of payload are recently supported by Gigabit and 10Gigabit interfaces from many Ethernet devices, but these have not become part of the official IEEE 802.3 Ethernet standard.

- *Frame Check Sequence (FCS; 4 bytes):* The FCS field provides a 32-bit cyclical redundancy check (CRC) calculation for error checking the content of the frame. The source station creates the MAC frame with user data, calculates the CRC value, and appends it to the MAC frame before transmitting the frame across the network. The CRC calculation is performed on all the fields from the destination MAC address to the Pad field excluding the preamble, start frame delimiter, and obviously the FCS to be calculated. The destination Ethernet device recalculates the CRC value on arrival of the transmitted frame and compares the CRC value with the one contained in the frame; if they do not match, then data corruption is detected and the arrived Ethernet frame is discarded.

The Ethernet frame provides a basic data packet encoding at Layer 2, the data link layer. It captures the fundamental communication needs at the data link layer with the source and destination addresses, actual data payload, and error detection fields. Various higher layer protocols like Transmission Control Protocol/Internet Protocol (TCP/IP), User Datagram Protocol (UDP), Address Resolution Protocol (ARP), Open Shortest Path First (OSPF), Resource Reservation Protocol (RSVP), and so on are commonly carried over Ethernet frames to deliver loss-free transport, signaling, routing, and other networking-specific data units.

10.2.4 Half-Duplex and Full-Duplex Ethernet

The half-duplex mode of Ethernet is the traditional CSMA/CD-based Ethernet that allows multiple Ethernet devices to share common media and uses collision detection and a backoff mechanism to revolve the connection and achieve communication among multiple Ethernet devices. In 1997, IEEE 802.3x introduced full-duplex Ethernet for point-to-point connections between two Ethernet stations. Because no contention exists, collision does not occur and CSMA/CD is not required. Full-duplex mode enables stations to simultaneously transmit and receive data over a single point-to-point link. As a result, the overall throughput of the connection is doubled.

Full-duplex operation includes an optional flow control mechanism using a PAUSE frame. If one station transmits a high volume of frames resulting in serious congestion at the other end, the receiving station can send back a PAUSE frame to ask the sourcing station to temporarily cease transmission, giving it the opportunity to recover from the congestion. A PAUSE frame contains the period of pause time requested by the congested destination; once the time period expires, the source station returns to normal operation to transmit frames. PAUSE frames are bidirectional, meaning that either of the peer stations on the link can issue them.

Ethernet Optical Transport Networks

10.2.5 ETHERNET REPEATERS AND HUBS

Ethernet repeaters are devices used to extend the length of Ethernet segments. Coaxial Ethernet segments had a restricted length for signal degradation and timing reasons. Ethernet repeaters are capable of forwarding Ethernet frames from one cable to another in wire speed. Repeaters that have multiple ports are also called Ethernet hubs, in which Ethernet frames from one port are multicasted to all other ports. If a collision is detected, a jam signal will be transmitted to all ports to ensure proper collision detection on all Ethernet segments attached. Because all extended Ethernet segments belong to one collision domain and every Ethernet device has to be able to detect a collision within a short time window, there is a limitation of how many Ethernet segments can be extended using repeaters. A general guideline is to make sure there are a maximum of five Ethernet segments between any two hosts, three of which can have attached Ethernet devices. Repeaters and hubs can also help isolate cable breakage problems by detecting a broken link and stopping forwarding frames from it, which allows other segments to continue working properly.

10/100Base-T Ethernet hubs are very cheap today, and computer professionals are getting used to add a small four-port or five-port Ethernet hub to share a twisted-pair Ethernet connection among multiple computers in homes and offices. Hubbed Ethernet networks still use half-duplex and CSMA/CD, and the entire network is one collision domain. Every packet is broadcast to all other ports on the hub with no security control. The total throughput of the hub is also limited to that of a single link and all links must operate at the same speed.

10.2.6 ETHERNET BRIDGES AND SWITCHES

Whereas Ethernet repeaters and hubs extend Ethernet physical links with little data processing, Ethernet bridges work one layer higher, at the data link layer, to replicate Ethernet frames and provide a more robust and flexible Ethernet network. Ethernet bridges are capable of learning MAC addresses of incoming Ethernet frames and associate them with the incoming Ethernet segments or ports automatically. They can make intelligent decisions to forward well-formed Ethernet frames to the destined Ethernet segments only. This greatly improves overall performance, provides collision and frame error isolation, and overcomes the limits on total segments between two hosts.

Ethernet bridging, often called transparent bridging, typically requires the setup of a spanning tree that allows every pair of bridges to reach each other through a unique low-cost path within the LAN. The Spanning Tree Protocol (STP), defined in IEEE Standard 802.1D, essentially provides a loop-free topology within which broadcast, multicast, and flooding and learning can all operate properly. Spanning trees are automatically set up by Ethernet bridges through exchange of Bridge Protocol Data Units (BPDUs) to determine a root bridge, a network-wide tree topology, and the forwarding state of each port. Forwarding tables in each switch are updated to reflect the tree topology, and once an Ethernet frame is switched onto a spanning tree, it follows the tree path to its destination device. The entire process of building a spanning tree

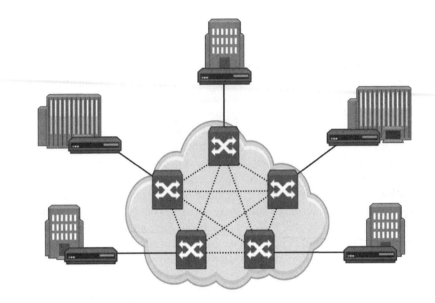

FIGURE 10.3 Switched Ethernet networks.

could take up to 30 to 50 seconds. A much faster protocol called Rapid Spanning Tree Protocol (RSTP) was introduced in IEEE 802.1w in 1998 and was refined in IEEE 802.1D-2004 to provide second-level convergence and reaction time. The automatic formation of a spanning tree enables almost zero configuration of switched Ethernet networks, as adding a new switch is essentially plug-and-play. The simplicity and low-maintenance characteristics of Ethernet allow it to prevail over other alternative LAN technologies like Token Ring and Fiber Distributed Data Interconnect (FDDI).

Early Ethernet bridges examined each packet one by one using software on a CPU and were slow compared with Ethernet hubs at forwarding traffic. Ethernet switches were invented to forward frames at full wire speed and benefitted from hardware implementation of transparent bridging. There are various techniques to switch Ethernet frames between ports; some switches buffer and perform a checksum on each frame before forwarding and others read MAC addresses only with no error checking. Functionally, Ethernet bridges and switches are equivalent, and these two terms are often used interchangeably.

Today, switched Ethernet (Figure 10.3) has become the dominant LAN technology and Ethernet switches are available in capacity from hundreds of megabits to multiple terabits, driven by exponential growth of LANs for e-commerce, Internet, video, storage area networks, and bandwidth-intensive applications.

10.2.7 Evolution of Ethernet Header: VLAN, PB, and PBB

As Ethernet takes off for business and residential LAN services, there is a growing demand for Ethernet to be carrier friendly. The Ethernet protocol has evolved from the original shared medium broadcast protocol with source and destination MAC addresses to Virtual LAN (VLAN), Provider Bridges (PBs), and Provider Backbone

Bridges (PBBs) to allow service providers to provide Ethernet services in a more secure and scalable way. These changes are reflected by the evolution of the Ethernet header (Figure 10.4).

10.2.7.1 Virtual LANs (IEEE 802.1 Q)

The simplicity and cost efficiency of Ethernet has made it the dominant network access technology in LANs. Service providers started to offer point-to-point Ethernet services to enable enterprise customers to connect their branch offices together as a single integrated Ethernet LAN. These applications revealed some deficiencies of the original shared-medium Ethernet as all devices from different customers have to share a single physical transport medium and there are no security and scalability measures in place to isolate and support hundreds of enterprise customers at the same time. VLAN was introduced in IEEE Standard 802.1Q to resolve these deficiencies by adding a 12-bit VLAN tag to the Ethernet header to create independent logical LANs with disjoint broadcast domains over a single physical network. With this definition, a VLAN tag would be added to arriving Ethernet frames by edge switches in the service provider network. VLAN-enabled switches would admit and forward frames only on ports that were configured with the same VLAN identifier. This effectively provides security over a shared physical network through logical traffic segregation and more efficient bandwidth utilization by limiting the scope of flooded broadcast traffic to a VLAN instead of the entire physical network. The 12-bit VLAN identifier is able to support up to 4,094 (Identifier 0 and 4095 were reserved) VLAN networks, enough for a small to midsize service provider network.

10.2.7.2 Provider Bridges (IEEE 802.1AD)

The 802.1Q VLAN capability has been leveraged by service providers to offer metro Ethernet services to enterprise customers. Some enterprise customers also deployed their own VLAN networks for controlling quality of service (QoS) and building a more scalable Layer 2 network. This created a conflict as service providers would have to honor the client VLAN IDs and coordinate among their customers to make sure there is no overlap, a task that sometimes is not achievable. IEEE Standard 802.1ad, now part of IEEE Standard 802.1Q-2005, resolved this problem by inserting a new service VLAN tag before the 802.1Q VLAN and now the 802.1Q VLAN tag can be used as the customer VLAN tag in the Ethernet frame. The service VLAN tags are added by provider edge bridges to customer Ethernet frames as they enter a service provider network. The tags are then removed from those frames by provider edge bridges as they leave the service provider network. The frames are switched based on the service VLAN within the provider network and the customer VLANs remain invisible to the service provider. The 802.1ad is often called Provider Bridges as it enables a service provider to uniquely identify the customer service instance and allows for the transparency of customer VLANs within the service provider network. For example, customer A may operate 20 VLANs using VLAN IDs 90 to 109, and customer B may operate 50 VLANs using VLAN IDs 100 to 149. Without the 802.1ad Provider Bridges support, it is difficult for a service provider to support these two customers at the same time, not to mention having a medium-size provider support thousands of customers. With the PB support, customer A can be assigned a service VLAN

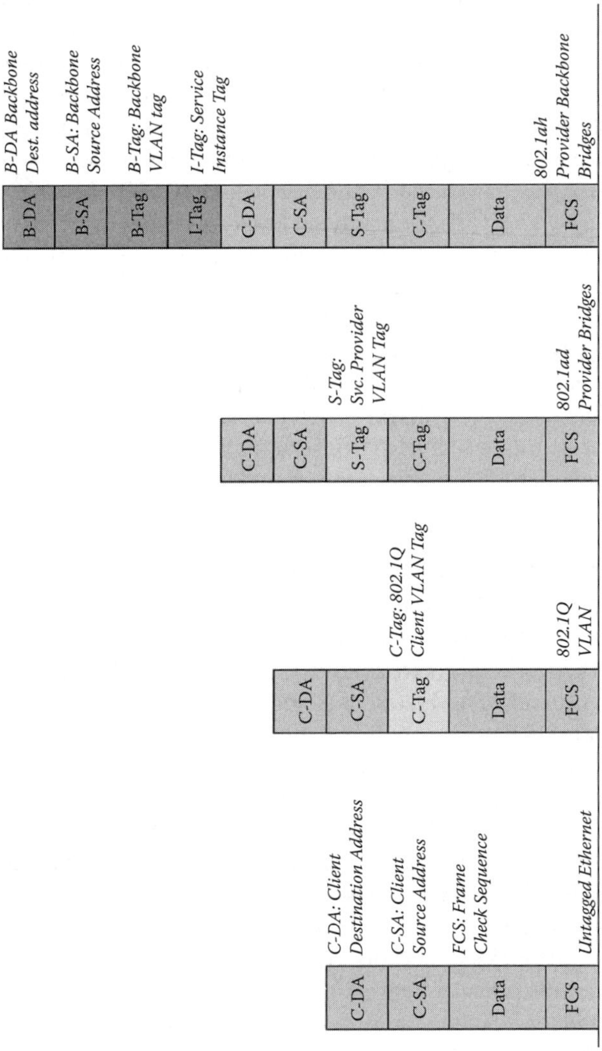

FIGURE 10.4 Evolution of Ethernet header.

ID of 10, and customer B can be assigned a service VLAN ID of 12. Their private customer VLAN space now can be kept intact and totally isolated from the service VLAN space. IEEE Standard 802.1ad is also called Q-in-Q with the outer 802.1Q ID being the service provider ID and the inner 802.1Q ID being the customer ID. The Q-in-Q term came from early proprietary implementations of PBs.

10.2.7.3 Provider Backbone Bridges

Although PBs solved the problem of multiple administration of the VLAN space, there were two limitations that still hindered the deployment of Ethernet private line services in service provider networks. First, the maximum number of service VLANs is only 4,094, a significant limitation for large service providers that have more than 4,094 clients. Second, customer MAC addresses still need to be learned and forwarded by service providers, which poses not only security issues but also scalability challenges, as there could be hundreds of thousands of MAC addresses a service provider switch has to handle. A typical state-of-the-art Ethernet switch today has a maximum MAC address forwarding table of 4,000 to 64,000 entries [2]. Most medium- and low-end Ethernet switches are not able to handle a large retail department store that might have 200 sites in the U.S. and 50 computers per site.

The IEEE 802.1ah Provider Backbone Bridges overcomes these two limitations by adding four additional fields into the Ethernet header to fully encapsulate customer Ethernet frame including MAC addresses into the service provider Ethernet frame, a technique often termed MAC-in-MAC. These four fields are a 48-bit Backbone Destination Address (B-DA), 48-bit Backbone Source Address (B-SA), 12-bit Backbone VLAN tag (B-VLAN), and 24-bit Service Instance tag (I-SID). These four fields are populated with service-provider-defined values when Ethernet frames ingress a PBB network and customer MAC addresses are hidden from the PBB core. Devices in the PBB core forward traffic based on backbone MAC addresses and have no knowledge of the customer MAC address space. This greatly enhances the scalability of the Ethernet transport solution.

The 12-bit backbone VLAN identifier (B-VLAN) enables service providers to partition their PBB network into different broadcast domains to improve network utilization especially for multicast applications. The PBB protocol uses Multiple Spanning Tree Protocol (MSTP) to establish rooted broadcast domains in the network, with each domain associated with a B-VLAN ID. Unknown MAC addresses are only broadcast within the given B-VLAN.

To overcome the customer service instance scalability limitation, the PBB frame format defines a 24-bit service identifier known as the I-SID. Each customer service instance would be assigned a unique I-SID value, which can scale to over 16 million instances, far more than the 4,094 instances offered by 802.1ad Provider Bridges. The I-SIDs are assigned and visible to backbone edge bridges only and are transparent to the devices in the core of the PBB network.

A PBB network operates in exactly the same manner as that of an 802.1ad (Q-in-Q) network. However, instead of operating on customer MAC addresses, the new backbone MAC addresses are used to forward traffic in the backbone core. The edge devices of the PBB network need to learn client MAC address plus VLAN IDs, and then add proper B-DA, B-SA, B-VLAN, and I-SID values to the Ethernet frame before forwarding it onto the core backbone nodes. The core backbone nodes do not even have to be PBB capable. PBB frames should be passed along transparently as long as the core nodes understand the IEEE 802.1ad Provider Bridges. For a service provider to upgrade existing 802.1ad networks to PBB networks, they need to upgrade only the edge devices to implement PBB functionality. The core of their 802.1ad networks could remain unchanged.

10.3 ETHERNET OPTICAL TRANSPORT TECHNOLOGIES

Few people disagree that the convergence of Ethernet and optics (EtherOptics) is one of the cornerstones of next-generation broadband networks. This convergence offers fixed-line and mobile network operators a powerful range of benefits, including optimization of packet transport and cost-efficient deployment of new revenue-generating services.

Although the industry has reached consensus that Ethernet will play a fundamental role in the evolution from circuit- to packet-based services, the real-world implementation is not always as simple as it seems. Ethernet will continue to be a key technology in access networks, but the existing core and regional service provider network infrastructure is still largely TDM (Time Dimension Multiplexing) centric. The benefits of Ethernet simplicity and low cost are unquestionably compelling; however, service providers managing end-to-end regional and core service networks are faced with other economic and practical considerations that weigh heavily as they cost-effectively balance TDM-to-Ethernet convergence and support for traditional revenue-generating SONET/SDH services. Key implementation considerations include carrier-class QoS; bandwidth optimization; standards-based interoperability; operations, administration, maintenance, and provisioning (OAM&P) robustness; and guaranteed service-level agreement (SLA) performance as the network scales.

Ethernet owes its phenomenal success in no small part to simplicity because it was designed for LAN connectivity in well-controlled private network environments supporting a finite number of users. The emergence of EtherOptics was born out of the desire to leverage Ethernet cost, simplicity, and efficiencies and to extend these attributes to metro and wide area (or long-haul) networks. Extending the LAN-oriented feature set of Ethernet to the WAN, the core of which is an optical infrastructure, poses more than a few interconnection challenges.

There is clearly more than one approach to extending Ethernet into a carrier's optical infrastructure. Although each approach has its advantages and disadvantages, ultimately the suitability of a specific approach will depend on, among other things, the types of services offered, the service provider's network infrastructure and operations, the deployment strategy (overlay vs. greenfield), the existing equipment, and the service provider's target customers. The following discussion highlights some of the more common approaches (Figure 10.5) that carriers can consider today for extending Ethernet services to the MAN (Metropolitan Area Network)/WAN environment.

10.3.1 ETHERNET OVER FIBER

In this approach, Ethernet interfaces on routers or switches are simply interconnected over dark fiber networks. Because no additional protocol or format conversion

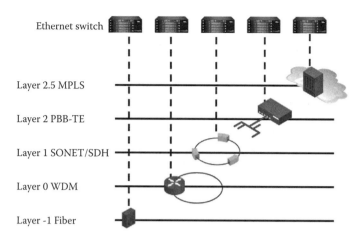

FIGURE 10.5 Ethernet transport approaches.

is required, this is the most straightforward method for extending Ethernet to the MAN. Large enterprises that have private fiber infrastructure have been among the first to adopt this option to transport Ethernet over fiber across multiple locations.

The primary drawbacks of this method are the same as the limitations of Ethernet for MAN/WAN applications outlined previously: carrier-class QoS, OAM&P robustness, network resiliency, interoperability, and so on. In addition, this approach is difficult to scale efficiently, as new optical interfaces and trunks need to be deployed as new nodes are added, even if the existing interfaces and trunks are not fully utilized.

10.3.2 Ethernet over WDM

This approach is similar to native Ethernet, except that the use of WDM (Wavelength-Division Multiplexing) provides additional reach and capacity in comparison to the standard Ethernet PHY. Only a simple wavelength conversion from the Ethernet PHY wavelength to the appropriate WDM wavelength takes place without any protocol change.

Depending on the number of channels and distance required, either Coarse WDM (CWDM) or Dense WDM (DWDM) can be deployed. CWDM is typically used for small-capacity access networks, whereas DWDM is used more in the long haul and core networks for much larger bandwidth needs and more expensive fiber routes. WDM can be used with both Gigabit Ethernet (GbE) and 10GbE. Optical layer protection and performance monitoring can be used to overcome the rudimentary, inherent limitations of Ethernet.

10.3.3 Ethernet over Next-Generation SONET/SDH (EoS)

In this approach, Ethernet is carried over a next-generation SONET/SDH network using standardized techniques such as Generic Framing Procedure (GFP), Virtual Concatenation (VCAT), and Link Capacity Adjustment Scheme (LCAS). VCAT provides a method for right-sizing the transport pipe, and LCAS provides a mechanism for in-service increase and decrease of transport capacity.

SONET/SDH networks have been the dominant transport network infrastructure in the past 20 years due to their standardization, high availability, 50-msec fast protection time, a rich set of performance monitoring capabilities, and comprehensive OAM functionalities. SONET/SDH networks were optimized to carry voice traffic at the beginning but have been enhanced to support Ethernet and data services lately with the ITU-T G.7041 GFP and ITU-T G.7042 VCAT/LCAS standards. The GFP protocol is a multiplexing technology that allows mapping of variable-length, higher layer client signals over a SONET/SDH transport network. The VCAT and LCAS protocols break the boundary of fixed SONET/SDH bandwidths (i.e., STS-1/STM-0, STS-3c/STM-1, STS-12c/STM-4, STS-48c/STM-16, etc.) and allow a connection that consists of multiple STS-1s, for example, to be created. This is especially important to transport 1GigE services efficiently as there is no standard SONET/SDH path that provides 1 Gigabit per second bandwidth and the best fit is a 2.5G STS-48c/STM-16 connection. Without VCAT and LCAS, the only choice is to place two 1GigE services into an STS-48c/STM-16 pipe, which would waste 500 Mbps of valuable transport bandwidth. VCAT supports connections that consist of X number of STS-1 or STS-3c paths and allows these paths to take different physical paths from the same source and destination pair. LCAS enables hitless bandwidth expansion and contraction. A typical configuration of 1GigE EoS service is to map the Ethernet traffic into a VCAT group of 21 SONET STS-1 or 7 SDH STM-1 connections, often termed STS-1-21v.

Widely deployed and proven, the primary advantages of EoS are its ability to leverage the existing MAN/WAN transport infrastructure and the superior and well-known protection, fault isolation, and OAM&P features of SONET/SDH.

10.3.4 Ethernet over MPLS and MPLS-TP

The Multi Protocol Label Switching (MPLS) protocol is a layer 2.5 protocol operating between layer 3 IP layer and layer 2 Ethernet layer. It is designed to overcome the connectionless deficiency of IP networks and provide a connection-oriented packet-switching network with a goal to support delay-sensitive real-time applications as well as virtual private networks (VPNs). A good analogy to understand the relationship between MPLS and IP would be thinking about a post office system. Imagine each letter is an IP packet and it has a destination address written on its envelope. Consider a letter from Boston to San Francisco and assume it needs to make two stops in Chicago and Salt Lake City along the way. At each intermediate post office, the letter's address will be scanned and it will be forwarded to the next stop based on a large forwarding table. A forwarding table entry at Chicago might be something like "For all letters destined to San Francisco, send to Salt Lake City." As one might be able to tell, the process is inefficient, as the address needs to be rescanned at each intermediate stop, a large forwarding table needs to be maintained to keep up with the growing number of destinations, and the table look-up through hundreds of thousands of entries could be time consuming. This post office system actually mimics an IP switched network very closely as, in the same manner, an IP router maintains a large routing table to route IP packets based on their destination IP addresses. Imagine an improvement to the system is to assign a label, let's say 100, to the letter and call the Chicago office to set up a rule: When receiving a letter with label 100, change the label to 105 and forward it to Salt Lake City. A similar label swapping rule can be established in the Salt Lake City office. In this way, letters would be processed more efficiently without address lookup and could reach their destination sooner. The MPLS technology works in the same way to provide efficient and connection-oriented services for both circuit-centric and packet-centric client applications. It adds an MPLS label to packet headers and leverages a set of signaling and routing protocols to set up label switched paths (LSPs) for fast packet forwarding. Over the past decade, this has evolved as the dominant connection technology in backbone IP networks around the globe.

Ethernet over MPLS is similar to Ethernet over SONET/SDH, except that it uses MPLS LSPs instead of virtually concatenated circuits. In this case, Ethernet frames are encapsulated in MPLS frames and sent over MPLS connections. Ethernet over MPLS is advantageous in converged multiservice networks as it overcomes the scalability constraint of the 12-bit VLAN identifier field from native Ethernet. The MPLS label is a 20-bit field that allows 1 million identifiers. MPLS also supports fast reroute capability for network restoration after a failure. Large service providers typically own two independent transport networks: a SONET/SDH-based TDM-centric transport network that supports voice and legacy TDM services, and an emerging MPLS-based packet-centric transport network that supports IP and Ethernet services. Ethernet over MPLS provides an option for service providers to carry Ethernet over their emerging IP networks. One drawback of this approach is that the cost is significantly higher than the Ethernet over SONET/SDH approach, as the label switched routers tend to be more expensive than the more economic SONET/SDH devices.

The latest development of the MPLS transport solution is the MPLS-Transport Profile (MPLS-TP) initiative (Figure 10.6), currently driven by a joint working group from the Internet Engineering Task Force (IETF) and the International Telecommunication Union (ITU-T). MPLS-TP is a cost-effective transport technology using a subset of MPLS standards with transport-oriented OAM capabilities.

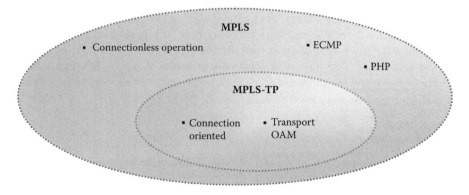

FIGURE 10.6 MPLS and MPLS-TP.

It removed connectionless functions such as Layer 3 capabilities, penultimate hop popping (PHP), LSP merging and Equal-Cost Multiple-Path (ECMP) routing because these are not needed in a connection-oriented environment. Once the standard is finalized and implemented by various vendors, carriers will have another potentially economical option to transport Ethernet over WAN networks, more specifically, over IP core networks.

10.3.5 Ethernet Transport over PBB-TE

Provider Backbone Bridging-Traffic Engineering (PBB-TE) is an initiative to provide carrier-class Ethernet transport capabilities (e.g., determinism, QoS, scale, 50-msec resiliency, carrier-class OAM&P tools, etc.). It is an extension of the IEEE 802.1ah Provider Backbone Bridges explained earlier in this chapter. The goal is to provide connection-oriented Ethernet (i.e., using Ethernet to transport Ethernet) by leveraging VLAN and MAC-in-MAC encapsulation defined in IEEE 802.1ah, but disabling flooding, spanning trees, and other non-connection-friendly Ethernet functions. In the initial proposal of PBB-TE, the forwarding table is not constructed automatically; rather, a centralized network management system is used to populate the forwarding entries manually. It provides more predictable and controllable behavior compared with PBB. PBB-TE is now being defined in a draft standard, IEEE 802.1Qay. PBB-TE also carries a comprehensive set of OAM capabilities from IEEE 802.1ag.

PBB-TE's connection-oriented features and behaviors, as well as its OAM approach, are inspired by SONET/SDH networks. PBB-TE can also provide path protection levels similar to the UPSR (Unidirectional Path-Switched Ring) protection in SDH/SONET networks. Although relatively new to the Ethernet optical transport landscape, PBB-TE adds to the existing set of options available to network operators. Limited from a deployment perspective, PBB-TE is essentially designed for manually provisioned point-to-point Ethernet (transport) connections. PBB, on the other hand, allows Ethernet transparent bridging and therefore enables automated establishment of Ethernet connectivity, but thus far it has not been deployed.

Recently, there has been intensive debate in the industry about the cost savings, OAM&P capabilities, and interoperability issues between PBB-TE and the incumbent MPLS transport technology. With carrier reaction mixed to date, and the initiatives tainted to a certain degree from excessive hype, it is premature at this time to fully assess the market impact these new transport options will have. Nevertheless, on the access side, PBB-TE might eventually evolve as a leading Ethernet access technology, serving to complement the various core transport network architectures and technologies required for end-to-end service performance.

10.3.6 Ethernet Optical Transport Standards Activities

Progress in the standards organizations and Ethernet-focused industry forums has played a critical role in advancing Ethernet optical applications and services. ITU standards—for example, GFP, VCAT, and LCAS—have been instrumental in enabling operators to transform existing SONET/SDH networks into a flexible and efficient transport vehicle for the IP traffic driving network growth. Meanwhile, because there is no universally accepted meaning for "Ethernet services," the Metro Ethernet Forum (MEF) has led a commendable effort to standardize service definitions to clearly communicate the features of various Ethernet services among service providers and end customers.

Based on the MEF service definitions, there are two broad categories of Carrier Ethernet services: point-to-point E-LAN services, and multipoint E-LAN services. These services are accompanied by five attributes that define carrier-class service: standardized services, scalability, reliability, QoS, and service management. More and more Ethernet access and transport devices have become MEF certified in the last couple of years.

On the transport side, the ITU-T OTN defines the next-generation framing protocol, G.709, to enable efficient transportation of 10GbE services. The 10.7-Gbit/sec throughput provided by G.709 overcomes the 9.9-Gbit/sec SONET/SDH throughput limitation and is designed to provide transparent transport for 10GbE services.

On the Ethernet standard front, a newly formed task force of the IEEE has begun to review technical proposals for a new standard for 40GbE and 100GbE. The IEEE established 100GbE as its next step in 2006 but later decided to include 40GbE in the same standard.

The need for 40G/100G Ethernet is growing as IP video and transaction-intensive Web applications are exploding across the Internet. Companies such as YouTube™ and Yahoo™ regularly add 10-Gbit/sec service pipes to meet growing demand, and 40/100GbE will truly unlock the potential of the transport network and enable service providers to aggregate 10GbE services much more efficiently.

On the control plane side, the Optical Interworking Forum (OIF) is finalizing the User Network Interface (UNI) 2.0 implementation agreement to allow Carrier Ethernet devices to request connections from transport networks in a dial-up fashion using the standard control plane capability. UNI 2.0 Ethernet Virtual Private Line (EVPL) services were successfully demonstrated during the 2009 OIF Worldwide Interoperability Demo.

10.3.7 Ethernet Transport with Intelligent Optical Layer

An intelligent optical control plane is the key to realizing the full potential of Carrier Ethernet services across multiple service provider networks that may comprise a variety of legacy transport equipment, heterogeneous administrative domains, and different levels of survivability mechanisms. The ITU-T Automatic Switched Optical Network (ASON) framework, embodied by OIF UNI/E-NNI, IETF GMPLS, and vendor-specific I-NNI signaling and routing protocols, provides the unified control plane capability to enable provisioning and operation of Carrier Ethernet services in a true multicarrier, multivendor environment.

Besides multivendor interoperability, the intelligent optical control plane enables new revenue-generating services such as bandwidth-on-demand Ethernet. Traditionally, it takes weeks if not months for a service provider to turn up an end-to-end Ethernet connection, and these service contracts typically require 12-month commitments from end customers. Enabled by end-to-end automatic service provisioning and termination, the optical control plane changes the entire service paradigm by introducing new short-term, high-speed Ethernet services to meet the growing demand of dynamic applications like video conferencing and IPTV distribution applications. These services can be offered on a daily or even hourly basis.

With the intelligent optical layer, Carrier Ethernet services become far more dynamic and richer in available service parameters than previous static point-to-point Ethernet connections. The 2009 OIF Worldwide Interoperability Demonstration showcased end-to-end provisioning of dynamic switched Ethernet services over multiple, control-plane-enabled intelligent optical core networks through the use of OIF's implementation agreement of UNI 2.0 and E-NNI.

The adoption of ASON-empowered optical switched networks has also contributed to one of the most compelling applications for carrier-class EtherOptics. ASON-empowered optical switched networks incorporate automatic network topology discovery, the ability to dynamically provision end-to-end SONET/SDH circuits across a variety of topologies (ring, mesh, linear), and flexible ways to assign specific protection or restoration characteristics (e.g., UPSR/SNCP (Subnetwork Connection Protection), BLSR/MS-SPRing (Bidirectional Line-Switched Ring/Multiplex Section-shared Protection Ring), dynamic path restoration, dynamic span restoration, etc.) to each constituent circuit.

These two features (flexible topology and tiered protection schemes) enable service providers to choose the transport mechanism best suited to the Ethernet service requirement based on the SLA. For example, a best-effort service for Internet access can use a bandwidth-efficient mechanism such as dynamic path restoration on a mesh topology, whereas a high-availability, delay-guaranteed service can use a BLSR protected ring circuit.

The combination of Ethernet transport protocols and an intelligent, dynamic optical layer provides a robust foundation for next-generation Ethernet services while lowering overall transmission costs and simplifying operations. Ethernet packet intelligence, for example, tightly integrates the service and transport layers of the network, which simplifies EoS provisioning, increases security for network partitioning, and enhances the manageability and resiliency of Ethernet transport. High-density aggregation and intelligent packet-to-circuit mapping routes Ethernet traffic flows over SONET/SDH networks more efficiently and provides a scalable, cost-effective alternative to packet-over-SONET/SDH (PoS) interfaces to accommodate ever-increasing data traffic.

Mapping Ethernet traffic to virtual concatenation groups (VCGs) based on Layer 2 label information (VLAN tag, MPLS label) extends the traffic engineering features of packet networks to the optical domain (Figure 10.7). Combined with support for VCAT, GFP, and LCAS standards, and robust control plane and management software, enhanced provisioning, flow control, and protection mechanisms simplify the management and customization of Ethernet services.

Ethernet packet intelligence supports diverse applications, such as connecting remote offices to a centralized resource or larger corporate facility, disaster recovery, or any situation that requires fully distributed and protected services.

Backed by significant technology advances, carrier-class Ethernet services will continue to proliferate in quantity, range of data rates, and sophistication of defined

Ethernet Optical Transport Networks

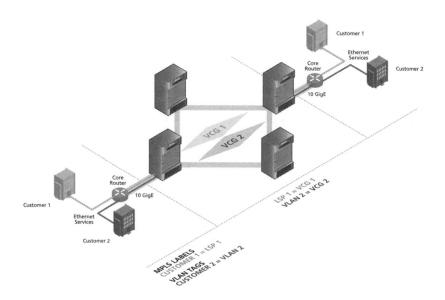

FIGURE 10.7 Ethernet packet intelligence.

service attributes. The critical relationship between Ethernet technology and the optical layer will continue to deepen and grow in importance as this market evolves. By combining the positive attributes of Ethernet with intelligent optical layer network elements, network operators will be well positioned to deliver a highly scalable, flexible, carrier-class transport infrastructure in a cost-optimized manner.

10.4 CARRIER ETHERNET FOR MOBILE BACKHAUL

Mobile networks have been experiencing exponential growth for the last decade due to advances of wireless technology and phenomenal penetration of mobile handsets in the consumer market. Not only are there almost 4 billion mobile phone subscribers worldwide today [3], but the bandwidth usage of each user is growing as mobile data applications are taking off with 3G and LTE (Long-Term Evolution) network deployments. All this growth is driving bigger and faster mobile backhaul networks that aim to support demanding mobile users. T1/E1 and ATM (Asynchronous Transfer Mode) connections are commonly used today for mobile backhaul solutions, and it is difficult and expensive to scale these TDM and ATM connections to support more backhaul capacity. Carrier Ethernet with its enhanced QoS, reliability, and service management is well posed as an economic mobile backhaul alternative to all mobile carriers.

10.4.1 CARRIER ETHERNET DEFINITION

Historically, Ethernet was designed and optimized for LAN deployment. It had huge success due to its low price, the abundance of components and devices, ease of installation, simplicity to operate, and low cost of ownership in general. However, unlike SONET/SDH connections, Ethernet was not intended to be a service provider interface that provides robust, manageable, and deterministic carrier-class services over WANs. For example, it does not have the performance monitoring capabilities in SONET/SDH networks that can track bit error rates closely and report to management modules when a user-configured threshold is crossed. When an Ethernet connection was lost, it might take hours to determine the root cause and recover the connection accordingly. All those deficiencies limited the potential of Ethernet beyond the LANs.

On the other hand, from the service provider perspective, the client service paradigm is shifting from TDM-based services to Ethernet-based services. More and more enterprises and end customers are willing to accept Ethernet client handouts instead of T1/E1, DS3, and STS-N/STM-M services for the simplicity and low cost of ownership of Ethernet. Delivering carrier-class Ethernet across metro and WANs poses a huge market opportunity, and standardization of Ethernet handout services becomes inevitable. Since 2001, the MEF has been focusing on defining Carrier Ethernet and promoting interoperability and deployment of Carrier Ethernet worldwide. Today, the MEF is made up of more than 150 organizations including telecommunication service providers, cable operators, MSOs (Multiple System Operators), network equipment vendors, testing equipment vendors, semiconductor vendors, networking labs, and software manufacturers [4].

The MEF has defined Carrier Ethernet as a ubiquitous, standardized, carrier-class service and network defined by five attributes that distinguish it from familiar LAN-based Ethernet:

1. *Standardized services:* Standardized point-to-point, point-to-multipoint, and multipoint-to-multipoint LAN services (Figure 10.8) that require no change to the existing LAN devices including TDM clients. Provide wide range of bandwidth and QoS options that facilitate building converged voice, video, and data networks.
2. *Scalability:* Ability to support millions of users with a diverse set of applications running at the same time. 1 Mbps to >10 Gbps bandwidth services in granular increments across metro and WANs over a wide range of transport network infrastructure from multiple service providers.
3. *Reliability:* Ability to recover from network failure automatically within a short period of time that could be as low as the well-known 50-msec SONET/SDH protection time. Provide high quality and availability of services that are comparable to the existing TDM services.
4. *QoS:* Provide a rich set of bandwidth and QoS options through SLAs provisioning that is based on Committed Information Rate (CIR), frame loss, delay, and delay variation characteristics.
5. *Service management:* Ability to rapidly provision end-to-end Ethernet services and provide full carrier-class OAM capabilities.

Carrier Ethernet is designed to be a commercial service offered by service providers and therefore needs to be associated with a clear definition of service parameters like legacy TDM services. Service availability, delay, jitter, and packet loss can all become part of SLAs. Historically, Ethernet is not a service interface but rather a networking solution for LANs and is short of meeting stringent SLAs. A new

Ethernet Optical Transport Networks

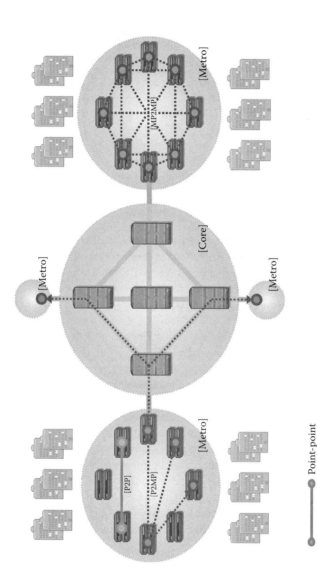

FIGURE 10.8 Carrier Ethernet connectivity.

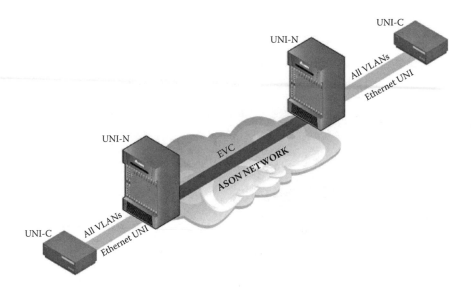

FIGURE 10.9 MEF UNI and EVC.

service architecture needs to be introduced to provide a clear demarcation point between user and network. The MEF specified two key service attributes for Carrier Ethernet services: the UNI and the Ethernet Virtual Connection (EVC). Coupled with the ITU-T ASON network architecture for intelligent optical transport, Carrier Ethernet can provide end-to-end Ethernet connections with well-defined service attributes (Figure 10.9).

The UNI is the interface that connects a subscriber to an Ethernet service provider. It consists of two network elements: the UNI-C element, client-side equipment that makes UNI requests and receives transport services from the network, and the UNI-N element, network-side equipment that takes UNI requests and finds a transport path that meets all requirements in the UNI request. The EVC associates two or more UNIs to provide Ethernet connections between subscriber client endpoints. There can be one or more subscriber Ethernet connections mapped to a single EVC. There are two fundamental rules governing the delivery of Ethernet frames over an EVC. First, an Ethernet frame must never be forwarded to the UNI where it is sourced; second, the entire Ethernet frame including MAC addresses needs to be transparent to EVC and kept unchanged. The MEF defined two types of EVCs: point-to-point and multipoint-to-multipoint. The two types of Ethernet connectivity are designed to provide point-to-point Ethernet Private Line (EPL) and any-to-any E-LAN services.

10.4.2 Carrier Ethernet for Mobile Backhaul: Motivations and Challenges

The mobile backhaul network connects mobile subscribers and wireless networks. It provides critical connections from remote cell towers to base station controllers (BSCs) for 2G mobile networks and to radio network controllers (RNCs) for 3G mobile networks. In a 2G mobile network, the primary application is voice, and a couple of T1/E1 lines with roughly 10 Mbp aggregated bandwidth from a cell power to a BSC should be sufficient. As broadband mobile handsets and applications take off with 3G networks, more backhaul capacity is required to carry mobile data applications like YouTube, video phone, wireless broadband to laptops, and so on. Wireless carriers are looking at hundreds of Mbps of bandwidth to reach a cell power. Many wireless carriers lease the last mile access lines from regional or local metro service providers, and it is very expensive to scale up the existing T1/E1 or ATM leased line services to meet the capacity needs. Keep in mind that wireless carriers are typically offering unlimited wireless data service to mobile subscribers for a flat monthly fee, and the increased broadband wireless data usage does not actually bring direct revenue to them. As a result, they need to be very cautious about how much to invest in the mobile backhaul capacity. For service providers who have access to a Carrier Ethernet network, mobile backhaul using Carrier Ethernet transport is an alternative solution because Ethernet is economical and more scalable than TDM connections, and in addition it is packet centric and more suitable to carry data traffic (Figure 10.10). The Ethernet optical transport technology outlined before offered various Carrier Ethernet access options to mobile backhaul networks; Ethernet over fiber, Ethernet over WDM, Ethernet over SONET/SDH, Ethernet over MPLS and MPLS-TP, and Ethernet over PBB-TE are all applicable depending on the location of the Ethernet access and the existing transport infrastructure. In addition, for mobile radio access networks, Carrier Ethernet can also be carried over copper cables, microwave radios, PDH (Plesiochronous Digital Hierarchy) connections like T1/E1/DS3 and Gigabit Passive Optical Network (GPON). Having listed all the possibilities, there are a number of technical challenges that Carrier Ethernet has to meet to achieve the current performance requirements of wireless backhaul networks.

10.4.2.1 Synchronization

Mobile cell sites require a very accurate reference clock to ensure licensed radio frequency accuracy to support a wide range of wireless handsets. Whereas some CDMA base station platforms include a global positioning system (GPS) clock, most of the installed base stations rely on an accurate (stratum-1) reference clock to be delivered from core networks through mobile backhaul connections. Failure to secure an accurate timing can result in dropping calls or adversely affecting the quality of voice and data services over wireless handsets. TDM connections like T1/E1 lines are designed to maximize delivery timing information over the transport channels, but this is not natural to Ethernet connections, which do not run over a synchronized network.

10.4.2.2 Delay and Jitter

Mobile backhaul connections have a very stringent delay requirement due to the nature of voice service and no-latency feeling of response time. A typical latency budget between a base station at cell sites and a BSC in the core is only a couple of milliseconds. Jitter is an unwanted variation brought to traffic due to transport distances and mechanism. The ITU-T G.823/G.824/G.825 defined E1/T1 jitter and

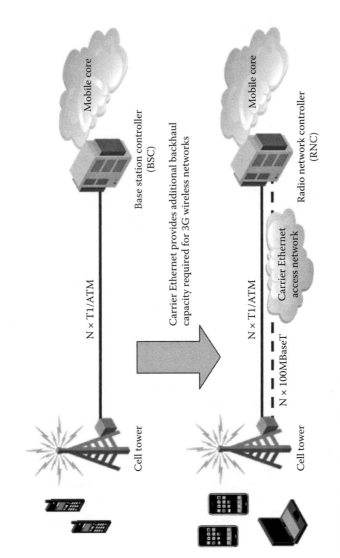

FIGURE 10.10 Carrier Ethernet for mobile backhaul.

wander impairments. Adhering to these standards not only ensures quality of service, but also avoids any significant upgrade of existing mobile backhaul infrastructure.

10.4.3.3 Reliability

The existing TDM-based mobile backhaul networks can achieve a "four nines" or "five nines" network availability budget as it is engineered to meet the performance of wire line SONET/SDH networks. It incorporates fast sub-50-msec protection mechanisms to keep service up during network failures. Ethernet, originated from the low-cost LAN architecture, was not initially designed to achieve high availability, but the Carrier Ethernet initiative is trying to change the perception of Ethernet services by making them more reliable and more suitable for wireless backhaul applications.

There are technologies and tools available to meet the preceding challenges for Ethernet mobile backhaul applications. As a result, Carrier Ethernet for mobile backhaul has been deployed in the United States as well as other parts of the world. To standardize these technologies and tools, the MEF has developed three technical specifications: MEF 3—Circuit Emulation over Ethernet Framework and Requirements, MEF 8—Circuit Emulation over Ethernet, and MEF 18—Abstract Test Suite for Emulated Circuits. Essentially, the circuit emulation over Ethernet definition provides a service interface carriers can rely on to draw TDM-like performance from Ethernet connections. This also facilitates the migration path as the existing base station and BSCs will continue to require TDM and ATM connectivity, and it is important to support legacy services with the new Ethernet backhaul solution. MEF is also working toward a Carrier Ethernet for Mobile Backhaul Interoperability Agreement aiming to specify a set of service parameters to be implemented in a standard manner to comply with the MEF Implementation Agreement. All these standards efforts will streamline the technology, facilitate interoperability among Ethernet system vendors, and make Carrier Ethernet more suitable for mobile backhaul applications.

REFERENCES

1. B. Metcalfe. (2008, April 24). Terabit Ethernet around 2015: Bob Metcalfe (Ethernet coinventor) gave a keynote speech, Toward Terabit Ethernet. [Online]. Available: http://www.netbigpicture.com
2. P. Wang, C. Chan, and P. Lin, "MAC address translation for enabling scalable virtual private LAN services," in *Proc. 21st International Conf. on Advanced Information Networks and Applications Workshop,* Vol. 1, 2007, pp. 870–875.
3. Worldwide mobile cellular subscribers to reach 4 billion mark late 2008. (2008, Sept. 25). [Online]. Available: http://www.cellular-news.com
4. About the MEF. (2009, Jan.). http://www.metroethernetforum.org

11 Optical Burst Switching
An Emerging Core Network Technology

Yuhua Chen and Pramode K. Verma

CONTENTS

11.1 Introduction 266
11.2 Optical Switching Technologies 267
    11.2.1 Optical Circuit Switching 267
        11.2.1.1 Optical Add/Drop Multiplexers 267
        11.2.1.2 Reconfigurable OADM 267
        11.2.1.3 Wavelength Router 268
    11.2.2 Optical Packet Switching 268
    11.2.3 Optical Burst Switching 269
11.3 Optical Burst Switching Fundamentals 269
    11.3.1 OBS Network Architecture 269
    11.3.2 OBS Signaling Protocols 270
    11.3.3 Burst Assembly 272
    11.3.4 OBS Router Architecture 273
11.4 Burst Scheduling 273
    11.4.1 Horizon Scheduling 273
    11.4.2 LAUC-VF 276
    11.4.3 Optimal Burst Scheduling 277
        11.4.3.1 Dual-Header OBS 277
        11.4.3.2 Practical Optimal Burst Scheduler 278
11.5 Supporting Quality-of-Service in OBS Networks 278
    11.5.1 Offset-Based QoS 278
    11.5.2 Proportional QoS 279
    11.5.3 Burst Early Dropping 279
    11.5.4 Wavelength Grouping 279
    11.5.5 Look-Ahead Window 280
    11.5.6 Contour-Based Priority 280
11.6 Security Considerations in OBS Networks 281
    11.6.1 Security Concerns 281
        11.6.1.1 Orphan Burst 281
        11.6.1.2 Malicious Burst Headers 282
        11.6.1.3 Network Security versus End-to-End Security 283

11.6.2 Embedded Security Framework in OBS Networks...........................283
11.6.3 OBS Secure Router Architecture..284
11.7 OBS in Metropolitan Area Networks 287
11.8 Conclusion 288
References..289

11.1 INTRODUCTION

Optical fibers and dense wavelength division multiplexing (DWDM) techniques are already widely deployed in commercial telecommunication networks. In addition, there is a continuing need to expand the optical technology toward the customer or business premises that are the sources and sinks of information. DWDM technology allows several hundred wavelength channels to be transmitted over a single optical fiber at a rate of 10 Gb/s per channel and beyond. This can lead to data rates reaching 10 Tb/s in each individual fiber.

Although DWDM is universally used in transmission, different switching technologies can be used to direct input data to the desired outputs at router nodes. Current switching technologies can be characterized into *electronic switching* and *optical switching* technologies based on how data is processed inside the router. Electronic switching converts DWDM optical signals to electronic signals using optical-to-electrical (O/E) conversion, and directs them electronically. The directed information is converted back to optical signals using electrical-to-optical (E/O) conversion, and sent on DWDM links leading toward the downstream routers. As the number of DWDM channels increases, O/E/O conversion required by electronic switching adds to the system cost significantly. For example, although it is technologically feasible to carry 512 wavelengths in a single optical fiber, it requires 512 O/E/O pairs in electronic routers just to terminate a single DWDM link.

The next-generation switching technology needs to be able to scale cost effectively to support a large number of wavelength channels. Optical switching technologies allow DWDM channels to directly pass the core router node without O/E/O conversion, greatly reducing the cost to deploy DWDM channels over existing network infrastructure.

Optical switching can be further divided into optical circuit switching, optical packet switching, and optical burst switching. *Optical circuit switching* makes switching decisions at the wavelength level and passes data through routers in pre-established lightpaths. Because switching is done at the wavelength level, this technology suffers from coarse granularity. *Optical packet switching* [1–3] can switch data at the packet level optically and has a fine granularity. However, optical packet switching is unlikely to be commercially available in the foreseeable future, largely due to the synchronization issues associated with the packet header and the payload and the lack of random access optical buffers. *Optical burst switching* (OBS) [4, 5] provides a granularity between optical circuit switching and optical packet switching. In many ways, it combines the benefit of both while obviating their shortcomings. Currently, OBS is considered the most promising optical switching technology [6, 7].

In this chapter, we describe the OBS technology in detail and position it as a viable and emerging core switching technology. We provide a brief overview of optical switching technologies. OBS fundamentals are then described. We also present an OBS security framework. Following that, OBS in metropolitan area networks is discussed.

11.2 OPTICAL SWITCHING TECHNOLOGIES

11.2.1 Optical Circuit Switching

Optical circuit switching is realized in practice in various forms such as Optical Add/Drop Multiplexers (OADMs), Reconfigurable Optical Add/Drop Multiplexers (ROADMs), and wavelength routers. This section briefly reviews these technologies.

11.2.1.1 Optical Add/Drop Multiplexers

The OADMs are the most widely used form of optical switching today. Such devices allow one or more wavelengths to be added or dropped at a node. In an OADM node, DWDM signals from the incoming fiber are demultiplexed into separate wavelengths. A small number of wavelengths are dropped at the node and the majority of the wavelengths bypass the node optically. The dropped wavelengths become available to the node to insert local traffic. To add wavelengths at the node, data to be inserted is carried in the optical wavelengths corresponding to the dropped wavelengths. The added wavelengths are combined with the bypassed wavelengths using an optical multiplexer onto the outgoing optical fiber. In OADMs, the wavelengths to be dropped or added at a node are fixed.

11.2.1.2 Reconfigurable OADM

ROADMs are reconfigurable versions of OADMs where the wavelengths to be added or dropped can be reconfigured, resulting in operational flexibility. In metropolitan area networks where OADMs have played a pivotal role, there is often a need to reprovision the network to address evolving customer needs. Setting up an OADM is often a complex and time-consuming task that cannot effectively function in a changing business environment. ROADMs give service providers the flexibility to reconfigure the wavelengths based on changing demand.

The reconfigurability of ROADMs can be provided in several ways. For example, each of the incoming wavelengths separated by a demultiplexer passes through an optical switch that could either divert the wavelength to the dropped wavelength bundle or retain it as part of the wavelengths that bypass the node. The setting of optical switches can be reconfigured to allow a different wavelength to be dropped. A ROADM can also be implemented using a tuning-based architecture, where a wavelength to be dropped or added can be selected by a wavelength tunable device such as a Bragg grating or a thin film filter. The ROADM allows different wavelengths to be added or dropped by reconfiguring the node and provides flexibility of handling varying network demands. Readers can find additional details in [8].

OADMs and ROADMs can reduce the cost of O/E/O conversion compared to electronic switching when only a few wavelengths need to be added or dropped at a node as most of the wavelengths continue onto the outgoing fiber. However, because OADMs or ROADMs only have the drop, add, and bypass options, they can only be used to construct linear or ring network topologies and are not suitable for building mesh network architecture.

11.2.1.3 Wavelength Router

A *wavelength router* can switch wavelengths among multiple incoming and outgoing fibers and can be used to build mesh networks. Wavelength routers can be either nonconfigurable or reconfigurable. In a nonconfigurable wavelength router, each incoming wavelength is routed to a fixed outgoing fiber. An incoming wavelength in a reconfigurable wavelength router can be reconfigured to be routed to a different outgoing fiber.

The wavelengths in each of the incoming fibers are first demultiplexed into separated wavelengths. In a nonconfigurable wavelength router, each individual wavelength is then connected to the respective optical multiplexer associated with the outgoing fiber depending on where that particular wavelength is required to be routed to. Routing of incoming wavelengths in a nonconfigurable wavelength router cannot be changed. A reconfigurable wavelength router uses space division optical switching fabric between the demultiplexers and multiplexers to direct any of the incoming wavelengths to any of the multiplexers based on the optical switching fabric configuration. Routing is determined by both the incoming wavelength and the configuration of the optical switching fabric. Reconfigurable wavelength routers are more flexible than nonconfigurable wavelength routers, as an incoming wavelength can be routed to a different fiber by reconfiguring the optical switching fabric.

Wavelength routers present a coarse level of granularity as routing is performed at the wavelength level, typically 10 Gb/s and up. The use of wavelength routers is largely restricted to core networks where optical connections are provisioned to last for months or years.

11.2.2 Optical Packet Switching

Optical packet switching makes switching decisions on a packet-by-packet basis and has the finest granularity among all forms of optical switching. In optical packet switching, the packet payload is buffered optically and the packet header is converted to electronic signals and processed electronically at each node.

When the packet header is being processed, the payload needs to be stored optically. In addition, when two packets contend for an output, one of them needs to be buffered optically. Both factors—storing optical payloads during header processing and buffering optical packets for contention resolution—require optical buffers. Unfortunately, random access optical buffers do not exist. The only way to provide limited delays in the optical domain is through *fiber delay lines* (FDLs). Time delays to the optical payload are achieved either through the use of FDLs of varying lengths or through the use of a fiber delay loop where the optical data to be stored circulate through the loop a number of times. Unfortunately, the lack of optical random access memory severely degrades the performance of optical packet switching. In addition, implementing FDLs at the scale needed is not practical. For example, a 1-microsecond delay would require over 1,000 feet of FDLs.

The synchronization between the packet header and its payload is another serious problem in optical packet switching. After the packet header is processed, it is converted to optical signals and needs to be combined with the optical payload. Synchronizing the header and the payload is a challenging task that has not been yet accomplished beyond research laboratories.

In short, due to the lack of random access optical buffers and the stringent synchronization requirements between the packet header and optical payload, optical packet switching is likely to remain in academic research laboratories, and its commercialization is probably decades away.

11.2.3 Optical Burst Switching

OBS is an emerging core switching technology that allows variable-size data bursts to be transported over DWDM links. To reduce the switching overhead at the core nodes, the ingress edge node aggregates packets from a number of sources destined for the same egress node dynamically to form a large data burst. An optical path is created on the fly to allow data bursts to stay in the optical domain and pass through the core routers transparently. It is achieved by launching a burst header on a separate control channel ahead of the data burst and setting up the optical path prior to the arrival of the data burst.

OBS effectively addresses the limitations of optical circuit and packet-switching technologies. Because the optical path in a core router is set up before the burst arrival, there is no need for optical buffers in OBS networks. In addition, there is no stringent synchronization requirement as burst headers and data bursts are sent on separate DWDM channels. OBS has received a considerable amount of attention within the last several years. It is likely that the first commercial realization of an all-optical core network will be based on the OBS technology due to its scalability and ability to handle dynamic traffic. The rest of the chapter focuses on the details of the OBS technology.

11.3 OPTICAL BURST SWITCHING FUNDAMENTALS

11.3.1 OBS Network Architecture

OBS is a core switching technology that can scale cost effectively to support a large number of DWDM channels. Figure 11.1 illustrates a core OBS network that consists of a set of OBS edge routers and core routers connected by DWDM links. At least one of the wavelength channels in a DWDM link is reserved as the *control channel* to carry burst headers. The rest of the channels are used as data channels to carry data bursts. OBS edge routers can interface with different types of networks such as Internet Protocol (IP) networks, Gigabit/10 Gigabit Ethernet, and wireless networks. Packets with the same egress edge router address are assembled into *data bursts* at ingress edge routers and are disassembled back to packets at the egress edge routers. QoS can be supported by assembling packets into bursts according to priorities.

Once bursts are formed at the ingress router, *burst headers,* also referred as *burst header cells* (BHC) [4, 9], or *burst header packets* (BHPs) [10], are generated and

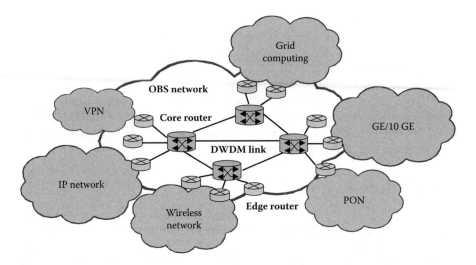

FIGURE 11.1 Optical burst switching network architecture (Copyright © IEEE 2008).

sent on a separate control channel ahead of the data bursts to set up optical paths. Data bursts are then transmitted on the data channels, usually without waiting for the acknowledgment. Once launched, the data burst traverses the OBS core network as an optical entity transparently without encountering O/E/O conversion. Once the burst passes through a core router, the optical path within the router will be released immediately, which allows the same wavelength to be used by other bursts.

In OBS networks, each intermediate core router only creates a transient optical path for the duration of the burst when the burst is actually passing through that particular router. Note that the burst assembly and disassembly functionality is only provided by the OBS edge routers. There is no burst reassembly in the OBS core network.

11.3.2 OBS Signaling Protocols

The OBS signaling protocol can be represented by the relationships between burst headers and their corresponding bursts using a simple timing diagram illustrated in Figure 11.2. The burst header specifies the *offset* time between the header and the burst, and the *length* (time duration) of the bursts. In addition, the burst header also carries the routing information and the wavelength on which the burst is to be launched, and possibly the QoS parameters. After a burst header is sent on the control channel, the corresponding data burst will wait for an offset time before being transmitted on one of the data channels.

At each OBS core router node, the burst header is converted to electronic signals and processed electronically. The OBS core router uses the offset and the length fields in the burst header to calculate the burst arrival time and makes a wavelength reservation for the burst accordingly. Once an outgoing wavelength is selected, the OBS router will update the offset time and the wavelength field in the burst header before forwarding the burst header to the downstream router. The OBS core router will configure the optical switching fabric before the arrival time of the burst to

Optical Burst Switching

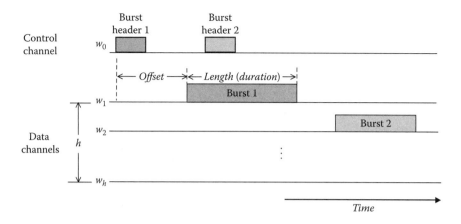

FIGURE 11.2 Burst and burst header timing relationships in OBS signaling protocol (Copyright © IEEE 2008).

direct the burst to the destined output port. The offset field in the burst header is adjusted at every hop, reflecting the header processing time at each individual core router node. If the offset time reduces to zero (or negative value) before the burst header reaches the egress edge router, no optical path will be set up at the core router when the burst arrives, and the burst will be lost.

The OBS signaling protocol just described was independently developed in [9] and [4] and, in principle, resembles the well-known Just-Enough-Time (JET) protocol [9]. Both protocols allow the OBS core routers to use delayed reservations by calculating the projected future burst arrival time and the burst release time using the information carried in the burst header. As a result, the OBS core router can manage the wavelength resources efficiently by reserving the wavelength only for the duration of the burst.

Although the JET-like protocol is the prevailing protocol used for OBS, the other notable OBS signal protocol is Just-In-Time (JIT) [12, 13]. In JIT, a setup signaling message is sent on a dedicated signaling path to set up the optical path. The core router reserves the wavelength when the setup message is received. After a short time, the actual data is transmitted on the data path, without waiting for the acknowledgment message. When the data is completely transmitted, a release message is sent to tear down the established optical path. Compared to JET, JIT reserves the channel bandwidth when a setup message is received, not when the data burst is to arrive, and, therefore, is less efficient in terms of bandwidth usage.

It was originally believed that the JIT protocol with instant reservations would be easier to implement than the JET protocol with delayed reservations. It has been demonstrated that several channel scheduling algorithms based on the JET-like protocols are very efficient. For example, *horizon scheduling* [4, 9], which makes burst reservations using the projected burst arrival and finish times, is extremely suitable for efficient implementation in hardware [14] and can achieve one burst reservation every two clock cycles, regardless of the total number of wavelength channels [15]. In addition, *optimal burst scheduling,* which overcomes the inefficiency of horizon scheduling, can also be realized in hardware with O(1) constant runtime complexity [14]. Therefore, the JET-like signaling protocol is the protocol of choice for OBS deployment.

11.3.3 BURST ASSEMBLY

The OBS ingress edge router aggregates packets into bursts before forwarding them to the core network. This process is called *burst assembly*. Bursts are assembled based on the destination edge routers. In a mixed burst assembly scheme, a burst is formed when either of the following conditions is met: (1) the size of the assembling burst reaches the burst length threshold, or (2) the timer expires. The timer is triggered by the arrival of the first packet in a burst. When the timer expires, the burst is formed even though it might not yet have reached the burst length threshold. The timeout value restricts the burst assembly latency and is especially useful under light traffic loads. Bursts usually reach the burst length threshold before the timer expires under heavy traffic load. Although the threshold-based burst assembly policy and the timer-based burst assembly policy have been discussed, the mixed burst assembly scheme described earlier works well under both light and heavy traffic loads and is easy to implement. Therefore, the mixed burst assembly scheme should be considered the default burst assembly scheme in OBS networks.

The burst length threshold and the timeout value in the aforementioned mixed burst assembly can be either *fixed* (*static burst assembly*) or *adaptive* (*dynamic burst assembly*). A dynamic burst assembly can adjust either the timeout value or the burst length threshold or both based on dynamic traffic conditions. Although a static burst assembly is easier to implement, a dynamic burst assembly can potentially bring additional performance benefits. Burst assembly algorithms are intrinsically distributed algorithms that are independently executed at individual edge router nodes. Each OBS ingress edge router has the freedom to choose a specific burst assembly algorithm to meet its own cost and performance objectives, subject to some networkwide parameter constraints.

To support QoS in OBS networks, additional criteria can be utilized in burst assembly. For example, priority burst assembly assembles bursts based on priorities. There are two major priority burst assembly schemes: (1) assembling packets with the same priority into a burst, and (2) assembling packets with different priorities into a burst. In the first scheme, each burst contains packets of the same priority. When there is a contention at the core router where overlapping bursts compete for an outgoing wavelength, the OBS core routers can make burst dropping decisions based on burst priorities. In the second scheme, packets with different priorities are aggregated separately before being assembled into bursts. When the burst is assembled, packets with the lowest priority are placed at the head of the burst and the packets with the highest priority are placed at the tail end of the burst. When contention occurs at the core router, the low-priority packets at the head of the contending burst can be dropped while the high-priority packets at the end of the burst are preserved, resulting in lower burst loss probability for higher priority packets [16, 17]. It is worth noting that although each ingress router can choose or adjust the basic burst assembly parameters such as the burst length threshold and the timeout value independently, a particular priority burst assembly scheme is meaningful only if the OBS core routers support such priority scheduling.

11.3.4 OBS ROUTER ARCHITECTURE

OBS routers are crucial to OBS deployment. In this section, we explain the practical realizations of the OBS edge router and the core router. Figure 11.3 illustrates the architecture of an OBS edge router. In the ingress direction, packets received on different line interfaces such as IP and Gigabit Ethernet (GE)/10 GE are sent to the *burst assembler*. The burst assembler classifies the data according to its destination and possibly the QoS levels and assembles data into different bursts. Once a burst is formed, the burst assembler generates a burst header, which is transmitted on the control channel. After holding the burst for an offset time, the burst assembler releases the data burst to be transmitted on one of the data channels. The control channel and the data channels are combined onto the outgoing DWDM link using a passive *optical multiplexer* (MUX). The outgoing DWDM link is connected to the OBS core router. In the egress direction, the wavelengths on the incoming DWDM link are separated using an optical demultiplexer (DEMUX). The burst headers received on the control channel and the data bursts received on data channels are forwarded to the *burst disassembler,* which converts bursts back to packets and forwards them to the appropriate line interfaces.

The architecture of an OBS core router is illustrated in Figure 11.4. The OBS core router consists of an optical data path and an electronic control path. The control channel on each DWDM link is tapped off at the core router and converted to electronic signals. The burst header sent on the control channel is processed electronically by the burst header processing unit. Based on the information carried in the burst header, the burst header processing unit selects an outgoing wavelength for the upcoming burst and configures the optical interconnects to allow the data burst to pass to the desired outgoing link optically. The burst scheduling algorithms and the QoS scheduling algorithms described later in this chapter are implemented in the burst header processing unit.

11.4 BURST SCHEDULING

In OBS networks, OBS core routers set up optical paths on the fly, based on the information carried in burst headers. This requires an online burst scheduling algorithm that can support a large number of wavelength channels. Major burst scheduling algorithms are discussed next.

11.4.1 HORIZON SCHEDULING

Horizon scheduling [4, 9] was the first burst scheduling algorithm to support delayed reservations in OBS networks and is very cost effective in terms of implementation. In horizon scheduling, the scheduler maintains a time horizon for each of the wavelength channels of an outgoing link. The horizon is defined as the earliest time after which there is no reservation made on that wavelength. When a burst header arrives

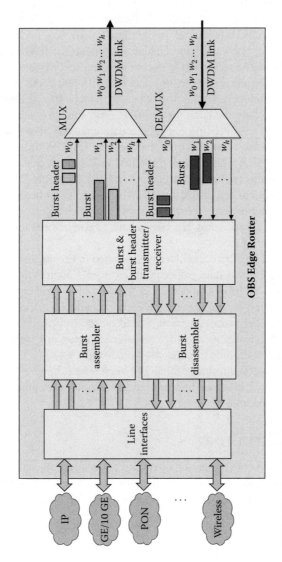

FIGURE 11.3 OBS edge router architecture.

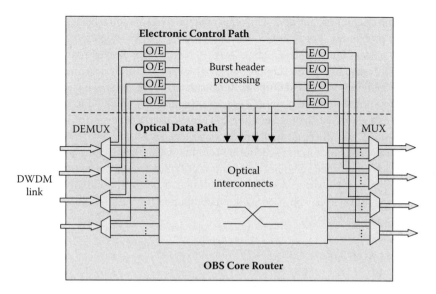

FIGURE 11.4 OBS core router architecture.

at the OBS core router, the horizon scheduler first calculates the projected burst arrival time based on the time the burst header is received, as well as the offset time carried in the burst header. The horizon scheduler selects the channel with the latest horizon that is earlier than the projected arrival time of the burst, if such a channel can be found. Otherwise, the burst scheduling request is rejected. After a burst is successfully scheduled, the ending time of the scheduled burst becomes the new horizon of the channel. An example of horizon scheduling is shown in Figure 11.5.

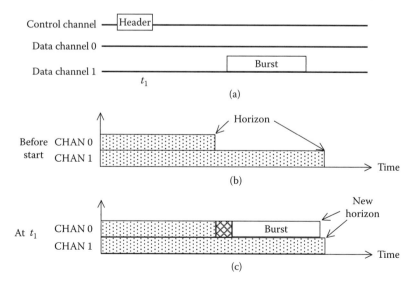

FIGURE 11.5 Example of horizon scheduling.

Horizon scheduling can work with other contention resolution schemes to reduce the burst loss probability. For example, deflection routing can be supported by forwarding the burst header of a failed reservation to a different port for another round of scheduling. Horizon scheduling can support burst segmentation [16] by selecting the channel with the smallest horizon and dropping the head part of the burst, which overlaps with the horizon of the selected channel, when no channel can be found to accommodate the burst as a whole. Horizon scheduling can also be used to manage the lane groups in *multilane OBS* [18] where no wavelength converters need to be deployed.

The drawbacks of horizon scheduling are that it only keeps track of a single status for each wavelength channel and cannot utilize the gaps created by previously scheduled bursts. However, horizon scheduling can perform optimally under some conditions [14] or it can be used as a subcomponent in building an optimal burst scheduler, described later.

11.4.2 LAUC-VF

The latest available unscheduled channel with void filling (LAUC-VF) [10] makes use of the voids (gaps) between the successive bursts scheduled on the channel to improve bandwidth efficiency. The LAUC-VF algorithm keeps track of the voids on each of the wavelength channels and tries to schedule a burst in one of the voids whenever possible. If more than one void can fit a burst, the wavelength channel with the latest available unused time before the arrival time of the burst is selected. Figure 11.6 illustrates how LAUC-VF works.

Because LAUC-VF utilizes the voids created by previously scheduled bursts, link utilization of LAUC-VF is in general higher than that of horizon scheduling. However, LAUC-VF has high computational complexity as the scheduler needs to

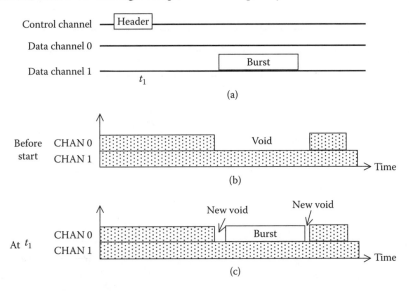

FIGURE 11.6 Example of LAUC-VF.

keep track of every void on each of the wavelength channels and identify a suitable void to accommodate the burst. The complexity of LAUC-VF is $O(m)$, where m is the total number of voids on all wavelengths. In addition, despite the ability to utilize the voids, LAUC-VF is not an optimal scheduling algorithm. Under certain burst header arrival sequences, LAUC-VF can result in burst loss, whereas an optimal burst scheduler would schedule all bursts successfully [19].

The *Minimum Starting Void* (Min-SV) [20] is a more efficient implementation of LAUC-VF. The Min-SV algorithm uses a geometric approach by organizing the voids into a balanced binary search tree and finding a void that minimizes the distance between the starting time of the void and the starting time of the burst. The complexity of Min-SV algorithm is $O(\log m)$, which is a significant improvement over LAUC-VF. However, Min-SV has a large memory access overhead for each burst scheduling request.

11.4.3 OPTIMAL BURST SCHEDULING

The aforementioned LAUC-VF algorithm increases bandwidth utilization by filling the voids on the channel. However, as explained earlier, even with void filling, LAUC-VF might not be able to produce optimal burst schedules. In fact, the voids are caused by the variable offset times between the burst headers and bursts. It turns out that if the variability in the offset can be removed, optimal burst scheduling can be achieved without the need for void filling. Such observations and corresponding treatments were first discussed in [9], and later in [19], and were independently approached as dual-header OBS in [21]. An $O(1)$ constant runtime optimal burst scheduler was described in [14].

11.4.3.1 Dual-Header OBS

Dual-header OBS [21] is a signaling protocol that removes the variability in the offset to achieve optimal scheduling. Dual-header OBS decouples the resource request from the resource reservation process by using two different types of control packets for each burst: the *service request packet* (SRP) and the *resource allocation packet* (RAP). The SRP is similar to a traditional burst header that contains the routing information, the offset time, and the length of the burst and is initiated by the OBS ingress edge router. The only difference between the SRP and a traditional burst header is that the SRP does not contain information about the physical wavelength on which the burst is to arrive. The OBS core router, on receiving the SRP, records the burst information carried in the packet and forwards the SRP immediately. This process is continued until the SRP reaches the egress edge router.

The wavelength reservation in dual-header OBS is made at a *functional offset* time (fixed for all burst scheduling requests) ahead of the burst arrival time, regardless of the initial offset carried in the SRP. After the wavelength reservation is made, a RAP is initiated by the core router to send the wavelength selection information to its adjacent downstream router. The downstream router will use the incoming wavelength information in the RAP to configure the optical switching fabric. Dual-header OBS alleviates the offset variation associated with different bursts and can achieve optimal scheduling in wavelength reservation.

11.4.3.2 Practical Optimal Burst Scheduler

Horizon scheduling can operate at a very high speed. This is especially important in a practical OBS router implementation. Unfortunately, horizon scheduling cannot utilize bandwidth efficiently if the variation in the offset is large. It was explained earlier that if the offset is fixed, optimal burst scheduling can be achieved using a simple horizon scheduler. A practical optimal burst scheduler called the *Constant Time Burst Resequencing* (CTBR) scheduler, presented in [14], can produce optimal burst schedules at a speed comparable to the horizon scheduler.

The idea behind the CTBR scheduler is that rather than processing bursts as soon as the burst headers arrive, the burst headers can be delayed and processed in the order of the expected burst arrival times. This can be achieved by passing the burst headers through a burst resequencer and holding them there until a fixed time unit before the expected burst arrive time. Once the burst headers are resequenced, they are processed by a horizon scheduler. To achieve an overall O(1) runtime complexity, the practical optimal burst scheduler uses a *timing wheel* structure, which allows the burst header to be appended to the timing wheel slot corresponding to the expected arrival time of the burst in a single list operation. The burst headers that are resequenced can be removed from the timing wheel and sent to the horizon scheduler for burst scheduling.

The CTBR scheduler described above was initially designed to work with OBS core routers with input FDLs. In fact, the CTBR scheduler can be extended to the traditional router architecture with no FDLs by incorporating dual-header OBS signaling. This can be achieved by placing a copy of the burst header in the timing wheel and releasing the original burst header immediately. After the burst is scheduled, a burst configuration packet carrying the selected wavelength is generated and sent to the adjacent downstream router. The CTBR scheduler with dual-header signaling will provide an optimal and practical solution to OBS deployment.

11.5 SUPPORTING QUALITY-OF-SERVICE IN OBS NETWORKS

11.5.1 OFFSET-BASED QOS

Offset-based QoS [22] was the first QoS scheme proposed for OBS networks. In the offset-based QoS scheme, higher priority bursts are assigned a larger offset time. More specifically, the burst header of a high-priority burst is launched ahead of the burst header of a low-priority burst, supposing the bursts of both priorities are to be transmitted at the same time. Because the burst header of the high-priority burst arrives at the OBS core router ahead of the low-priority one, it has a better chance to reserve a wavelength successfully.

Although the offset-based QoS scheme has lower burst loss probability for high-priority bursts, it has the following shortcomings. First, offset-based QoS can only be supported by burst scheduling algorithms with void-filling capability. As mentioned earlier, void-filling algorithms such as LAUC-VF suffer from high computational complexity and are not optimal. Second, because a larger offset time is associated with high-priority bursts, offset-based QoS increases the end-to-end latency for high-priority bursts. In addition, it has been shown that offset-based QoS favors bursts with shorter lengths [23] as shorter bursts are more likely to fit in the voids.

11.5.2 Proportional QoS

A proportional differentiation service model provides quantitative QoS differentiation between service classes. In proportional QoS, a particular QoS metric can be quantitatively adjusted to be proportional to some service differentiation factors. Suppose q_i and s_i are the performance metric and the differentiation factor for priority class i, respectively. The proportional differentiation model requires the following relationship to be satisfied for all pairs of service classes:

$$\frac{q_i}{q_j} = \frac{s_i}{s_j} \quad (i, j = 0, \cdots, N).$$

To support proportional QoS in OBS networks, the Intentional Burst Dropping scheme [23] keeps track of the proportional burst loss probabilities among different priorities. A low-priority burst is intentionally dropped if the proportional differentiation relationship just described is violated. By intentionally dropping low-priority bursts, more outgoing link capacity is freed up for high-priority bursts to be scheduled successfully. This quantitative approach provides the network operators with better controllability over different service classes. However, this scheme encounters higher overall burst loss probability due to intentional dropping.

11.5.3 Burst Early Dropping

The burst early dropping scheme [24] is a probability-based mechanism in which low-priority bursts are intentionally dropped with some probability to avoid potential contention with high-priority bursts. Each OBS core router node measures the burst loss probability for each priority class over a fixed window. Burst early dropping is triggered when the measured burst loss probability of the high-priority class exceeds some predefined burst loss threshold. At this time, the core router computes an early dropping probability for the low-priority class and starts to drop the low-priority class bursts in a probabilistic matter to meet the worst-case burst loss guarantee for the high-priority class. Because low-priority bursts might be dropped even when there is no contention, the early dropping approach also experiences high overall burst loss probability.

11.5.4 Wavelength Grouping

Absolute service differentiation provides worst-case quantitative service guarantees for each priority class. *Wavelength grouping* [24] provides service differentiation in OBS networks by guaranteeing the worst-case burst loss probability for each service class. In wavelength grouping, the wavelengths at the outgoing link are divided into wavelength groups. Each priority is limited to use no more than a maximum number of wavelengths computed for that priority class based on the target burst loss probability. There are two ways to assign wavelength groups. In *static wavelength grouping* (SWG), a fixed wavelength group is assigned to each priority class. In *dynamic wavelength grouping* (DWG), bursts are allowed to use wavelengths dynamically, as long as the total number of wavelengths occupied by a particular priority class does not exceed the predefined value.

Wavelength grouping guarantees the predefined burst loss probability for each priority class at the expense of elevated overall burst loss probability, as statistical multiplexing performance of a large number of channels is much better than that of a collection from a small number of channels.

11.5.5 LOOK-AHEAD WINDOW

The *look-ahead window* (LaW) [25, 26] resolves burst contention by constructing a window of W time units and makes collective decisions on which burst should be dropped. The LaW allows the core router to collect burst information from multiple burst headers before making a scheduling decision. This is possible because of the offset time between the burst header and the burst. Because the burst header needs to be delayed by W time units, the offset time has to be larger than W. If the original offset time is to be maintained, FDLs will be used at each core router node to delay data bursts by W time units. The look-ahead contention resolution algorithm has three phases: (1) collecting burst headers for an output port and creating an LaW of size W, (2) determining the contention regions, and (3) applying heuristics to select the contending bursts within the window to be dropped.

The contention is resolved by constructing an auxiliary directed graph representing all bursts in the LaW. The start time and the end time of each burst in the window are represented as vertices in the graph with a directed edge between them. By assigning different weights to the edges in the directed graph, service differentiation can be realized. A set of zero-weight directed edges between adjacent nodes are added under certain conditions to make the directed graph connected, before shortest-path algorithms such as Bellman-Ford can be used to identify the bursts to be dropped.

Unlike some previously mentioned QoS algorithms that intentionally drop bursts, LaW only drops a burst when contention is identified. However, the algorithm has high computational complexity as the complexity of Bellman-Ford algorithm is $\Theta(|V| \bullet |E|)$, where the number of vertices $|V|$ is twice the number of bursts in the LaW, and the number of edges $|E|$ is at least equal to the number of bursts in the window due to the added zero-weight directed edges.

11.5.6 CONTOUR-BASED PRIORITY

The *contour-based priority* (CBP) algorithm [27] achieves priority scheduling by maintaining a contour of each priority and discarding a low-priority burst only if scheduling such a burst will result in a high-priority burst being discarded. A contour function $C_i(t)$, shown in Figure 11.7, represents the number of pending bursts of priority i. *Pending bursts* are the bursts with burst headers that have arrived but have not been scheduled. A rising edge in the contour corresponds to the start of a burst, and a falling edge in the contour corresponds to the end of a burst. Because of the unique properties of the contour as discussed in [27], a complete contour can be represented by a set of contour lists, which can be maintained in a set of First-In First-Out (FIFO) buffers.

Optical Burst Switching

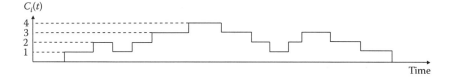

FIGURE 11.7 Contour of priority i (Copyright © IEEE 2007).

There are three operations associated with the CBP algorithm: inserting a burst into the contour, removing a burst from the contour, and scheduling a burst. Adding and removing a burst to or from the contour is a single FIFO write/read operation. During the scheduling phase, the scheduler only needs to access the first element in the contour lists. Therefore, the complexity of scheduling a burst is independent of the number of pending bursts in the contour. Because the variation in offset is automatically removed at the time of scheduling, a simple horizon-based scheduler can be used to produce optimal burst schedules. In other words, the CBP algorithm inherits the optimal burst scheduling properties described earlier. The CBP algorithm has $O(1)$ runtime complexity and is suitable for practical realization of priority scheduling in OBS networks.

11.6 SECURITY CONSIDERATIONS IN OBS NETWORKS

Network security is a growing concern in all applications. There are several aspects of security, but from a networking point of view, integrity, confidentiality, and authentication are potentially the most important [28]. *Integrity* would imply that the data entrusted to the network at the point of ingress has not been modified in any way before it is delivered to its destination. *Confidentiality* would mean that any exposed data cannot be meaningfully interpreted by an adversary. *Authentication* would imply that the sender of the data is verified for the receiver by the network. Because an OBS network provides an environment where its boundaries are well defined, a security umbrella over an OBS network can be effectively created at a relatively low cost. Implementing a security framework within the OBS boundary is also desirable as OBS networks are vulnerable due to their characteristics. The following sections discuss the security issues in OBS networks and present means of implementing an embedded OBS security framework (Copyright © IEEE 2008) [29, 30].

11.6.1 SECURITY CONCERNS

Potential security threats that are unique in OBS networks are presented here and the need to bring security measures to OBS networks is discussed. The examples illustrated here are not intended to be comprehensive.

11.6.1.1 Orphan Burst

In OBS networks, each valid burst is associated with a burst header, which is sent ahead of the data burst on a separate control channel. The burst header carries the control information and is responsible for making the wavelength reservation for its

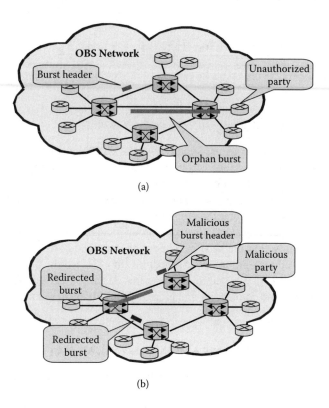

FIGURE 11.8 (a) Example of orphan burst. (b) Example of malicious burst header and redirected burst (Copyright © IEEE 2008).

corresponding burst. If the scheduling request is rejected at one of the OBS core routers, there will be no valid optical path set up for the arriving burst. Because the burst has been launched, it is going to arrive at the input of the core router in any case. At this point, the burst is no longer connected with its burst header and becomes an *orphan burst* as illustrated in Figure 11.8a. Depending on the configuration of the switching fabric at the time of the burst arrival, the orphan burst can take some unpredictable path and reach some unpredictable destination. As a result, orphan data bursts can be tapped off by some undesirable party, compromising their security.

11.6.1.2 Malicious Burst Headers

An active attack can be launched by injecting malicious burst headers into the OBS network. In an OBS network, the data burst bears no routing intelligence to the destination router and will follow the optical path set up by its associated burst header. If a burst header is injected into the network by a malicious party at an appropriate time, an optical burst can be misdirected to an unauthorized router. Because the OBS routers have no way of telling the authenticity of the burst headers, any active data that appears on the input channels can be misdirected. Figure 11.8b shows security compromises caused by a malicious burst header masquerading [28] as a legitimate one.

Optical Burst Switching

11.6.1.3 Network Security versus End-to-End Security

The preceding scenarios show the security vulnerability of the OBS networks. Other attacks are possible. For example, a *replay attack* [28] can be made using a legitimate but expired burst. It is, of course, entirely possible that part or all of the user data that formed the lost or misdirected data burst was source encrypted for decryption by the receiving user or user agent. In this case, its capture by an undesirable party would not compromise the confidentiality of communication. However, for several applications, end-to-end encryption might not be feasible or desirable. The all-optical part of the network in that case bears responsibility for managing security of the customer data. A network-based scheme for securing data within the OBS network can be efficiently implemented by embedding security as part of the OBS native network architecture, as opposed to a layer on top of it.

11.6.2 Embedded Security Framework in OBS Networks

An OBS security framework will provide the following services: (1) key distribution, (2) authentication of burst headers, and (3) confidentiality of data bursts. The security framework described here will work with various routing schemes in OBS networks (e.g., static routing, deflection routing, and dynamic load balancing). The secure OBS security framework illustrated in Figure 11.9 includes four components: (1) data burst encryption at ingress edge routers, (2) data burst decryption at egress edge routers, (3) per hop authentication of burst headers, and (4) key distribution mechanisms. The rationale behind the framework is explained as follows.

In OBS networks, data bursts assembled at an ingress edge router stay in the optical domain in the OBS core network and are only disassembled at the destination egress edge router. Because data bursts are transparent to OBS core routers, encryption and decryption of data bursts is only needed between a pair of ingress and egress edge routers. On the other hand, burst headers are converted back to an electronic form and are processed electronically at every OBS core router along

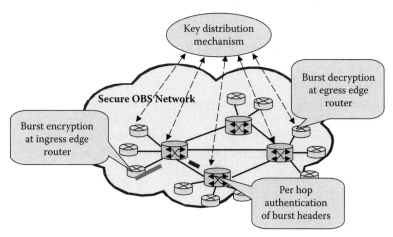

FIGURE 11.9 OBS security framework (Copyright © IEEE 2008).

the path. Therefore, per hop burst header authentication is needed to ensure that no malicious burst headers can alter the route of optical data bursts. Because data bursts are encrypted at ingress edge routers and decrypted at egress edge routers, keys for encrypting and decrypting data bursts only need to be distributed between pairs of ingress and egress routers.

Because burst headers need to be authenticated on a per hop basis, keys for burst header authentication need to be distributed between (1) the ingress edge router and the first hop core router, (2) any connected core router pairs, and (3) the last hop core router and the egress edge router. Although a different key distribution mechanism can be used for each of these scenarios, a network-wide unified key distribution mechanism is more desirable in terms of network management.

11.6.3 OBS Secure Router Architecture

The security framework described above can be embedded into the OBS edge router and core router architecture. Figure 11.10 illustrates the architecture of a secure OBS edge router. After a burst is formed, the burst assembler generates a burst header that carries information about the burst. The assembled bursts and their corresponding headers are encrypted before transmission onto the optical link. The burst header is encrypted and transmitted on the control channel for authentication purposes at the next hop. After holding the burst for an offset time, the burst assembler releases the encrypted data burst to be transmitted on one of the data channels. At egress, the received burst headers are authenticated before their corresponding bursts are decrypted. Decrypted data bursts are forwarded to the burst disassembler, which converts bursts back to packets and forwards them to the appropriate line interfaces. The key management block is responsible for key distribution and periodic updates. The flow chart shown in Figure 11.11 details the operations performed at the ingress edge router for secure transmissions across the OBS network.

The architecture of a secure core router is illustrated in Figure 11.12. The burst headers sent on the control channel are authenticated in the burst header authentication block. Only the authenticated burst headers are processed by the burst header processing unit for burst scheduling. Based on the information carried in authenticated burst headers, the burst header processing unit allocates outgoing wavelength for the incoming data bursts and configures the optical interconnects to allow data bursts to pass to the desired outgoing link optically. The burst header is encrypted again for the next hop authentication. The control channel and the data channels are then combined onto the DWDM link at the output. The key management block in the core router maintains and updates proper keys for authenticating the headers. The detailed operations in a secure OBS core router are explained in a flow chart in Figure 11.13.

This section has presented an embedded security framework for OBS networks. The security framework is based on the unique architecture of OBS networks and can be extended across islands of OBS networks using techniques that are based on mutual trust.

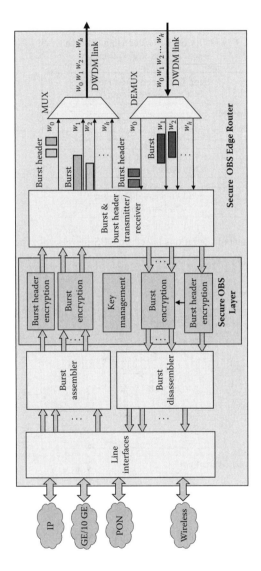

FIGURE 11.10 Secure OBS edge router architecture (Copyright © IEEE 2008).

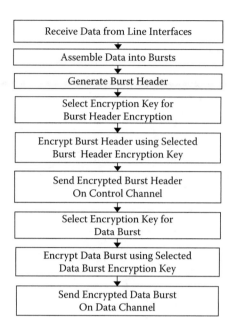

FIGURE 11.11 Flow chart of operations in secure OBS ingress edge router.

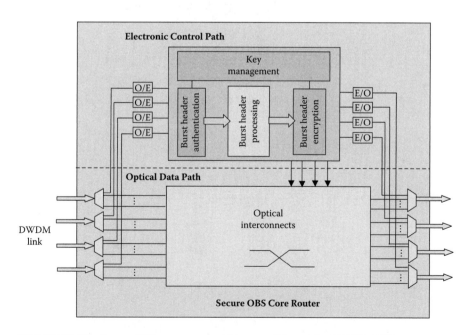

FIGURE 11.12 Secure OBS core router architecture (Copyright © IEEE 2008).

Optical Burst Switching

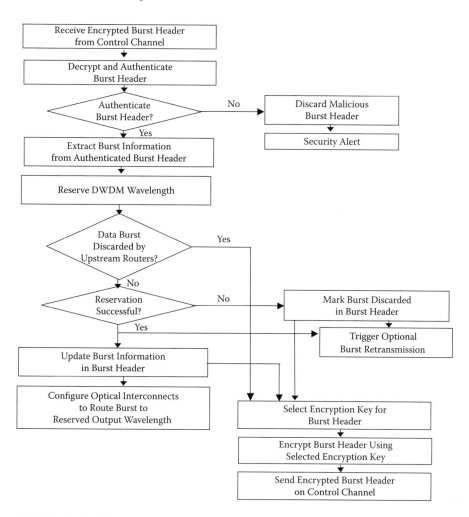

FIGURE 11.13 Flow chart of operations in secure OBS core router.

11.7 OBS IN METROPOLITAN AREA NETWORKS

Metropolitan area networks (MANs) carry increasing amounts of traffic. Because *local area networks* (LANs) form the access networks for MANs and the prevalent technology used by LANs is Ethernet, supporting Ethernet in MANs is a natural choice. There is a fair amount of global activity driven by the Metropolitan Ethernet Forum (MEF) [31] to make low-cost, high-bandwidth Ethernet services available beyond the LANs, which have been Ethernet's mainstay in the past. The MEF believes that its mission is driven by the opportunity to make ubiquitous and seamless services available in MAN as well as wide area network (WAN) environments. The objective of this section is to show that OBS provides better scalability and bandwidth efficiency in MANs compared to the currently used underlying transmission structure.

Metropolitan area Ethernet networks currently use either Synchronous Optical Network (SONET) or WDM as the underlying transmission facility. SONET can provision varying levels of fixed bandwidth between any two endpoints, such as a pair of endpoints on a ring. However, once provisioned, the bandwidth remains fixed and does not adapt to the varying traffic demands. The WDM allocates the entire bandwidth associated with a wavelength between the two endpoints, thus providing little flexibility, and requires a considerable amount of provisioning effort when a new endpoint is added or dropped.

Compared to SONET and WDM bandwidth provisioning, OBS can offer better utilization of the bandwidth by interconnecting its endpoints on demand. It offers advantages in two dimensions. First, because statistical multiplexing is an inherent part of OBS networking, it leads to a better utilization of resources. Second, the OBS networking offers flexibility in terms of accommodating a number of new endpoints, which is particularly desirable where the demand for connectivity between any pair of endpoints is varying and the number of endpoints is relatively large. In other words, OBS networks provide a finer granularity of transmission compared to the WDM technology.

The light paths created by the OBS technology do not constitute a permanent allocation of resources between the two endpoints. The OBS technology dynamically allocates the transmission resources to closely match the actual traffic load, along with the QoS levels required by the applications, with a relatively low overhead. As the ubiquity of Ethernet grows, static wavelength provisioning between any possible pair of endpoints becomes more and more unrealistic. The combination of scalability and efficiency leads to the superiority of the OBS technology in the metropolitan area environment. There is little doubt that the dynamic resource allocation capability makes OBS a strong candidate for MANs. As the upgrade in speed materializes over a wider communications infrastructure, the OBS technology will likely become more appealing beyond the core infrastructure of the network.

11.8 CONCLUSION

This chapter has discussed the technology of OBS, a flexible switching technology that has the advantages of both optical circuit and packet-switching technologies while obviating their limitations. Optical bursts are formed at the ingress node and delivered to the egress node while retaining their optical modality in the OBS core network. The OBS signaling protocols allow the burst header traveling on a separate control channel to set up an optical path for its corresponding burst on the fly, allowing the optical burst to pass OBS core routers without buffering, thus retaining its transparency. We described the architecture and the building blocks of the OBS routers with a focus on implementation aspects. We also addressed security considerations and presented a mechanism that can be deployed to retain the integrity of data within the OBS environment. Although positioned as a core network switching technology, OBS technology can be readily extended to a metropolitan area environment. It is expected that OBS will prove to be a commercially viable technology that will find increasing usage in the core network and beyond.

REFERENCES

1. D. Blumenthal, P. Prucnal, and J. Sauer, "Photonic packet switches: Architectures and experimental implementation," in *Proc. IEEE,* Vol. 82, 1994, pp. 1650–1667.
2. P. Gambini, M. Renaud, C. Guillemot, F. Callegati, I. Andonovic, B. Bostica, D. Chiaroni, G. Corazza, S. L. Danielsen, P. Gravey, P. B. Hansen, M. Henry, C. Janz, A. Kloch, R. Krahenbuhl, C. Raffaelli, M. Schilling, A. Talneau, and L. Zucchelli, "Transparent optical packet switching: Network architecture and demonstrators in the KEOPS project," *IEEE Journal on Selected Areas in Communication,* Vol. 16, pp. 1245–1259, Sept. 1998.
3. D. K. Hunter, M. C. Chia, and I. Andonovic, "Buffering in optical packet switches," *IEEE/OSA Journal of Lightwave Technology,* Vol. 16, pp. 2081–2094, Dec. 1998.
4. J. S. Turner, "Terabit burst switching," *Journal of High Speed Networks,* Vol. 8, pp. 3–16, 1999.
5. C. Qiao and M. Yoo, "Optical burst switching (OBS)—A new paradigm for an optical internet," *Journal of High Speed Networks,* Vol. 8, pp. 69–84, 1999.
6. Y. Chen, C. Qiao, and X. Yu, "Optical burst switching: A new area in optical networking research," *IEEE Network,* Vol. 18, pp. 16–23, May–June 2004.
7. M. J. O'Mahony, C. Politi, D. Klonidis, R. Nejabati, and D. Simeonidou, "Future optical networks," *IEEE/OSA Journal of Lightwave Technology,* Vol. 24, pp. 4684–4696, Dec. 2006.
8. B. Mukherjee, *Optical WDM Networks.* New York: Springer, 2006.
9. Y. Chen and J. Turner, "WDM burst switching for petabit capacity routers," in *Proc. IEEE Military Communications Conf.,* 1999, pp. 968–973.
10. Y. Xiong, M. Vanderhoute, and H. C. Cankaya, "Control architecture in optical burst-switched WDM networks," *IEEE Journal on Selected Areas in Communications,* Vol. 18, pp. 1838–1854, Oct. 2000.
11. M. Yoo and C. Qiao, "Just-enough-time (JET): A high speed protocol for bursty traffic in optical networks," in *Proc. IEEE/LEOS Tech. Global Information Infra.,* 1997, pp. 26–27.
12. J. Y. Wei and R. I. McFarland, "Just-in-time signaling for WDM optical burst switching networks," *IEEE/OSA Journal of Lightwave Technology,* Vol. 18, pp. 2019–2037, Dec. 2000.
13. I. Baldine, G. N. Rouskas, H. G. Perros, and D. Stevenson, "JumpStart: A just-in-time signaling architecture for WDM burst-switched networks," *IEEE Communications Magazine,* Vol. 40, no. 2, pp. 82–89, 2002.
14. Y. Chen, J. Turner, and P. Mo, "Optimal burst scheduling in optical burst switched networks," *IEEE/OSA Journal of Lightwave Technology,* Vol. 25, pp. 1883–1894, Aug. 2007.
15. Y. Chen, J. S. Turner, and Z. Zhai, "Design and implementation of an ultra fast pipelined wavelength scheduler for optical burst switching," *Photonic Network Communications,* Vol. 14, pp. 317–326, Dec. 2007.
16. V. M. Vokkarane and J. P. Jue, "Burst segmentation: An approach for reducing packet loss in optical burst switched networks," *SPIE/Kluwer Optical Networks,* Vol. 4, pp. 81–89, 2003.
17. V. M. Vokkarane and J. P. Jue, "Prioritized routing and burst segmentation for QoS in optical burst switched networks," in *Proc. IEEE/OSA Optical Fiber Communication Conf.,* 2002, pp. 221–222.
18. W. Tang and Y. Chen, "Methods for non-wavelength-converting multi-lane optical switching," U.S. patent application #12/268199, Nov. 2008, and U.S. provisional patent application #60/986818, Nov. 2007, patent pending.

19. J. Li, C. Qiao, J. Xu, and D. Xu, "Maximizing throughput for optical burst switching networks," in *Proc. IEEE Infocom,* 2004, pp. 1853–1863.
20. J. Xu, C. Qiao, J. Li, and G. Xu, "Efficient burst scheduling algorithms in optical burst-switched networks using geometric techniques," *IEEE Journal on Selected Areas in Communication,* Vol. 22, pp. 1796–1881, Nov. 2004.
21. N. Barakat and E. H. Sargent, "Separating resource reservations from service requests to improve the performance of optical burst switching networks," *IEEE Journal on Selected Areas in Communications,* Vol. 24, pp. 95–107, 2006.
22. M. Yoo and C. Qiao, "QoS performance of optical burst switching in IP-over-WDM networks," *IEEE Journal on Selected Areas in Communications,* Vol. 18, pp. 2062–2071, 2000.
23. Y. Chen, M. Hamdi, and D. H. K. Tsang, "Proportional QoS over OBS networks," in *Proc. Globecom 2001,* Vol. 3, 2001, pp. 1510–1514.
24. Q. Zhang, V. Vokkarane, B. Chen, and J. P. Jue, "Early drop and wavelength grouping schemes for providing absolute QoS differentiation in optical burst-switched networks," in *Proc. Globecom 2003,* Vol. 5, 2003, pp. 2694–2698.
25. F. Farahmand and J. P. Jue, "Look-ahead window contention resolution in optical burst switched networks," in *Proc. Workshop on High Performance Switching and Routing 2003 (HPSR 2003),* 2003, pp. 147–151.
26. F. Farahmand and J. P. Jue, "Supporting QoS with look-ahead window contention resolution in optical burst-switched networks," in *Proc. IEEE Globecom 2003,* Vol. 5, 2003, pp. 2699–2703.
27. Y. Chen, J. Turner, and Z. Zhai, "Contour-based priority (CBP) scheduling in optical burst switched networks," *IEEE/OSA Journal of Lightwave Technology,* Vol. 25, pp. 1949–1960, Aug. 2007.
28. W. Stallings, *Cryptography and Network Security,* 4th ed. Upper Saddle River, NJ: Prentice Hall, 2006.
29. Y. Chen and P. K. Verma, "Secure optical burst switching (S-OBS)—Framework and research directions," *IEEE Communications Magazine,* Vol. 46, pp. 40–45, Aug. 2008.
30. Y. Chen and P. K. Verma, "Methods and apparatus for securing optical burst switching (OBS) networks," U.S. provisional patent application #61/055696, May 2008, patent pending.
31. http://www.MetroEthernetForum.org

12 Intelligent Optical Control Planes

James J. Zik

CONTENTS

12.1 Introduction 291
12.2 Optical Control Plane Architectures 292
    12.2.1 Control Plane Layers 292
    12.2.2 Optical Control Plane Implementation 293
    12.2.3 Optical Control Plane Interactions 294
12.3 The ASON Optical Control Plane Model...294
    12.3.1 Optical-User Network Interface 296
    12.3.2 External-Network Network Interface 298
    12.3.3 Internal-Network Network Interface 298
    12.3.4 Network-Network Interface-Based Control Plane Operation...........298
12.4 Optical Control Plane Elements 299
    12.4.1 Neighbor Discovery 299
    12.4.2 Signaling...299
    12.4.3 Routing and Path Computation 301
        12.4.3.1 Bundling..303
12.5 Optical Control Plane Functionality 304
12.6 Optical Control Plane Standards 305
    12.6.1 ASON Standards 308
    12.6.2 GMPLS Standards...308
    12.6.3 Metro Ethernet Forum 310
    12.6.4 TeleManagement Forum 310
    12.6.5 OIF... 310
12.7 Optical Control Plane Usage in the Scientific Community 312
12.8 Conclusion 312
References... 313

12.1 INTRODUCTION

The optical control plane is a management layer that operates between the network management system (NMS) for the network and the data plane. It is defined as the software that controls the configurable features of a network. Its function is to automate the network by discovery of network elements, ports, and connectivity between ports in a network, creating inventory information, disseminating this information to all network elements, calculating and creating optimal paths, and providing network-level protection restoration. Essentially, the optical control plane is the central nervous system and information repository of the network and its topology. The optical control plane was invented in the late 1990s by Ciena and other companies to address issues with ring-based Synchronous Optical Network/Synchronous Digital Hierarchy (SONET/SDH) architectures such as manual provisioning, manual database entry, single node failure protection, and so on by providing network automation to reduce operational expenditures and incorporating advanced protection and restoration mechanisms to create efficient mesh architectures that reduce the number of ports and links needed for redundancy, resulting in decreased capital and operational outlays for building and operating a network. The optical control plane–enabled mesh network also supports high network availability.

Adding intelligence to the optical network results in additional benefits that have led the world's largest service providers to deploy intelligent optical control planes in continental, subsea, and global networks. Intelligent optical control planes enable networks to scale globally without performance degradation. They automate time-consuming, error-prone manual management functions, including provisioning, inventory, resource management, and path computation. Control plane data can be used to create graphical real-time charts for network analysis, including parameters such as node utilization and link utilization, with unsurpassed accuracy. In addition, they enable new services including dynamic connections for short-term services, real-time applications, on-demand bandwidth creation, and modification and class-of-service bandwidth reservation. They enable highly resilient survivable mesh networks that improve network utilization over traditional ring topologies.

12.2 OPTICAL CONTROL PLANE ARCHITECTURES

Intelligent optical control planes come in different types and operate with different interactions. The type of control plane is most significant because it affects the performance and scalability of a network. However, first it is important to understand where an optical control plane operates versus nonoptical control planes that have been created in the past for layers such as Asynchronous Transfer Mode (ATM) and multiple protocol label switching (MPLS), and to make clear the distinction between optical transport in general and photonic (all-optical) transport, which has its own special set of control plane issues.

12.2.1 Control Plane Layers

Control planes exist at different layers in the network and have been deployed on ATM and MPLS networks for over a decade. Internet Protocol (IP) routers employ MPLS to control the way Layer 3 data flow (IP traffic) is mapped onto Layer 2 traffic between adjacent network nodes. This is known as an IP/MPLS control plane. Optical transport represents a number of different technologies, ranging from Carrier Ethernet, to SONET/SDH, to the Optical Transport Network (OTN), and to all-optic networks that transport and switch multiple wavelengths through the use of Dense Wave Division Multiplexing (DWDM) and reconfigurable optical add-drop multiplexer (ROADM) technology without electrical regeneration. These technologies are

Intelligent Optical Control Planes

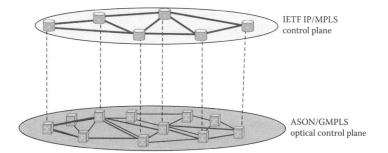

FIGURE 12.1 Control plane layers.

often layered in a hierarchical fashion, and connections at one layer can be used as a path to establish connections at a higher layer. This is an extension from classical IP routing, which assumes a flat underlying Layer 2 transport. This hierarchical structure has implications for both signaling and routing. In today's networks, the IP/MPLS control plane overlays onto the optical control plane as shown in Figure 12.1. The control plane is segmented into an IP domain and an optical domain; the router is a client of the optical domain and has no knowledge of the optical topology [1].

Optical control planes employ an Automatically Switched Optical Network (ASON) model or Generalized Multi-Protocol Label Switching (GMPLS) protocols, which can manage and route traffic that does not require IP routing. As much as 80% of IP traffic through core IP routers is transit traffic that does not require routing. Switching traffic at the lowest layer necessary maximizes the efficient use of IP/MPLS network capabilities for traffic that requires IP routing. The ASON/GMPLS model allows interoperability between network operators and equipment-enabling networks to globally scale. More detail on ASON and GMPLS is provided in subsequent sections where standards for optical control planes are addressed.

12.2.2 Optical Control Plane Implementation

An optical control plane can either be centralized or distributed as shown in Figure 12.2. A centralized control plane resides on the NMS and employs a hierarchical

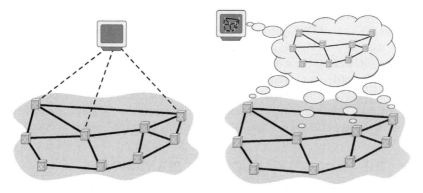

FIGURE 12.2 Centralized versus distributed control plane types.

decision-making process. This means it must scale exponentially as the network grows linearly to maintain performance such as subsecond protection and restoration in core backbone networks. Loss of connectivity to the management plane causes the control plane to lose its real-time control over the networks, which can potentially result in a loss of protection or restoration capabilities. However, centralized control planes can serve a useful purpose in access and small metro networks where ring topologies are typically employed with a limited number of network elements and the need to scale the network is capped due to its limited geography. Consequently, centralized control planes can offer a simpler alternative in these networks. For large backbone networks, however, centralized control planes typically cannot quickly restore circuits because the NMS sequentially processes new path information, creating a bottleneck, instead of parallel processing the information as in a distributed control plane. A distributed control plane resides on the network elements, employing them to distribute the decision-making process. Each device knows the network topology, allowing fast decision making that scales linearly with the network, unlike a centralized control plane that needs to grow exponentially as the network itself grows linearly to handle the exponential growth in possible paths. With a distributed control plane, the control plane continues to function when connectivity to the management plane is lost.

12.2.3 Optical Control Plane Interactions

The initial drivers for deploying dynamic intelligent optical networks were automation and reliability, but the growing demand for new end-to-end broadband and bandwidth-on-demand (BoD) are becoming main drivers as well. These new services are mainly based on Ethernet as the service interface. The current SDH/SONET network architecture with its inherent resilience provides the basis for the service-level agreements to manage these new services. Dynamic bandwidth allocation is enabled by means of SDH/SONET networks and their efficient mapping of Ethernet packets into the SDH/SONET data stream. The basis to those enhanced data-handling capabilities are three technologies—Generic Framing Procedure (GFP), Link Capacity Adjustment Scheme (LCAS), and Virtual Concatenation (VCAT)—that encapsulate and frame data traffic for transport over the SONET/SDH infrastructure. However, the lack of transparency of SONET/SDH and the need to transport full-rate 10 Gigabit Ethernet (10GbE) services, which cannot be done with the 10G SONET/SDH OC-192/STM-64 rates, has driven the creation of OTN infrastructures that enable transparent end-to-end service delivery as shown in Figure 12.3. Consequently, dynamic intelligent optical control planes are now available that operate on OTN infrastructures that can also carry legacy SONET/SDH traffic. Optical control planes can also operate at the Ethernet, or even the photonic layer supporting wavelengths; however, standards, as well as the development of these control planes, are still in development. The ability to encapsulate service payloads in the different layers enhances interoperability between carriers or domains.

12.3 THE ASON OPTICAL CONTROL PLANE MODEL

The ASON optical control plane model is a control plane model driven by telecom carriers, who were the first to implement optical control planes in their networks.

Intelligent Optical Control Planes

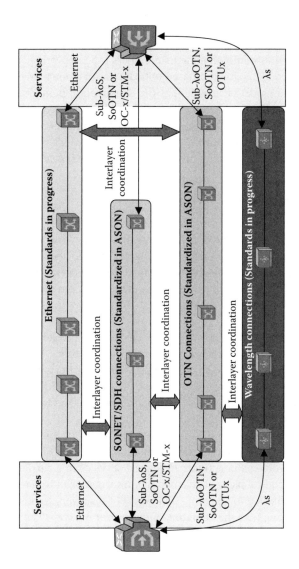

FIGURE 12.3 Control plane coordination with a connection-oriented transport network.

Because carriers operate large continental-sized networks and undersea networks, they promoted a model that consisted of several domains to handle growth to global networks and the ability to interface with other networks operated by other carriers or other domains in their own network. As defined by the Optical Internetworking Forum (OIF) [2], a control domain is "an architectural construct that provides for encapsulation and information hiding. The characteristics of the control domain are the same as those of its constituent set of distributed architectural components." Essentially, a domain is defined as a set of network resources that is administered as a group where full information among its members (i.e., network elements) within the domain is shared. This domain-based optical control plane model is illustrated in Figure 12.4. This model clearly defines the demarcation points between different domains as determined by geography, carrier policies, or technology (e.g., new equipment vs. legacy equipment). This model limits the amount of information and control across domains and easily scales to handle global carrier networks. It is defined by three main interfaces as described in the following subsections.

12.3.1 Optical-User Network Interface

With the development of optical multiplexing, a new optical layer was integrated into the telecommunication network. This optical network layer provides transport services to interconnect clients such as IP routers, MPLS label switching routers, Ethernet switches, SONET/SDH/OTN equipment, Multi-Service Provisioning Platforms (MSPPs), and other similar networking elements. Consequently, a UNI was created to define the service control interface between the transport network and client equipment.

Signaling over the UNI is how an edge device can invoke services that the transport network offers to clients. It is the signaling interface between the client (service consumer) and the network (service provider or carrier). It offers the ability to the client to dial up an end-to-end service such as a 100-Mbps Ethernet connection through a carrier's network to another device on the far end with another UNI. Because these UNIs are almost always "optical," they will be referred to as the Optical-User Network Interface (O-UNI) in the remainder of this chapter. The OIF has specified an O-UNI signaling implementation agreement that allows various devices from numerous manufacturers to request services in a carrier network that employs an optical control plane. The amount of information (attributes) that crosses over the O-UNI is limited to signaling and neighbor discovery to prevent the client device from interfering with the operation of the carrier network. The basic function of the O-UNI is primarily signaling as follows:

- Connection establishment (signaling).
- Connection deletion (signaling).
- Connection modification (signaling).
- Status exchange (signaling).
- User data (traffic).

Intelligent Optical Control Planes

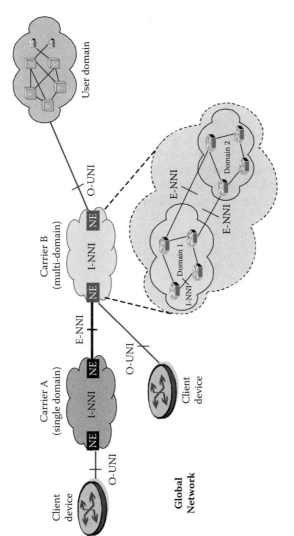

FIGURE 12.4 ITU-T ASON/GMPLS domain-based optical control plane model.

12.3.2 EXTERNAL-NETWORK NETWORK INTERFACE

The External-Network Network Interface (E-NNI) is how carriers or domains within a carrier network, who employ different control plane protocols within a domain, interface with each other. It provides an interface between network elements (NEs) belonging to different domains, which limit the information shared with its neighbors. In order to support deployment of an optical control plane into a heterogeneous environment, the E-NNI provides a necessary boundary between network domains and specifies how signaling and routing information are exchanged between such domains.

The E-NNI reference point is defined to exist between control domains. The OIF states that "the nature of the information exchanged between control domains across the E-NNI reference point captures the common semantics of the information exchanged amongst its constituent components, while allowing for different representations inside each control domain."

A carrier will typically define a domain by geography, technology, or equipment manufacturer, and when a carrier hands off traffic to another carrier, each will have its own domain to maintain control of its own network. The E-NNI differs from an O-UNI in that the E-NNI adds routing capability and therefore is not limited to signaling of connection requests. The primary functions of the E-NNI are as follows:

- Connection establishment (signaling).
- Connection deletion (signaling).
- Status exchange (signaling).
- Connection modification.
- User data (traffic).
- Path protection and restoration (routing).

12.3.3 INTERNAL-NETWORK NETWORK INTERFACE

The Internal-Network Network Interface (I-NNI) is the signaling and routing specified within a domain. The OIF does not support implementation agreements at the I-NNI because this is within a domain and not considered an internetworking issue. Within a domain, control planes users do not require interoperability between different manufacturers equipment regarding the I-NNI. This concept is similar in nature to SONET/SDH four-fiber and two-fiber line-switched rings. Even though SONET/SDH specifies how rings operate, there is no interoperability or internetworking between rings created by different manufacturers. Consequently, equipment manufacturers of optical control planes are allowed flexibility for signaling and routing within the I-NNI, enabling them to offer advanced features including mesh topologies with fast, efficient protection and restoration capabilities. The I-NNI can be based on standards including ASON and GMPLS or proprietary standards.

12.3.4 NETWORK-NETWORK INTERFACE-BASED CONTROL PLANE OPERATION

Policy, scalability, and performance needs usually dictate that a network be structured into different domains, separating the client and network at the O-UNI. With this

Intelligent Optical Control Planes 299

model, the client device at the O-UNI requests a connection to another client device but receives no detailed topology information about which networks the connection must traverse. The connection may traverse single or multiple domains. The client device is aware only of its attachment to the ingress/egress NE of the optical network. However, each node within a domain is aware of the local topology and the clients attached to its local domain. The nodes within a domain are given limited information about the topology and clients of other domains through the E-NNI routing protocol.

12.4 OPTICAL CONTROL PLANE ELEMENTS

The optical control plane consists of three main elements: neighbor discovery, routing that includes path computation and link bundling, and signaling, which is also known as connection management. These three features are the enabling technologies that provide network automation for today's high-capacity networks.

12.4.1 NEIGHBOR DISCOVERY

Neighbor discovery allows an NE to automatically determine its connectivity to adjacent NEs and is one of the most useful features of the optical control plane. Without neighbor discovery, the interconnection information must be configured manually for each port and path, an error-prone, time-consuming process. Updates are automatic as changes occur in the connectivity between NEs. Neighbor discovery information is used to form a topology map by which each node contains the information of how each NE is to be connected to each other. The discovery information includes not only the NE itself, but knowledge of each port, the connectivity between ports, and the available services that can be provisioned on the port. The information gathered through neighbor discovery is used for the inventory knowledge base, connection provisioning, and to verify that link parameters are configured in a consistent manner at both ends of a link as shown in Figure 12.5.

The communication for neighbor discovery can be in-band or out-of-band. In-band communication for SONET/SDH is handled via the data communication channel (DCC) or the J0 byte in the Section (SONET) or Regenerator Section (SDH) overhead. For OTN-based control planes, in-band communication is handled through the general communications channel (GCC) or through the trail trace identifier (TTI). For Ethernet frames, in-band communication is typically handled through the Link Layer Data Protocol (LLDP) frame and for the physical layer through the optical service channel (OSC), also known as the telemetry channel. Portions of neighbor discovery can also be implemented via out-of-band communications through the Data Communication Network (DCN), although some in-band component is required to validate discovery and detect connection miswiring.

12.4.2 SIGNALING

Signaling, also known as connection management, is the control messaging between switching entities in the network. It is used to create, modify, tear down, and restore connections between the domains through the network interfaces as shown in Figure 12.6, and

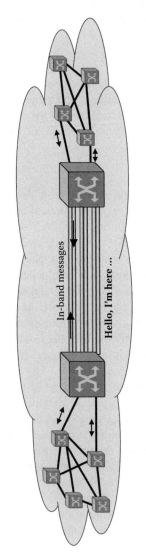

FIGURE 12.5 Neighbor discovery.

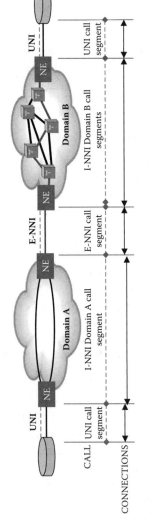

FIGURE 12.6 Signaling (connection management).

Intelligent Optical Control Planes

TABLE 12.1
Mapping of ASON DCM Attributes to RSVP-TE Objects

Attributes		RSVP-TE
Identity attributes	A-end user name	SOURCE_TNA
	Z-end user name	DESTINATION_TNA
	Connection name	SESSION + SENDER_TEMPLATE
	Call name	CALL_ID
Service attributes	SNP/SNPP ID	GENERALIZED_LABEL, other LABEL objects
	Directionality	(Implied by UPSTREAM_LABEL)
Policy attributes	CoS	SERVICE_LEVEL
	Explicit resource list	EXPLICIT_ROUTE
	Security	INTEGRITY or IP layer security (IPsec)
Additional protocol attributes	Implied layer info	GENERALIZED_LABEL_REQUEST, SENDER_TSPEC/FLOWSPEC
	For disabling monitoring	ADMIN STATUS
	For protocol robustness	HELLO_REQUEST, HELLO_ACK, MESSAGE_ID, RESTART_CAP
	For status/error codes	ERROR_SPEC
	For optional confirmation	RESV_CONFIRM
	Protocol-specific attributes	STYLE, TIME_VALUES

provides automated connection management for protection and restoration. Signaling enables different classes of service to be offered, allowing the service provider to offer numerous service types to provide differentiation from their competitors.

Connection provisioning requires the identification of two connection endpoints and the connection attributes that are characteristics of the connection. For example, the name of a connection or the name of the endpoints are connection attributes. Table 12.1 lists ASON-defined connection attributes for optical control planes for mapping to Resource Reservation Protocol with Traffic Engineering (RSVP-TE) objects for Distributed Call and Connection Management (DCM). In addition, the signaling also carries messages between the network elements. Table 12.2 lists several of the International Telecommunication Union (ITU)-defined messages for setup, teardown, query, and notification messages for ASON-based optical control planes. These messages are transmitted through the same in-band or out-of-band communication links as described earlier in neighbor discovery through the O-UNI, E-NNI, or I-NNI interfaces.

12.4.3 ROUTING AND PATH COMPUTATION

Routing is the process whereby a node determines the topology within the domain and allows the node to compute one or more paths to possible destinations in a network. Using network topology information, paths from a source to a destination can be computed using path computation algorithms based on link characteristics that include resource availability, route cost, route diversity (for protection and restoration

TABLE 12.2
ITU-Defined UNI/E-NNI Messages

Message Type	RSVP-TE
Setup messages	Path
	Resv, PathErr
	ResvConf
Release messages	Path or Resv (with D & R bits set)
	PathErr (Path_State_Removed flag) or Path Tear
Query messages	Path (implicit in RSVP-TE via periodic refreshes)
	Resv (implicit in RSVP-TE via periodic refreshes)
Notification message	Notify

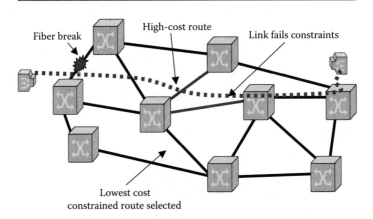

FIGURE 12.7 Routing and path computation.

paths), link protection capabilities, and other parameters. The optimal path is selected based on a set of chosen link weights as shown in Figure 12.7. A connection is then established along the computed route by activating the cross-connects in the appropriate NEs. The routing protocol distributes network topology and state information among all nodes within a local domain so the nodes can determine optimal routes for a particular service class specified by the user. The protocol supports typical routing functionality between all nodes and bandwidth advertisements (including automatic flooding), capturing changes in available resources. A single path computation command to the source node initiates a signaling sequence through the network to create a light path, or subnetwork connection (SNC), between any two nodes.

Similar to connection management, routes have ITU-defined attributes associated with them. These attributes include characteristics such as endpoints, capacity, link cost, protection, and others, as shown in Table 12.3. As numerous paths are set up in a network, the number of attributes to manage can grow quite large, requiring large computational processing capability. However, given the physical and practical constraints in a real network, the processing power can be limited. Consequently,

Intelligent Optical Control Planes

TABLE 12.3
Path Link Attributes

Link Attribute	Capability	Usage
Local link endpoint	Mandatory	Mandatory
Remote link endpoint	Mandatory	Mandatory
Layer-specific characteristics	Signal types supported	
	Link weight or cost	
	Resource class	
	Available capacity	
	Link protection characteristics	
	Link diversity or SRLG information	

Note: Examples of scope of connection control limited to a single call segment. The service is realized in different ways within each domain. Separate address spaces are used within each domain. There is a trust boundary. There is independence of survivability (protection/restoration) for each domain.

the concept of bundling links together was created to limit the amount of computations necessary for managing a network.

12.4.3.1 Bundling

Bundling of links improves the scalability of routing protocols by factors up to 20 times. In typical large networks, neighboring nodes are often interconnected by a large number of parallel links (fibers). Bundling allows parallel links to be handled as one entity, enabling scalability of routing protocols by eliminating unnecessary computational load and superfluous control traffic on the NEs as shown in Figure 12.8. Some control planes apply this same bundling concept to protection also for the same reason. Network survivability can be enhanced by applying the concept of link bundling to protection bundling using shared risk link groups (SRLGs) for mesh protection and restoration. Each element within an SRLG would typically share the same structure (i.e., conduit, cable, amplifier, etc.) as links that are likely to fail at the same time. SRLGs enable mesh networks to minimize the computational load and reduce event generation in a line failure which can create alarm storms. Mesh restoration utilizing SRLG information enables subsecond protection times while enhancing network utilization and providing greater network availability by increasing the number of potential alternate paths. Use of mesh restoration can improve network availability by a factor of up to 10 times.

FIGURE 12.8 Link bundling.

12.5 OPTICAL CONTROL PLANE FUNCTIONALITY

An optical control plane allows transport networks to automate and distribute many functions formerly performed through the combination of centralized management systems and manual processes. As a result of this automation, the network reacts flexibly and supports rapid deployment of new services. On the basis of the technologies presented here, the optical control plane provides a number of operational advantages in service-provider networks, including the following:

- Flexible network architectures for rapid provisioning of new services along with rapid restoration mechanisms.
- Dynamic and efficient bandwidth allocation.
- Different classes of service, which enables service providers to offer their customers differentiated services (service-level agreements).
- Resource tracking and allocation, as well as detailed real-time network-control information (inventory, topology, etc.).
- Autodiscovery, which provides an accurate picture of network topology and resource availability to make the network the database of record.
- Autoprovisioning, which increases the speed of service setup and teardown.

Intelligent control plane functionality can be combined with link-based functionality to enable further enhancements to network survivability and class of service. Optical control plane–enabled networks today support fast mesh restoration, SONET/SDH line protection mechanisms, and OTN protection mechanisms. Link-based linear and ring protection can be implemented concurrently with fast mesh restoration to provide the highest levels of service availability. Fast mesh mechanisms restore connections on an end-to-end basis and operate independently of underlying linear and ring protection schemes. Therefore, in addition to 50 msec Automatic Protection Switching (APS) on the port or client level, the network also provides a framework for a multitiered protection hierarchy that uses the control plane to recover the network from catastrophic failures (which often cause simultaneous multiple network failures) such as earthquakes, hurricanes, and widespread power failures. An illustration of multiple failure mesh networking is shown in Figure 12.9. When a fiber cut occurs in one link, an alternative path is immediately selected. When another failure occurs and could occur simultaneously, the control plane automatically calculates another route, restoring traffic in millisecond time frames. Recent high-profile multiple simultaneous undersea cable cuts and Hurricane Katrina have illustrated the vulnerability of traditional ring-based architectures against multiple failures. Other enhancements to optical control planes include the following:

- Link aggregation allows multiple links between two nodes to be managed as a single link. Link aggregation supports intelligent network scaling, faster restoration times, and network simplification.
- Local Span Mesh Restoration (LSMR) supports faster restoration times in an intelligent network. In response to a failure event such as fiber breakage,

Intelligent Optical Control Planes

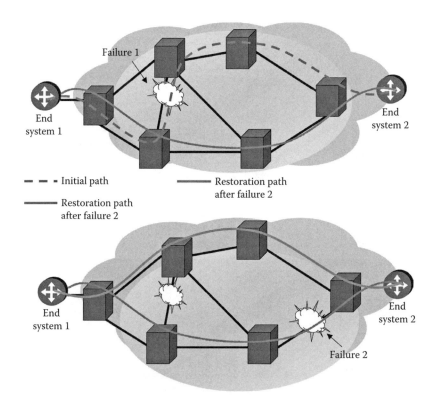

FIGURE 12.9 Multiple failure scenario in intelligent optical control plane–based mesh networks.

the SNCs on that fiber normally are restored from their originating nodes using signaling functionality across the network. LSMR allows the SNCs to restore locally, between the nodes adjacent to the break, increasing the speed of service restoration.

- Crankback provides continuous SNC restoration retries if the initial attempt is blocked. Each retry will seek alternative paths in the network using any available capacity, allowing restoration to function even in the presence of rapidly changing conditions and multiple failures.

12.6 OPTICAL CONTROL PLANE STANDARDS

Several organizations have been working on optical control plane standards and implementation agreements for the last decade, namely ITU-T standardization sector, the Internet Engineering Task Force (IETF), the TeleManagement Forum (TMF), and the OIF. As shown in Figure 12.10, there are a multitude of standards that can define an optical control plane. These standards cover architecture, autodiscovery, link management, signaling, routing, DCN, and management of the control plane. From a high level, it appears that the ITU-based ASON control plane is competing with the IETF GMPLS control plane; however, this is not the case. The ITU

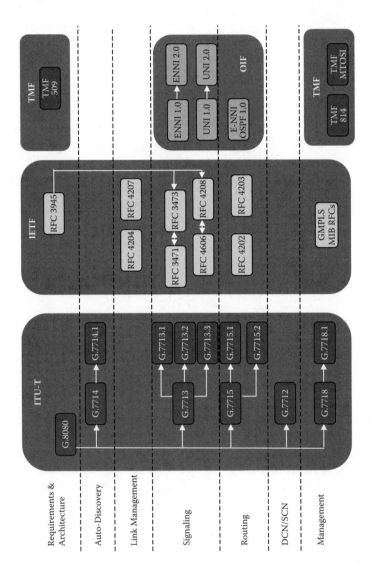

FIGURE 12.10 Optical control plane standards overview.

produces requirements and architectures based on the requirements of its members, many of which are large service providers and suppliers to them. Meanwhile, the IETF produces protocols in response to general industry requirements, potentially including those coming out of the ITU. Therefore, ASON and GMPLS can be and often are complementary to each other.

ASON was developed by the ITU-T, Study Group 15 [3], which is the group within the ITU-T that focuses on transport technologies. The work was initiated in response to demands from its members to create a complete definition of the operation of automatically switched transport networks including the management of the data plane. ASON is actually not a protocol but an architecture that defines the components in an optical control plane and the interactions between those components. In addition, it identifies which of those interactions will occur across a multivendor divide and therefore require standardized protocols. Other areas are intentionally not standardized to allow equipment manufacturers to provide differentiation with value-added features. ASON was developed in a top-down fashion, starting with a full and explicit list of requirements, beginning with high-level architectures and then moving down to individual component architectures. Because ASON is a collection of protocols, any protocol that fits the requirements of the component architecture can potentially be approved as an ASON standard.

The ITU created the ASON architecture based on transport network requirements. Because ITU members primarily come from a telecom background, ASON standards were based on concepts from protocols used heavily in telecom transport and switching networks, such as SONET/SDH. Consequently, the creators of ASON intended it to be complete, future-proof, highly scalable, and highly resilient to faults, specifically targeted at transport networks, used by service providers and carriers throughout the world.

GMPLS, on the other hand, was developed in a community strongly associated with IP-based data networks and evolved out of a set of existing protocols used in IP networks. The IETF introduced GMPLS employing a suite of protocols that in effect take packet-based technology from the IP domain and apply it to the optical domain. Based on the signaling and routing protocols of the GMPLS optical control plane, the various network elements of an optical network can communicate and work with each other. Among these elements might be SDH/SONET or OTN gear, DWDM equipment or switching elements (i.e., wavelength, fiber, time-division multiplex, or packet-aware switches).

GMPLS grew out of MPLS, a packet-routing technology designed to improve the efficiency of IP data networks. A flavor of MPLS, known as MPLS-TE, provided for provisioning of end-to-end connections using signaling with constraint-based routing. Consequently, it was felt that it should be generalized and extended to cover circuit-oriented optical switching technologies such as time-division multiplexing (TDM) and DWDM. Within the IETF, the MPLS protocol development was carried out by the MPLS Working Group, whereas GMPLS was defined in the Common Control and Measurement Plane (CCAMP) Working Group [4]. As an IETF protocol, GMPLS uses an IP-based control plane.

12.6.1 ASON STANDARDS

The ASON-related standards are shown in Figure 12.10 and Table 12.4 with the main protocols listed here:

- Architecture for Automatically Switched Optical Networks (G.8080).
- Distributed Call and Connection Control (G.7713), which covers signaling.
- Architecture and Requirements for Routing in the Automatic Switched Optical Networks (G.7715).
- Generalized Automated Discovery Techniques (G.7714).

In addition, various protocols have been integrated into the ASON architecture alongside the core ASON specifications, which include the following:

- PNNI-based signaling (G.7713.1).
- Generalized RSVP-TE-based signaling (G.7713.2).
- Generalized CR-LDP-based signaling (G.7713.3).
- Discovery for SONET/SDH, incorporating some aspects of LMP (G.7714.1).

As illustrated here, there are many ASON-compliant protocols, some of which are redundant. However, interoperability within a domain is not required, allowing flexibility of control implementation within the domain. However, once the network interfaces to another domain, internetworking issues are critical, which is where the OIF plays an important role.

12.6.2 GMPLS STANDARDS

The protocols defined by the IETF for GMPLS [5] include the following:

- Generalized RSVP-TE for signaling.
- Generalized Constraint Based Label Distribution Protocol (CR-LDP) for signaling.
- Open Shortest Path First (OSPF) with Traffic Engineering (TE) extensions for intra-area routing.
- Intermediate System to Intermediate System (IS-IS) with TE extensions, also for intra-area routing.
- Link Management Protocol (LMP) and LMP-WDM for assorted link management and discovery functions.

Like ASON, GMPLS also includes redundant standards. RSVP-TE and CR-LDP are alternative protocols that effectively do the same thing and were inherited from MPLS-TE. The IS-IS and OSPF extensions are also functionally equivalent. Although the IETF has agreed to limit future specification activity to RSVP-TE and

TABLE 12.4
Optical Control Plane Standards List

ITU-T	Title
G.7712	Architecture and Specification of Data Communication Network
G.7713	Distributed Call and Connection Management (DCM)
G.7713.1	Distributed Call and Connection Management (DCM) Based on PNNI
G.7713.2	Distributed Call and Connection Management: Signaling Mechanism Using GMPLS RSVP-TE
G.7713.3	Distributed Call and Connection Management: Signaling Mechanism Using GMPLS CR-LDP
G.7714	Generalized Automatic Discovery for Transport Entities
G.7714.1	Protocol for Automatic Discovery in SDH and OTN Networks
G.7715	Architecture and Requirements for Routing in the Automatically Switched Optical Networks
G.7715.1	ASON Routing Architecture and Requirements for Link State Protocols
G.7715.2	ASON Routing Architecture and Requirements for Remote Route Query
G.7718	Framework for ASON Management
G.7718.1	Protocol-Neutral Management Information Model for the Control Plane View
G.8080	Architecture for Automated Switched Optical Network (ASON)

TMF	Title
509	Network Connectivity Model
814	Multi-Technology Network Management Solution Set
MTOSI	Multi-Technology Operations System Interface

IETF RFC	Title
3471	Generalized Multi-Protocol Label Switching (GMPLS) Signaling Functional Description
3473	Generalized Multi-Protocol Label Switching (GMPLS) Signaling Resource Reservation Protocol-Traffic Engineering (RSVP-TE) Extensions
3945	Generalized Multi-Protocol Label Switching (GMPLS) Architecture
4202	Routing Extensions in Support of Generalized Multi-Protocol Label Switching (GMPLS)
4203	OSPF Extensions in Support of Generalized Multi-Protocol Label Switching (GMPLS)
4204	Link Management Protocol (LMP)
4207	Synchronous Optical Network (SONET)/Synchronous Digital Hierarchy (SDH) Encoding for Link Management Protocol (LMP) Test Messages
4208	Generalized Multiprotocol Label Switching (GMPLS) User-Network Interface (UNI): Resource Reservation Protocol-Traffic Engineering (RSVP-TE) Support for the Overlay Model
4606	Generalized Multi-Protocol Label Switching (GMPLS) Extensions for Synchronous Optical Network (SONET) and Synchronous Digital Hierarchy (SDH) Control

OIF	Title
OIF-E-NNI-Sig-01.0 & 02.0	Intra-Carrier E-NNI Signaling Specification
OIF-ENNI-OSPF-01.0	External-Network Network Interface (E-NNI) OSPF-Based Routing – 1.0. (Intra Carrier) Implementation Agreement
OIF-UNI-01.0 & 02.0	User Network Interface (UNI) Signaling Specification (1.0, 2.0)

cap development of CR-LDP, both IS-IS and OSPF extensions continue to be developed for GMPLS.

12.6.3 METRO ETHERNET FORUM

The Metro Ethernet Forum (MEF) [6] was formed to create technical specifications and implementation agreements on Carrier Ethernet and to promote interoperability and deployment of Carrier Ethernet worldwide. Carrier Ethernet is defined as a ubiquitous, standardized, carrier-class service and network defined by several attributes that distinguish Carrier Ethernet from familiar LAN-based Ethernet. Essentially, it is a way to make Ethernet "carrier grade" like SONET/SDH networks for robust transport with the benefit of using the Ethernet cost model to lower network cost. Many of the MEF standards are still in progress, but Ethernet characteristics have been supported in the OIF UNI-2.0 implementation agreement. The ultimate goal is for optical control planes to also operate over Carrier Ethernet in a similar fashion as they operate now over SONET/SDH and OTN protocols.

12.6.4 TELEMANAGEMENT FORUM

The TMF [7] is a large industry association involved in the communication, online, and entertainment industries. It creates open standards for management systems including Network Management Systems and Operating Support Systems (OSS), which interface to optical control planes. TMF 814 defines the Multi Technology Management (MTMN) Solution Set, which defines how an OSS can interact with, and fully manage, the entire intelligent optical network from a single interface. Multi-Technology Operations System Interface (MTOSI) is a standard for implementing interfaces between OSSs because many large carriers use multiple OSSs to manage complex networks. TMF-509 is an interface agreement for management of SONET/SDH transport networks. Figure 12.10 illustrates how the TMF standards and agreements interact with other optical control plane standards.

12.6.5 OIF

The OIF is responsible for creating implementation agreements (IAs) that subset standards as needed for a specific set of features and are interoperability tested to allow internetworking. Its mission is to promote the development and deployment of interoperable networking solutions and services through the creation of IAs for optical networking products, network processing elements, and component technologies. IAs are based on requirements developed cooperatively by end users, service providers, equipment vendors, and technology providers, and aligned with worldwide standards, augmented if necessary. The OIF is critical to the success of optical control plane interoperability. It employs the applicable GMPLS, ITU, and MEF standards create the UNI and E-NNI implementation agreements that allow interoperability between the client interface to the network and between networks or network domains as shown in Figure 12.11. Furthermore, it sets up interoperability demonstrations approximately every 2 years to test the effectiveness of the

Intelligent Optical Control Planes 311

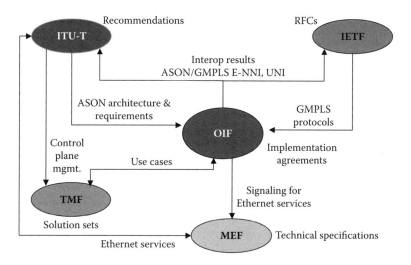

FIGURE 12.11 Optical control plane standard's body interactions.

implementation agreement and to sort out any ambiguities in the agreements. The recent demonstrations were as follows:

- 2004: 1st UNI 1.0 and E-NNI interoperability demonstration of the signaling of SONET/SDH connections from network edge to network edge at SuperComm 2004 in Chicago.
- 2005: 2nd UNI 1.0 and E-NNI interoperability demonstration of client-to-client Ethernet-over-SONET/SDH signaling Ethernet at SuperComm 2005 in Chicago.
- 2007: 3rd interoperability demonstration but with UNI 2.0 and E-NNI, which supported MEF-based Ethernet services over SONET/SDH networks at ECOC 2007 in Berlin, Germany.
- 2009: 4th interoperability demonstration enabling broadband on-demand services through dynamically switched Ethernet Virtual Private Line (EVPL) services over multiple, control-plane enabled intelligent optical core networks via the OIF's UNI 2.0 and E-NNI 2.0 implementation agreements.

The last OIF demonstration was a worldwide demonstration that took place from March 2009 to June 2009 in major carrier laboratories in Asia, Europe and the United States with seven major carriers and ten leading equipment manufacturers participating. The demonstration successfully demonstrated:

- Data plane interoperability of Ethernet Virtual Private Line (EVPL) services over Next Generation (NG) transport network switching equipment of different technology types, including Provider Backbone Bridging Traffic Engineering (PBB-TE), MPLS Transport Profile, OTN and SONET/SDH
- Data and control plane interoperability of on-demand EVPL and SONET/SDH services including multi-domain service restoration

These interoperability demonstrations have been critical to illustrating how on-demand world-wide services can be implemented through different control plane domains operated by different carriers.

12.7 OPTICAL CONTROL PLANE USAGE IN THE SCIENTIFIC COMMUNITY

Intelligent optical control planes are deployed in many of the world's largest networks. In the research and education community, optical control planes are used to create networking solutions that form a truly dynamic environment to support the most demanding applications known today. Scientific experiments including the large hadron collider (LHC) experiment at the European Organization for Nuclear Research (CERN) require large temporary data transfers, which contain very large data flows, to research and education institutes worldwide. Some of this traffic will be transported over Internet2's network.

The Dynamic Resource Allocation via GMPLS Optical Networks (DRAGON) project [8, 9] created optical control plane software that enables users of the Internet2 network to dynamically provision wave services of up to 10 Gbit/sec, allowing researchers to share large amounts of data through dedicated connections. This optical control plane allows very large data flows and deterministic quality of service (QoS). It also allows Internet2 to take large data flows off the normal IP network, avoiding saturation of the IP network.

The optical control plane–enabled network is highly dependable and highly resilient, providing carrier-class reliability and supporting both automatic protection switching and mesh-based network restoration for greater survivability in catastrophic failures. This capability ensures uninterrupted capacity to the research and education community's critical global experiments and collaborations. Furthermore, with the optical control plane–enabled network's inherent flexibility and scalability, the network allows the ability to perform in-service enhancements and upgrades without disrupting network traffic.

12.8 CONCLUSION

Today's mission-critical networks, which include those at educational entities, large enterprises, multinational corporations, medical services, and large financial institutions, are in need of near-zero downtime. Operators rely on global infrastructures to sustain and grow high-margin revenue streams, such as those associated with global voice and data services. An intelligent control plane expands carrier-class operator networks to help operators thrive; scale new service; lower both capital and operational costs; and—most important—survive single, multiple, or catastrophic failures.

Intelligent optical control plane–enabled networks are becoming a major tool to capture time-sensitive customers by turning networks into programmable service delivery engines. In addition, the resulting network automation enables service providers to offer new BoD services, and the ability to offer differentiated services can meet the variety of needs that customer have. As optical control planes move to the Ethernet and photonic layers coupled with the continued advancement in interoperability, optical control planes are fast becoming the cornerstone of communication networks, allowing profitable service revenue generation.

REFERENCES

1. G. Bernstein, B. Rajagopalan, and D. Saha, *Optical Network Control: Architecture, Protocols and Standards*. Reading, MA: Addison-Wesley Professional, 2003, pp. 131–163.
2. OIF UNI Signaling Specification and E-NNI OSPF-Based Routing Implementation Agreement OIF, Standards, 2007.
3. ITU-T G.7712, G.7713, G.7713.1–3, G.7714, G.7714.1, G.7715, G.7715.1–2, G.7718, G.7718.1, G.8080 ITU-T Recommendations, 2003–2007.
4. The IETF Common Control and Measurement Plane (CCAMP). [Online]. Available: http://www.ietf.org/html.charters/ccamp-charter.html
5. IETF RFCs 3471, 3473, 3945, 4202, 4203, 4204, 4206, 4207, 4208, Request for Comments, 2003–2005.
6. Metro Ethernet Forum, MEF Overview. [Online]. Available: http://www.metroethernetforum.org
7. Tele Management Forum, TMF 509, 814, MTOSI, TMF Standards, 2001–2007.
8. DRAGON (Dynamic Resource Allocation via GMPLS Optical Networks). [Online]. Available: http://dragon.east.isi.edu
9. T. Lehman, X. Yang, C. P. Guok, N. S. V. Rao, A. Lake, J. Vollbrecht, and N. Ghani, "Control plane architecture and design considerations for multi-service, multi-layer, multi-domain hybrid networks," presented at the IEEE High Speed Networks Workshop, 2007.

Index

1xEVDO
   TCP-impacting characteristics, 104–105
   TCP throughput, simulated scenario, 121
   web page transfer time, simulated scenario, 122
2.4 GHz frequency direct sequence spread spectrum, 27
2.4 GHz frequency hopped spread spectrum, 27
3G UMTS network architecture, 132–133
3G wireless networks
   poor indoor signal penetration, 128
   slow development, 142
3GPP/3GPP2, 90
10/100Base-T Ethernet hubs, 245
40 MHz channel, 30
   in 802.11n standard, 41–42
40 MHz coexistence, 48–49
40 MHz mixed mode preamble, 42
802.11a standard
   channels used in, 41
   introduction of OFDM in, 27–28
   packet structure review, 38, 39
802.11g standard, 28
802.11n standard, 27–33
   11a packet structure review, 38
   antenna spacing, 36
   coverage area increases, 44
   CSI information use, 45–46
   explicit feedback, 46
   frame aggregation, 32
   future standard developments, 50–51
   Greenfield preamble, 42–43
   guard intervals, 41
   high throughput provisions, 40–43
   implicit feedback, 46
   low-density parity check codes, 47
   MAC enhancement features, 31
   MAC overview, 47–50
   mandatory and optional PHY features, 30
   MIMO and SDM basics, 33–37
   mixed format high throughput preamble, 38, 40
   multipath fading environment with, 36
   partial overlapping channel conditions, 49
   PHY interoperability with legacy devices, 37–40
   protection mechanisms, 49–50
   receive diversity in, 44
   robust performance features, 43–47
   space-time block coding (STBC), 44–45
   spatial expansion, 44
   transmit beamforming, 45–47
   use of weights to improve reception, 45
802.16 standards
   advanced technologies, 76
   control channels, 74–75
   frame structure, 74
   Mobile WiMAX and, 54–55
   pilot overhead, 75
   variable frame sizes, 55
802.16 standards, frame structure, 73
806.16c standard, 54
802.16e standard, 55
911 emergency service, 140–141

A

Absolute service differentiation, 279
Accelerometer sensor, 9
Access deferral, 49
Access network delayering, 182–183
Access networks, 182
ACK packet congestion, 105, 106
Adaptive burst length threshold, 272
Adaptive modulation, 54
Adaptive switching, 68
Adaptive mode selection, 61–62
Advanced Antenna Subsystem (AAS) modes, 54
Advanced Metering Infrastructure (AMI), 202
   utility benefits from, 228
Advertisements, preinsertion, 86
Affordability, BPL communications, 198
Air interface standards, broadband satellite systems, 157
Alamouti scheme, 44, 162–163
Aloha protocols, 210–211
Alternative energy sources, 2, 8
Ambient-to-ambient communications, 17
Amplitude and phase shift keying (APSK), in satellite communications, 156
Antenna spacing, 802.11n standard, 36
Application-Specific Integrated Circuit (ASIC) design, 205
Application widgets, 96
ARABSAT, 148
Arbitration-based protocols, QoS and, 210, 211
ASON optical control plane model, 294, 296
   domain-based model, 297
   external-network network interface, 298
   internal-network network interface, 298

mapping of ASON DCM attributes to RESP-TE objects, 301
network-network interface-based control plane operation, 298–299
optical-user network interface, 296
ASON standards, 307
Attenuation
  as limiting factor for BPL networks, 212
  over power lines, 207
Authentication, 16
  and OBS networks, 280
  services considerations, 91–92
Authorization, services considerations, 91–92
Automated management functions, 292
Automatic Protection Switching (APS), 304
Automatically Switched Optical Network (ASON) model, 293. *See also* ASON optical control plane model
Autonomic systems, 1, 15, 17, 20
  technology evolution, 13–15
Aware environments, 22

B

B-PON standard, 175
Backbone VLAN identifier (B-VLAN), 249
Backhaul
  BPL for wireless, 202
  femtocell solutions, 128
Backhaul communications, 229–230
Balance checks, 85
Bandwidth
  drop in value, 4
  entertainment screens, 11
Bandwidth Delay Product (BDP), 107
Bandwidth estimation techniques, 112
Bandwidth measurement, for BPL, 213
Bandwidth oscillation effect, 102
  spurious TCP retransmission due to, 103
Bandwidth variations, in BPL networks, 212, 220
Base station coordination, in 802.16m standard, 76
Bidirectional transmission, with GPON, 181
Bill of material (BOM) weight, 187
Bindings, 87
Biometrics, in cell phones, 10
Bit transfer rate, technology evolution, 6
Block acknowledgment (BA) mechanism, 802.11n standard enhancements, 33, 48
Block error rate (BLER) performance, Mobile WiMAX, 64
Blogs, 83
Bose–Chaudhuri–Hocquenghem (BCH) codes, 156
BPL broadband access, 218
  PON-BPL architecture, 220, 224–225
  using passive optical network (PON), 218, 220

BPL communications, 195
  advantages over wireless, 235
  as backhaul for WiMAX, 202
  broadband access, 218–225
  for broadband access, 195, 200–202
  coupling issues, 209
  developing world benefits, 198–200
  historical perspective, 196–198
  home Internet access, 217
  for home networking, 195, 200, 214–218
  for meter communications, 203
  networking standards, 232–234
  noise problems, 208–209
  PHY and MAC layer technologies for, 210–211
  signal characteristics, 206
  smart power grids for, 195, 202–205, 225–232
  technology concept, 197
  for wireless backhaul, 202
BPL gateway
  implementation example, 222
  MDU Internet access using, 219
BPL network devices, 211–212
BPL network management, 212, 214
BPL networking standards, 232
  CEPCA standards, 233
  ETSI PLT standards, 233
  HomePlug Powerline Alliance standards, 233–234
  IEEE standards, 232
  research programs, 234
  UPA standards, 232
BPL networks, 199
  advantages over wireless home networks, 215
  bandwidth variations in, 212
  limiting factors, 212
BPL signal characteristics, 206
BPL technology, 205
  bandwidth measurement, 213
  frequency modes, 208
  network devices, 211–212
  network management technologies, 212, 214
  overcoming power lines' limitations with, 207–210
  PHY and MAC layers, 210–211
  power grid architecture, 205, 207
  signal characteristics, 206
Broadband access, BPL-enabled, 195, 200–202
Broadband Global Area Network (BGAN) service, 157
Broadband over Power Line (BPL) communications, 195. *See also* BPL communications
Broadband proliferation, and femtocell vision, 126
Broadband satellite communications, 156–158
  air interface standards, 157
Broadband services, satellite-based, 155–156

Index

Broadcast services, in 802.16m standard, 76
Broadcasting-satellite services (BSS), 147
Bundling, in optical control planes, 303
Burst assemblers, 273
Burst assembly, 272–273
Burst disassemblers, 273
Burst early dropping, 279
Burst header packets, 269
Burst headers, 269
    authentication of, 283
    signaling by, 270
Burst length threshold, 272
Burst priority, 272
Burst scheduling, 273
    horizon scheduling, 273, 275–276
    LAUC-VF, 276–277
    optimal, 277–278
Business ecosystems, technology evolution, 18–21
Business models, service layer considerations, 96–97
Buying behavior, 82

C

Capacity
    adding through femtocell technology, 126
    femtocell solutions, 127, 128–129, 141
    GPON vs. WDM-PON case, 191
    MIMO-SISO system comparison, 37
    WDM-PON vs. GPON, 178–179
Capital investment, in deployment of next-generation access networks, 168
Car manufacturing companies, as business ecosystems, 19
Carrier Ethernet for mobile backhaul, 239, 240, 257
    connectivity scheme, 259
    defined, 257–260
    delay and jitter issues, 261, 263
    motivations and challenges, 260–261
    reliability issues, 263
    schematic, 262
    synchronization challenges, 261
Carrier Sense Multiple Access (CSMA)
    for BPL systems, 211
    with collision detection, 240
Carrier to interference ratio (CIR), 105
Case studies
    next-generation access architectures and technologies, 189–192
    next-generation digital media services, 92–96
    novel TCP performance-enhancing proxy, 114–122
cdma2000 1xRTT, 101–102
cdma2000 Evolution Data-Only (1xEVDO), 104
CDWM wavelength comb, 176

Cell phones
    1-TB capacity, 7
    contextualization, 22–23
    as control devices, 25
    data generation by, 12
    detecting biometric parameters with, 10
    embedding sensors in, 9
    as lens for information, 25
    and limitations of telephone networks in developing countries, 198
    mobile network shift, 15
Cell splitting
    increasing capacity by, 127
    through femtocell technology, 126
Cellular radio, TCP-impacting characteristics, 101–102
Centralized control, demise of, 14
Centralized control planes, 293–294
Channel quality information (CQI), 54
    satellite communication requirements, 161
Channel quality information channel (CQICH), 58
Channel state information (CSI), uses in 802.11n standard, 45–46
Channel switching behavior, in TCP, 102
Chip production plants, investment required for, 2
Churn, femtocell solutions, 128–129
Clarke, Arthur C., 146
Clean energy applications, with BPL networks, 204
Clear channel assessment (CCA), 48
Client service paradigm, shift to Ethernet-based services, 258
Client-side demand management, with BPL, 203
Closed femtocells, 140
    911 issues, 141
Closed loop beamforming, 69
Closed loop multiple antenna technologies, 62
    maximum ratio transmission, 62–63
    statistical Eigen beamforming, 63
Cloud computing paradigm, 9, 13
Cluster paradigm, 17
Coarse WDM (CWDM) standard, 174
Code Division Multiple Access (CDMA), 90, 100
    in BPL communications, 210
Code multiplexing, 105
Coexistence rules, in 802.11n standard, 49
Collaboration, health care industry example, 97
Collision avoidance, for BPL systems, 211
Collision detection, 240
    for BPL systems, 211
Color noise, over BPL systems, 209
Colored ONU optical interfaces, 174
    architectures eliminating need for, 176
Colorless ONUs, 176, 177
Common Management Information Protocol (CMIP), 14
Communication service providers (CSPs), 80

Communications, as fabric of connection, 25
Communications models, technology evolution, 22–25
Competing uplink and downlink flows, 106
Competing uplink flows, 105
Competition, 22
Complementary code keying (CCK), 27
Complementary cumulative distribution function (CCDF), 37
Composite services, 81
Computer industry, as business ecosystem, 19
Confidentiality
 of data bursts, 283
 in security context, 280
Congestion control capability, 114
Congestion window (CWND), 100, 101
Connection management, 299, 300. *See also* Signaling
Connectivity
 charges for, 20
 disappearance of importance, 20
Constant Time Burst Resequencing (CTBR) scheduler, 278
Consumer demand, *vs.* spectral efficiency improvement, 127
Consumer Electronics Power Line Communication Alliance (CEPCA), BPL networking standards, 233
Consumer service providers (CSPs), 83
Consumer services trends, 80–81
 context-aware services, 81–82
 service hierarchy, 81
 social networking, 83
 user-generated content, 82
Content producers. *See also* User-generated content
 consumers as, 81
Content upload, 93–94
Contention-based protocols, higher data rates with, 210
Context-aware objects, 10
Context-aware services, 81–82
Contextualization, 5
Contour-based priority, 280–281
Control channel coverage, 74
 Mobile WiMAX, 58–59
Control channel transmission, Mobile WiMAX, 75
Control message sizes, Mobile WiMAX, 57
Control planes. *See* Intelligent optical control planes
Control points, 18
Convergence, 80
 and need to improve TCP, 100
 and satellite communications, 156
Convolution Turbo Code (CTC), 54
Cooperative diversity techniques, satellite communications, 161–164

Copper-based access network
 connection limitations, 167
 evolution to fiber-based architecture from, 167
 as passive structure, 169
 physical signal separation via, 168
 topology, 169
Core routers, in OBS networks, 270, 273, 275
Corinex management software, 214
Coupling, BPL issues, 209
Coverage
 capacity, churn (3Cs), 128–129
 of satellites systems, 148, 149
CPE Wide Area Network (WAN) Management Protocol, 138
Crankback, 305
Critical Peak Pricing (CPP) programs, 228
Customer care, paradigm shifts, 21–22
Customer premises equipment (CPE) devices, 127, 137. *See also* Femtocell base stations
 in BPL networks, 218
CWDM PICs, cost behavior, 189
CWDM SFP, cost comparison with PIC couple, 189
Cyclic shift transmit diversity, 61
Cyclical Array Wave Guides (AWGs), 177

D

Data aggregation
 in 802.11n standard, 47
 in PON-BPL networks, 201, 220
Data bursts, 269
Data centers, shrinking numbers, 16
Data flow, 7
Data rates
 with 802.11n standard, 28
 with contention-based protocols, 210
 efficiency issues with 802.11n standard, 31
 femtocell *vs.* UMA, 131
 increases with MIMO, 40
 with MIMO/SDM systems, 34
 over BPL systems, 205
Data synchronization, 16
Datacentric service, 80
Dead spots, in femtocell applications, 140
Decentralization, enabling in power generation, 203
Delay, in carrier Ethernet, 261, 263
Demand response (DR), in BPL networks, 227
Demand-side power management, in BPL networks, 229–230, 2332
Democracy, contextualization issues, 5
Democratization of energy, through BPL technology, 235
Dense Wavelength Division Multiplexing (DWDM), 171, 266, 292
 support by OBS, 269
Destination MAC address field, 242

Developing world, BPL benefits for, 198–200
Digital cameras, resolution, 11
Digital locker service, 82, 93
    case study, 93–94
Digital media services, service layer for, 79–80
Digital storage, technology evolution, 6–8
Displays, 1
    technology evolution, 10–12
Disruption threshold, 4
Distributed control, 14
Distributed control planes, 293–294
Distributed data banks, 13
Distribution cost, service domain, 4
Distribution network, leveraging, 83
Diversity gain
    with Alamouti scheme, 163
    in satellite communications, 162
Downlink spectral efficiency, 68
    comparison with HTTP traffic, 68
Dual-header OBS, 277
DUPACKs, additional data transmission on, 115
DVB project, 156
DVB-return channel by satellite (DVB-RCS), 157
DVB-S2, 156, 161
DVB-SH, 161
Dynamic frequency selection (DFS) rules, 50
Dynamic Resource Allocation via GMPLS
    Optical Networks (DRAGON), 312
Dynamic routing, 84
Dynamic service discovery and invocation, 85
Dynamic sharing, 105
Dynamic wavelength grouping, 279

E

E-PON standard, 175
E-textiles, 9
Eastbound interfaces, 88
Ecosystems
    loose relationships in, 18
    value added, 19
Edge routers, in OBS networks, 273, 274
Educational capital availability, as barrier to
    evolution, 5
Effective bandwidth per user, WDM-PON *vs.*
    GPON architectures, 179
Effective SINR, 65, 67
EFLOPS, 8
Eigen beamforming. *See* Statistical Eigen
    beamforming
Electronic watches, evolution of, 4
Embedded intelligence, 21
Emergency calls, femtocell issues, 140–141
Energy consumption
    as bottleneck to technological evolution, 2
    growth of, 2, 3
    reduction per GFLOP, 8

enhanced Telecom Operations Map (eTOM), 88
Enterprise Application Integration (EAI), 87
Enterprise environments, 28
Enterprise Service Bus (ESB), 87
Error rates, with power line data transmission, 208
Error recovery, improving in TCP, 110
Ethernet bandwidth growth, 241–242
Ethernet-based PON, 173
Ethernet bridges, 245–246
Ethernet frame, 242, 244
    Ethernet Type/Length field, 242
    frame check sequence, 244
    MAC client data and pad field, 242, 244
    preamble, 242
    source MAC address field, 242
    start frame delimiter field, 242
Ethernet header evolution, 246–247, 248
    PBBs, 249–250
    PBs, 247–249
    VLANs, 247
Ethernet hubs, 245
Ethernet LAN, shared medium, 241
Ethernet MAC frame, 243
Ethernet optical transport networks, 239–240
    carrier Ethernet for mobile backhaul, 257–263
    technology overview, 240–250
Ethernet optical transport standards activities,
    254–255
Ethernet optical transport technologies, 250
    Ethernet optical transport standards
        activities, 254–255
    Ethernet over fiber, 250–251
    Ethernet over MPLS and MPLS-TP, 252–254
    Ethernet over next-generation SONET/SDH
        (EoS), 251–252
    Ethernet over WDM, 251
    Ethernet transport over PBB-TE, 254
    Ethernet transport with intelligent optical layer,
        255–257
Ethernet over fiber, 250–251
Ethernet over MPLS and MPLS-TP, 252–254
Ethernet over next-generation SONET/SDH (EoS),
    251–252
Ethernet over WDM, 251
Ethernet packet intelligence, 256, 257
Ethernet repeaters, 245
Ethernet services
    client service paradigm shift to, 258
    extending to MAN/WAN environment, 250
Ethernet switches, 245–246
Ethernet technology overview, 240
    bandwidth growth, 241–242
    Ethernet bridges and switches, 245–246
    Ethernet frame, 242, 244
    Ethernet header evolution, 246–250
    Ethernet repeaters and hubs, 245

half-duplex and full-duplex Ethernet, 244
historical development, 240–241
Ethernet transport approaches, 251
Ethernet transport over PBB-TE, 254
Ethernet transport with intelligent optical layer, 255–257
ETSI PLT, BPL networking standards, 233
European Organization for Nuclear Research (CERN), 312
European Telecommunications Standards Institute (ETSI), 156
    BPL networking standards, 232
EUTELSAT, 148
Exabyte capacity, 7
Explicit Congestion Notification, 107
Explicit feedback
    in 802.11n standard, 46
    weights requirements, 47
Exponential Effective SIR Mapping (EESM), 64
Extensible Markup Language (XML) schemas, 86
Extension LTFs, use in 802.11n standard, 46
Extremely high-voltage (EHV) power lines, 205

F

Fabry–Perot Laser Diode (FPLD), 176
Facebook, 83
Fast link adaptation, 33
Fast mesh restoration, 304
Fast Retransmission, compromises to, 102
Fast scheduling, in 802.16 standard, 54
Federated architecture, with SOA approach, 85
Femto Forum, 138
Femto Gateway, 133, 134
FEMTO solution, 15, 16
Femtocell access point (FAP), 129
Femtocell base stations, 125–126
    3C challenges, 128–129
    3G UMTS network architecture, 132–133
    backhauling over IP broadband links, 128
    business drivers, 130
    concept and vision, 126–132
    cost advantages, 129–130, 139
    cost per bit issues, 141
    emergency calls, 140–141
    evolution of network architecture, 133
    future developments, 141–143
    ideal indoor ratio conditions, 129
    in 802.16m standard, 76
    integrating with existing core networks, 133
    licensed spectrum, 138–139
    LTE femtocells, 141–142
    Iuh network architecture, 136
    minimization of radio interference by, 129
    network architecture, 132–135
    network interface protocols, 136–137
    number of calls supported, 129
    open *vs.* closed access, 140
    operation in licensed spectrum, 131
    ping-pong effect, 140
    rationale, 126–128
    remote device management and control, 137–138
    revolutioning network rollout, 142–143
    RF interference management with, 139–140
    security aspects, 138
    as self-organizing networks, 138–141
    signaling plane, 135–138, 137
    SIP/IMS-based architecture, 135
    sniffer capabilities, 139
    standalone *vs.* integrated, 130
    star rating comparisons with UMA, 131
    tunneled Iub network architecture, 136
    UMA-based architecture, 134–135
    UMTS-centric network architectures, 133–134
    Uu interface protocols, 136
    vs. UMA, 130–132
Fiber-based architecture, transition from copper-based, 167
Fiber delay lines, 268
Fiber deployment, in Europe, America, Asia, 168
Fiber to the BPL (FTT-BPL) broadband access networks, 224
Fiber to the building (FTTB) architecture, 170, 218. *See also* FTTB architecture
Fiber to the cabinet (FTTCab) architecture, 171
Fiber to the curb (FTTC) architecture, 171, 218. *See also* FTTC architecture
Fiber to the home (FTTH) architecture, 170, 218. *See also* FTTH architecture
Fiber to the node (FTTN) architecture, 171, 200. *See also* FTTN architecture
Fiber utilization, WDM-PON *vs.* GPON, 181
Fixed burst length threshold, 272
Fixed mobile convergence (FMC), 130
Fixed module substitution (FMS), 130
Fixed-satellite services (FSS), 147, 148
Flash memory, 6
Flexibility, WDM-PON *vs.* GPON, 181
Flexible resource allocation, Mobile WiMAX, 59
Forward error correction
    in 802.16 standards, 54
    modal dispersion limitations, 180
Forward links, 156
    in satellite communications, 147
Foundation services, 80
Fractional frequency reuse, 69
    in Mobile WiMAX, 59–60
    Mobile WiMAX, 69–70
Frame aggregation, 31
    throughput *vs.* PHY data rate with, 32
Frame control header (FCH), 56
Frame error rate (FER), Mobile WiMAX, 59

Index

Frame structure
   802.16m, 73, 74
   Mobile WiMAX, 72
Frequency bands, satellite communications, 147–150, 149, 150
Frequency-diverse subchannelization schemes, 54
Frequency modes, for BPL, 208
Frequency shift keying (FSK), 210
Frequency-specific subchannelization schemes, 54
FTTB architecture, 170
   cost structure, 184
   GPON vs. WDM-PON deployment cost comparison using, 191
   installation cost structure, 184
FTTC architecture, 170
FTTH architecture, 170
   cost structure, 184
   installation cost structure, 184
FTTN architecture, 170
FTTx architecture, 190
Full-duplex Ethernet, 244
Full Service Access Network (FSAN) working group, 173
Full usage subchannelization (FUSC), 56
Functional offset time, in dual-header OBS, 277

G

Gain
   with MIMO systems, 37
   receive diversity, 44
General Packet Radio Service (GPRS), 103
Generalized Multi-Protocol Label Switching (GMPLS), 293. *See also* GMPLS standards
Generic Access Network Controller (GANC), 134
Generic Framing Procedure (GFP), 251, 294
Geostationary orbit (GSO) satellite, 146, 152. *See also* GSO satellite
GEPON standard, 173, 174
Gigabit-class equipment, 184
Gigabit Ethernet standard, 241
Gigabit PONs (GPONs). *See also* GPON architecture
   in BPL networks, 224
Global positioning systems (GPSs), femtocell uses, 139
Global System for Mobile Communications (GSM), 90, 100
   indoor coverage, 128
Glocal innovation, 4–6
Glocalization, 5
GMPLS standards, 307, 308, 310
Google™ Maps, 19
GPON architecture, 177–178
   bidirectional transmission, 181
   capacity, 178–179
   capacity per user, 179
   cost comparison with WDM-PON for different numbers of users, 192
   cost determinants, 184
   deployment cost comparison with WDM-PON, 191
   fiber utilization, 181
   flexibility, 181
   integrated optics approach, 185
   micro-optics approach, 185
   North American use, 173
   optical link budget, 179–180
   optical transmission infrastructure and protocols, 181
   performance comparison with WDM-PON, 178–181
   security and unbundling, 180–181
   standard wavelength allocation, 178
   triplexer functional scheme, 185
GPON optical interfaces, 185–188
GPON processor, cost *vs.* volumes, 187
GPON standard, 175
GPON triplexer
   cross-section, 186
   top view, 186
GPRS/GSM, TCP-impacting characteristics, 103–104
Greenfield environments, 30, 33, 49–50
Greenfield preamble, in 802.11n standard, 42–43
Grid infrastructure upgrades, 203
Group Special Mobile, 15
GSO satellite, equatorial plane, 147
Guard intervals
   in 802.11n standard, 41
   in Mobile WiMAX, 56

H

Half-duplex Ethernet, 244
Handheld terminals, mobile satellite broadcasting to, 158–159
Handoff issues, femtocell to macrocellular networks, 135
Handset availability, femtocell *vs.* UMA, 131
Handset battery life, femtocell *vs.* UMA, 131
Hard drives, technology evolution, 6
HD television, 11
Health care industry, services collaboration in, 97
Hierarchy control, absence in Internet, 14
High Speed Downlink Packet Access (HSDPA), 104
High throughput (HT)
   and 40 MHz channel, 41–42
   in 802.11n standard, 27–33
   802.11n standard provisions, 40–43
   protection mechanisms in 802.11n standard, 49–50

High Throughput task group, 28
High-voltage (HV) power lines, 205
HNB Gateway, 134
Home area communications, 230
Home base stations, 125–126. *See also* Femtocell base stations
Home eNodeB (HeNB), 141
Home Internet access, BPL-enabled, 217
Home networking, 214–215
  BPL-enabled, 195, 200, 214, 216, 220
  in multidwelling units, 215, 218
Home NodeB (HNB) luh, 134
Home plugs, in BPL networks, 201
HomePlug Powerline Alliance, BPL networking standards, 233–234
Horizon scheduling, 271, 273, 275–276
Hotspot environments, 28
HSPDA, TCP-impacting characteristics, 104–105
Hybrid Automatic Repeat Request (HARQ) techniques, 54
Hybrid networks
  PON-BPL, 200
  satellite-based, 159–160
  TCP performance enhancement in, 99–100
Hybrid systems, in satellite communications, 160

I

I-TCP, 110–111
Icons
  as centers of future communication, 23
  contextualization, 24
Identity management, 16
IEEE 802.1p, 212
IETF RFC standards, 309
IMMARSAT, 148, 157
Impedance variation, 208
  over power lines, 207
Implicit feedback
  calibration requirements, 47
  in 802.11n standard, 46
Impulse noise, over power lines, 209
In-home wireless traffic, routing through broadband connections, 127
In-shoring, 5
Inclination angle, of satellite orbits, 153
Indoor environments
  femtocell coverage in, 126
  poor 3G wireless penetration, 128, 142
Information elements (IEs), in Mobile WiMAX, 57
Information sharing, control of, 24
Infrared, in 802.11 standard, 27
Infrastructure aging, BPL solutions, 204
Iniewski, Krzysztof, vii
Initial Congestion Window (ICW), 10

Innovation
  global market for, 5
  glocal, 4–6
  pursuing to increase efficiency, 18
Institute of Electronic and Electrical Engineers (IEEE), BPL networking standards, 232
Integrated FAP, 130
Integrated optical components, 187
Integrated optics, GPON approach, 185
Integrated satellite networks, 159–160
  with cooperative diversity, 162
Integrity, in security context, 280
Intelligence
  at network edges, 14
  outside network, 15
  shift from network to edges, 15
Intelligent optical control planes, 291–292
  architectures, 292–294
  ASON model, 294–299
  centralized *vs.* distributed, 293
  control plane coordination with connection-oriented transport network, 295
  control plane interactions, 294
  elements, 299–303
  functionality, 304–305
  implementation, 293–294
  ITU-defined UNI/E-NNI messages, 302
  layers, 292–293
  link bundling in, 303
  multiple failure scenario, 305
  neighbor discovery in, 299
  path link attributes in, 303
  routing and path computation in, 301–303
  signaling in, 299–301
  standards, 305–312
  standards body interactions, 311
  usage in scientific community, 312
Intelligent state prediction, 114
Intentional Burst Dropping scheme, 279
Interactive objects, 22
Interactive satellite broadcasting, 156–158
Interactive services, REST for, 87
Interference
  BPL network advantages, 202
  BPL prevention methods, 209
  femtocell *vs.* UMA, 131
  to macrocell networks from MDU-based femtocells, 139
  management in femtocells, 139–140
  minimization by femtocell technology, 129
  RRM algorithm requirements for femtocells, 132
  with satellite communications, 149

Index

Intermediate nodes, in FTTx networks, 170
International Telecommunication Union (ITU), 154, 159, 173
   G.hn study group, 234
   recommendation and reports classification, 155
   satellite study groups, 154
Internet
   absence of hierarchy of control, 14
   with embedded intelligence in objects, 21–22
   resistance from telecom engineers, 14
   resistance to local faults, 14
Internet Engineering Task Force (IETF), 305
Internet Protocol over Satellite (IPoS) standard, 157
Internet Protocol Television (IPTV), 82
Interoperability
   Ethernet over fiber issues, 251
   Ethernet over PBB-TE, 254
   with SOA approach, 85
INTERSPUTNIK system, 148
Invertible channel matrix, with 802.11n standard, 36
IP DiffServ, 212
IP/MPLS control plane, 292, 293
IP Multimedia Subsystem (IMS), 88
iPhone, 9, 13
   tiny apps, 7
iPod, 20
IPsec, use in femtocell applications, 138
iTouch, 9
ITU-T Automatic Switched Optical Network (ASON) framework, 255, 307
   standards, 309

J

Jitter, carrier Ethernet issues, 261, 263
Just-Enough-Time (JET) protocol, 271

K

Kepler's laws, and satellite communications, 150–151
KOREASAT, 148
Ku/Ka bands, 156

L

Landline service providers, 80
Large hadron collider (LHC) experiment, 312
Last mile solutions, 218
   BPL communications, 196, 207
Latency, 113
   reducing for Mobile WiMAX, 73
   with TCP over WLANs, 105

Latest available unscheduled channel with void filling (LAUC-VF), 276
LAUC-VF, 276–277
   high computational complexity, 278
Law of equal areas, 151
Law of periods, 151
Layered coding, for satellite communications, 164
LDPC coding, in 802.11n standard, 43
Legacy devices
   coexistence of 802.11n with, 30
   in future standards, 50–51
   Mobile WiMAX, 72
   PHY interoperability under 802.11n standard, 37–40
Licensed spectrum, femtocell operation in, 131, 138–139
Limited Transmit, 107
Line-of-sight (LOS) environments
   806.16c standard and, 65
   importance for satellite communications, 146
   requirements for satellite quality, 149
Link aggregation, 304
Link budget classes, 178
Link bundling, in optical control planes, 303
Link Capacity Adjustment Scheme (LCAS), 294
Link layer latencies, with GPRS, 104
Link performance
   Mobile WiMAX, 64
   TRL-PEP, 115
Link simulators, 63
Link-to-system mapping, Mobile WiMAX, 65–67
Load control, in BPL networks, 228
Local data exchange, 7
Local span mesh restoration (LSMR), 304–305
Locking wavelength, 177
Long Term Evolution (LTE), 12
Look-ahead window, 280
Loose relationships, within ecosystems, 18
Loosely coupled reusable services, 84
Loss-based TCP variants, 108
   enhancements to, 109
Loudspeakers, 23
Low-density parity check codes, 31, 54
   in 802.11n standard, 47
   in satellite communications, 156
Low Earth orbits (LEOs), 152
Low voltage (LV) power lines, 205, 225
LTE femtocells, 141–142
   20 MHz-wide channel requirements, 142
   citywide and nationwide, 142
Luh network architecture, 134
   in femtocells, 136

M

MAC layer
  802.11n standard enhancements, 31
  and 40 MHz coexistence in 802.11n, 48–49
  BPL communications technologies, 210–211
  efficiency enhancements, 47–48
  fixed and dynamic access schemes, 210
  overview of 802.11 features, 47
  resource sharing at, 105
  throughput vs. PHY data rate, no changes to, 29
MAC protocol data unit aggregation (A-MPDU), 47
MAC service data unit aggregation (A-MSDU), 47
Malicious burst headers, 282
Many-to-many conversations, 83
Mashup applications, 19, 84
  case study, 94–96
Mashup assembler, 94
Mashup platform, 90
Matrix B scheme, 61
Maximum ratio combining (MRC) receiver, 58
Maximum ratio transmission, 62–63
Media access control (MAC) layer, negative impact on TCP throughput, 101
Media-enabled service bus, 85–87
Medium access control (MAC) layer, 28. *See also* MAC layer
Medium Earth orbits (MEOs), 152
Medium voltage (MV) power lines, 205
  gateway, 230
MEF specifications, 263
MEF UNI and EVC, 260
Message transformation, 85
Messaging properties, 85
Metadata, 13, 17
  managing for digital content, 86
Metaservice generation, 17
Meter communications, 230
  BPL for, in smart power grids, 203
  BPL implementation example, 231
Metro access networks, 182
  optical burst switching in, 287–288
Metro core networks, 182
  European vs. American, 183
Metro Ethernet, 250, 259
Metro Ethernet forum, optical control plane standards, 310
Metropolitan area networks (MANs), 287–288
Metropolitan Ethernet Forum (MEF), 287
Micro-optics, GPON approach, 185
Microelectronics, 187
  GPON interfaces, 186
Microprocessing, 8
MIMO equalizer, 35
MIMO receiver, 38, 40
MIMO systems, 55
  and 802.11n standard, 33–37

capacity comparison with SISO system, 37
femtocell applications, 141
mathematical model, 35
receiver block diagram, 40
schematic, 34
MIMO technologies, 802.16m standard, 75–76
Minimum Mean Square Error (MMSE) design, 61
Minimum Starting Void (Min-SV), 277
Mixed format HT preamble, 37, 42
  in 802.11n standard, 38, 40
MMS, case study, 93–94
Mobile backhaul, carrier Ethernet for, 257–263
Mobile base stations, migration to Ethernet backhauling, 181
Mobile-satellite services (MSS), 147, 148
Mobile service providers, 80
Mobile traffic, carrying over public Internet via femtocells, 138
Mobile WiMAX, 53–54
  802.16m frame structure evolution, 73
  802.16m MIMO technologies, 75–76
  802.16m physical resource allocation, 73–74
  and 802.16 standards, 54–55
  cell coverage, 59
  control channel coverage, 58–59
  control channels, 74–75
  downlink spectral efficiency, 66, 68
  downlink system performance summary, 67–69
  evolution, 53–54, 71–76
  fractional frequency reuse, 59–60
  with fractional frequency reuse, 69–70
  frame structure, 56–58
  IEs and sizes, 57
  link performance summary, 64
  link-system mapping methodology, 67
  link-to-system mapping, 65
  multiple-antenna technologies, 60–63
  number of OFDM symbols for bearer data, 58
  pilot overhead vs. 802.16m standard, 75
  relative improvements of fractional frequency reuse, 70
  system performance, 63–71, 68
  system simulation parameters, 64
  TDD frame structure, 56
  uplink performance, 66
  uplink performance with multiple receive antennas, 69
  VoIP performance, 71
Modal dispersion, in GPON architectures, 180
Modulation and coding scheme (MCS), 33
Moore's law, 4, 129
MPEG-21 Digital Item Declaration (DID), 86
MPLS, 252–254, 307
MPLS-TP, 252–254
MPLS-Transport Profile (MPLS-TP), 253
Multi Protocol Label Switching (MPLS), 252
Multiantenna operations, in 802.16 standards, 54

Index

Multicarrier operation
   in 802.16m standard, 76
   satellite technologies, 160–161
Multicast services, in 802.16m standard, 76
Multidwelling units (MDUs)
   BPL home networking in, 215, 218
   extension of home networking to, 200
   femtocell applications, 138
   Internet access with BPL gateway, 219
Multihop relays, in 802.16m standard, 76
Multilane OBS, 276
Multimedia message service (MMS), 80
Multipath fading environment, 36
   indoor environments, 128
Multiperson chat, 83
Multiple-antenna technologies
   closed loop, 62–63
   in Mobile WiMAX, 60
   open loop, 60–62
Multiple failure scenario, intelligent optical control plane mesh networks, 305
Multiple-input, multiple-output (MIMO) systems, 30. *See also* MIMO systems
   and 802.11n standard, 33–37
Multiple System Operators (MSOs), role in future femtocell technologies, 141
Multiuser data transmission, Mobile WiMAX, 76
MySpace, 83

N

Narrowband background noise, over BPL systems, 209
Neighbor discovery, 299
NettGain, 112
Network effects, 90
Network management system (NMS), interface with optical control plane, 291
Networking paradigms, technology evolution, 15–18
NewReno TCP, 101, 108
Newton's theory of gravitation, 150
Next-generation access, 167–169
   access network delayering, 182–183
   access segment after delayering, 183
   architectures, 169–174
   architectures performance comparison, 178–181
   case study, 189–192
   copper-based access network topology, 169
   cost per user, 191
   cost structure, FTTB and FTTH installations, 184
   CWDM wavelength comb and G.709 fiber attenuation profile, 176
   effects of volumes on costs, 187
   Fiber to the building (FTTB) architecture, 170
   Fiber to the cabinet (FTTCab) architecture, 171
   Fiber to the curb (FTTC) architecture, 171
   Fiber to the home (FTTH) architecture, 170
   Fiber to the node (FTTN) architecture, 171
   fiber utilization comparisons, 181
   flexibility comparisons, 181
   FTTx architectures, 170
   future-proofing infrastructure, 168
   GPON architecture, 177–178
   GPON optical interfaces, 185–188
   hybrid TDM/WDM schemes, 174
   key technologies, 184–189
   operational cost pressures, 168
   optical link budgets, 179–180
   passive optical network topology, 171
   security and unbundling, 180–191
   TDM-PON functional scheme, 173
   three-layer metro and access network, 182
   WDM-PON functional scheme, 172
Next-generation digital media services, 79–80
Next-generation networks (NGNs), future satellite systems for, 145
Next Generation Operational Support Systems (NGOSS), 88
NodeBs, 132. *See also* Femtocell base stations
Noise
   as limiting factor for BPL networks, 212
   over power lines, 207, 208–209
Non-IP communications, 10
Non-line-of-sight (NLOS) conditions
   and 802.16 standards, 54
   role in 802.11n standard, 36
Northbound interfaces, 88
Null Data Packet, use in 802.11n standard, 46

O

O/E.O conversion, 267
Objects
   as distribution and access points, 25
   embedded intelligence in, 21–22
   interactive, 22
   transformation into communicating devices, 10
OBS core router architecture, 275
   operations flow, 287
   security in, 286
OBS edge router architecture, 274
OBS ingress edge router, operations flow, 286
OBS networks, embedded security framework, 283–284
OFDM symbols, availability in Mobile WiMAX, 58
Offset-based QoS, 278
Offshoring, 5
OIF standards, 309, 310–312
OLED technology, 22
OMA, 90
   standards activities, 91

Radio resource management (RRM) algorithms, 132
Radio resource management time scale, 113
Rain attenuation
  mitigation by ACM mode, 161
  in satellite communications, 149
Rapid Spanning Tree Protocol (RSTP), 246
Rate-based control, 112
Rayleigh fading channel, 37
Real-time performance data, 85
Receive antennas, 44
  in 802.11n standard, 35
Receive diversity, in 802.11n standard, 44
Receive Time Gap (RTG), 56
Receiver block diagram, MIMO systems, 40
Receiver Window Size limit, 107
Reconfigurable OADM, 267–268
Regional Bell operating companies (RBOCs), 183
Registry lookup, 85
Regulatory issues
  femtocell classification as wireless base stations, 139
  femtocell *vs.* UMA, 131
Reliability
  with BPL networks, 204
  with carrier Ethernet, 258
  carrier Ethernet issues, 263
Remote device management
  in BPL networks, 214
  with femtocells, 137–138
Remote locations, BPL advantages for, 196
Remote procedure call (RPC) communications, 87
Replay attacks, 283
Representational State Transfer (REST), 85, 87
Residential environments, 28
Residential femtocell gateways, 129
Resolution, display devices, 11
Resource allocation packet, in OBS networks, 277
Resource sharing, at MAC/link layer, 105
Retailers, cell phone opportunities, 25
Retransmission, on partial acknowledgments, 116, 118
Retransmission timeout
  on partial acknowledgments, 115–116
  standard TCP reaction to, 101
Return links, 156
  in satellite communications, 147
RF data collectors, for BPL networks, 211
RFID technology, 9–10
R0GigE/100GigE standard, 242
RIFS bursting, 33, 49
Right-of-way issues, BPL advantages, 202
Robustness
  Ethernet over fiber issues, 251
  in 802.11n standard, 43–47
  STBC *vs.* SM-MIMO, 62
Roomba, 14, 19

Round-trip time (RTT)
  enhancements by WCDMA/UMTS, 104
  GPRS, 103
Routing
  of digital media, 85, 86
  in optical burst switching, 273
  in optical control planes, 301–303

S

SACK option, 114. *See also* TCP selective acknowledgment (SACK)
  vs. NewReno, 107
Satellite audio broadcasting, to mobile receivers, 158
Satellite communications
  altitude considerations, 148–149
  broadband satellite communications, 156–158
  c bands, 149
  centripetal and centrifugal forces, 145, 146
  circular orbit classification by altitude, 152
  cooperative diversity techniques, 161–164
  current technology status, 155–159
  diversity gain, 162
  downlink strategies to improve performance, 161
  forward and return links in, 147
  future technologies, 159–164
  GSO satellite, equatorial plane, 147
  history, 146
  hybrid and integrated networks, 159–160
  interactive broadcasting, 156–158
  ITU-R study groups, 154
  and Kepler's laws, 150–151
  layered coding, 164
  MBMS, 164
  mobile broadcasting to handheld terminals, 158–159
  multicarrier techniques, 160–161
  for NGNs, 145
  operational principles and characteristics, 145–147
  orbital periods and velocities, 147, 152
  orbits and spectrum resources, 154–155
  orbits by altitudes and characteristics, 153
  polar orbits, 153
  principles, 145–155
  propagation loss, 149
  rain attenuation, 149
  S-DMB system configuration using CDM, 159
  satellite orbits, 150–153
  satellite system and components, 148
  services and frequency bands, 147–150
  standardization issues, 154
  sun-synchronous orbits, 153
  time synchronization problems, 164
Satellite digital audio radio service (S-DARS), 158

Index

Satellite frequency bands, 147–150
Satellite orbits
   classification by altitude, 152
   classification by inclination angle, 153
Satellite services, 147–150
   classification methods, 147, 148
   to handheld user terminals, 158
   niche services, 155
Satellite systems, 147, 148
   efficiency, 156
   initial launch costs, 149
   TT&C stations in, 147
Satellite TV services, to fixed receivers, 158
Scalability
   with BPL networks, 204
   with carrier Ethernet, 258
Scalable OFDMA, 55
   in 802.16e standard, 55
SDM basics, and 802.11n standard, 33–37
SDM system, 34
   mathematical model, 35
Secure OBS edge router architecture, 284, 285
Security
   with BPL networks, 204, 215
   embedded security framework in OBS networks, 283–284
   femtocell-related, 138
   malicious burst headers, 282
   network *vs.* end-to-end, 283
   in OBS networks, 281–287
   OBS secure router architecture, 284
   orphan burst, 281–282
   with WDM-PON, 180
   WDM-PON *vs.* GPON, 180–181
Seeding wavelength, 177
Selective acknowledgments, 107
   intelligent use by TRL-PEP, 116–118, 117
Sender Window Size limit, 107
Sensors, 1
   decreased production costs, 8
   technology evolution, 9–10
   in terminals, 22
Serial Advanced Technology Attachment (SATA) standard, 6
Service delivery platform (SDP), 83–84, 88–89
   eastbound interfaces, 88
   northbound interfaces, 88
   service-oriented architecture, 84–87
   southbound interfaces, 88
   Web 2.0 and telecom, 89–90
   westbound interfaces, 88
Service enablers, 80
Service hierarchy, 81
Service layer
   authentication and authorization considerations, 91–92
   business model considerations, 96–97
   consumer services trends, 80–83
   illustrative call flows, 92–96
   next-generation digital media services, 79–80
   service delivery platform, 83–90
   standards, 90–91
Service-level agreements (SLAs), 97
Service management, with carrier Ethernet, 258
Service-oriented architecture (SOA), 84–85
   media-enabled service bus, 85–87
   service delivery platform, 88–89
   SOAP and REST technologies for, 87
   Web 2.0 and telecom, 89–90
Service reliability, with BPL networks, 202
Service request packet, in OBS networks, 277
Service virtualization, 85
Services
   in aggregator places, 17
   context-aware, 81–82
   local perception, 16
Set-top boxes (STBs), 129
Shannon capacity formula, 34, 35
Shared risk link groups (SRLGs), 303
Signal fading, in satellite communications, 149
Signal processing techniques
   advantages for BPL communications, 198
   overcoming power line problems with, 207
Signal separation
   in BPL networks, 212
   with copper-based networks, 168, 169
Signal-to-noise ratio, maximizing with MRT, 62
Signaling, 300
   in optical control planes, 299, 301
Signaling plane, in femtocells, 135–138
SIM cards, 10
Simple Network Management Protocol (SNMP), 214
Simple Object Access Protocol (SOAP), 85, 87
Single frequency network (SFN) concepts, 55
Single-input, single-output (SISO) systems, 33
Single-user data transmission, Mobile WiMAX, 76
Sirius satellite constellation, 158
Situational applications, 94
Smart materials, 12
Smart meters, 203, 227
Smart power grids, 195, 204, 225, 227
   BPL for, 202–205
   demand-side management programs, 229–230, 232
   outage management in, 228–229
   time-of-use power metering, 227–228
Smart sockets, 203
Sniffer capabilities, in femtocell applications, 139, 140
Snoop agents, 111
Social networking, 81, 82, 83
Solid state memory, technology evolution, 6

SONET/SDH, 242
    Ethernet over next-generation, 251–252
    for MANs, 288
    neighbor discovery, 299
    optical control plane solutions to, 292
Sound-beaming devices, 23
Southbound interface, 88
    protocol standards, 88
Space-time block coding (STBC), 31, 54, 60
    adaptive mode selection, 61–62
    in 802.11n standard, 44–45
    in satellite communications, 161, 162, 163
    transmit power penalty, 44
Spanning Tree Protocol (STP), 245
Spatial Channel Models (SCMs), 64
Spatial diversity, 58
    CSTD vs STBC, 61
Spatial division multiple access (SDMA), 55
Spatial division multiplexing (SDM), 31. *See also* SDM basics
    and 802.11n standard, 33–37
Spatial expansion, in 802.11n standard, 44
Spatial multiplexing MIMO (SM-MIMO), 61
Spatial stream, 30
Spectral efficiency
    downlink performance, Mobile WiMAX, 66, 68
    frequency reuse patterns and, 67
    increasing through femtocell technology, 126
    traffic model factors, 67
    vs. consumer demand, 127
Spectrum usage efficiency, 9
Split connection approaches, 110. *See also* Cell splitting
    nontransparent, 110–111
    transparent, 111–112
Spurious retransmissions, in TCP, 102
Standalone FAP, 130
Standardized services, with carrier Ethernet, 258
Standards
    collaborative nature of, 91
    Ethernet optical transport-related, 254–255
    major organizational activities, 91
    for satellite communications, 154–155
    service layer, 90–91
Statistical data analyses, 1
    technology evolution, 12–13
Statistical Eigen beamforming, 62, 63
Sticker communication paradigm, 17
Storage, 1
    technology evolution, 6–8
SUB-DL-UL-MAP, 58
Subcarriers
    with 40 MHz waveform design, 42
    in 802.11n standard, 40–41
Sun-synchronous orbits, 153
Supercrunchers, 8

Supervisor Host, 111
Switched Ethernet networks, 245–246
Synchronization, 17
    carrier Ethernet issues, 261
System performance, Mobile WiMAX, 63–64

T

TCP Eifel, 110
TCP Forward Acknowledgment (FACK), 108
TCP/IP Header Compression, disabling, 108
TCP Jersey, 109
TCP modifications, 108–110
TCP performance enhancement
    case study, 114–122
    and cdma2000 1xRTT, 101–102
    and cellular radio limitations, 101–102
    enhancing wireless technologies for, 113
    and GPRS/GSM limitations, 103–104
    in hybrid networks, 99–100
    intermediary solutions, 110–113
    introductory concepts, 100–101
    and packet-switched shared channel technologies limitations, 104–105
    PILC working group recommendations, 107–108
    possible solutions, 105
    proprietary solutions, 112–113
    reaction to random packet loss and retransmission timeout, 101
    split connection solutions, 110–113
    and TCP-impacting characteristics of wireless access networks, 101–106
    TCP modifications for, 108–110
    TCP replacement solutions, 112–113
    TCP selective acknowledgment (SACK) option, 106–10
    throughput behavior comparisons, loss-based and delay-based variants, 109
    and WCDMA/UMTS limitations, 104
    WLAN impediments, 105–106
TCP performance enhancing proxy
    additional data transmission on DUPACKs, 115
    architectural considerations, 114–115
    case study, 114
    intelligent use of selective acknowledgments by, 116, 118
    retransmission on partial acknowledgments by, 115–116
    simulation results, 118–119
    TCP state prediction by, 118
    TRL-PEP performance enhancement techniques, 115–118
TCP performance-enhancing proxy (PEP), 100
TCP retransmission, caused by bandwidth oscillation effect, 103
TCP selective acknowledgment (SACK), 106–107

Index

TCP Snoop, 111
    scalability disadvantages, 112
TCP state prediction, 118
TCP Vegas 0, 108
TDM-PON, 172
    functional scheme, 173
    standards and main optical characteristics, 175
Technology evolution
    autonomic systems, 13–15
    business ecosystems, 18–21
    digital storage, 6–8
    displays, 10–12
    educational capital availability barriers, 5
    energy as bottleneck to, 2
    future communications models, 22–25
    and glocal innovation, 4–6
    Internet with objects, 21–22
    limits of, 1–4
    networking paradigms, 15–18
    processing, 8–9
    sensors, 9–10
    statistical data analyses, 12–13
Telco service widgets, 96
Telecom services providers, 89–90
    competition with ISPs, 89
    distribution network reliability, 96
    evolution from bit-pipe to service providers, 80
    mashup application example, 94–96
    network-level services, 96
    perpetual beta syndrome, 90
    trusted customer relationships, 96
Telecommunications industry, 79
Telecommunications penetration, 4
TeleManagement Forum (TMF), 88, 305
    optical control plane standards, 310
Telemetry, tracking, and control (TT&C) stations, 147
Telephone networks
    problems in developing world, 196, 198
    stalls due to cellular telephony developments, 200
Telepresence, 6
Terabyte capacity, cell phones, 7
Throughput
    edge cell, 59
    as key metric for 802.11n, 28
    TCP
        over simulated cdma2000 1xRTT link, 119
        over 1xEVDO RA-100 scenario, 121
    TCP, WCDMA static scenario, 121
    TCP problems, 100
    for VHTSG, 50
    vs. PHY data rate with frame aggregation, 32
    vs. PHY data rate without MAC changes, 29
Throughput behavior, comparisons between loss-based and delay-based TCP variants, 109

Tiered protection schemes, Ethernet transport with intelligent optical layer, 256
Time division duplex (TDD) transmissions, 55
Time Division Multiplexing PON (TDM-PON), 172
Time-of-use power metering, 227–228
Timestamps Option, 108
Timing wheel, in OBS scheduling, 278
TMF, 90
    optical control plane standards, 309
    standards activities, 91
To-do lists, 24
TR-069 specification, 138
Traffic bundle characteristics, 190
Trail trace identifier (TTI), 299
Transaction-oriented traffic, 7
Transcoding, 86
Transformation, of digital media, 85, 86
Transimpedance Amplifier (TIA), 185
Transmission and distribution (T&D) networks, 202
Transmission Control Protocol (TCP), 100
Transmit antennas, in 802.11n standard, 35
Transmit beamforming, 31, 33, 43
    in 802.11n standard, 45–47
Transmit Time Gap (TTG), 56
Transparent bridging, 245
Transport layer functions, pushing upward in proprietary TCP solutions, 112
Triple-play technology, in BPL networks, 235
Triplexers, 185, 188
    cost *vs.* volumes for microelectronics GPON processor, 187
    cross-section, 186
    top view, 186
TRL-PEP, 114
    aggressive link utilization by, 115
    algorithmic improvements, 117
    network topology, 119
    performance-enhancement techniques in, 115–118
    vs, SACK-enabled standard TCP, 122
Tunnel-Iub architecture, 134, 137
Tunneled Iub network architecture, in femtocells, 136
Typical-Urban (TU) model, 64

U

UMA network architecture, in femtocells, 137
UMA Network Controller (UNC), 135
Unacknowledged data, 100
Unbundling, 181
    with copper-based networks, 169
    GPON limitations, 181
    WDM-PON *vs.* GPON, 180–181
UNI/E-NNI messages, 302

Unidirectional transmission, 181
  in GPON architecture, 177
  in WDM-PON scheme, 172
Universal Mobile Telecommunications System (UMTS) standard, 100
Universal Power Line Association (UPA), BPL networking standards, 232
Unlicensed Mobile Access (UMA), 130
  as femtocell architecture, 134–135
  star rating comparisons with femtocell, 131
  vs. femtocells, 130–132
Unlimited Mobile TV, 158
Uplink capacity, 7
Uplink performance
  Mobile WiMAX, 66
  Mobile WiMAX with multiple receive antennas, 69
Uplink spectral efficiency, Mobile WiMAX, 70
User Datagram Protocol, TCP proprietary solutions using, 112
User-generated content, 82
User profiles, 82
User value
  shift from connectivity to service, 20
  shifts in customer care, 22
Uu interface protocols, 135
  and femtocells, 136

V

Value chains modeling, 18, 19
Value tracking/sharing, 17
Valuets, 7
Van Allen radiation belts, 152–153
VDSL2, 169
Very High Throughput Study Group (VHTSG), 50
Video services, as drivers of fiber-based architecture, 167
Viral marketing, 83
Virtual Concatenation (VCAT), 294
Virtual concatenation groups (VCGs), mapping Ethernet traffic to, 256
Virtual LANs, 246, 247
Virtual Local Area Network (VLAN) assignment. *See also* VLAN technology
  in BPL networks, 212
VLAN technology, in BPL networks, 225
Voice communication, 80
Voice over IP (VoIP), 127. *See also* VoIP
  in BPL networks, 214
Voice quality, femtocell *vs.* UMA, 131
VoIP
  performance of mobile WiMAX systems, 71
  persistent scheduling for, 72

W

Walled garden approach, 83, 90
Watermarking, 86
Wavelength Division Multiplexing PON (WDM-PON), 172
Wavelength grouping, 279–280
Wavelength routers, 268
WCDMA/UMTS
  indoor coverage problems, 128
  TCP-impacting characteristics, 104
  TCP throughput over simulated scenario, 121
WDM-PON, 190
  architecture, 174–177
  capacity, 178–179
  capacity per user, 178
  cost comparison with GPON, 191, 192
  dedicated wavelengths, 180
  fiber utilization, 181
  flexibility, 181
  functional scheme, 172
  optical interface as determinant of cost, 184
  optical link budget, 179–180
  performance comparison with GPON, 178–181
  protocol transparent transmission infrastructure, 181
  security and unbundling, 180–181
WDM-PON optical interfaces, 188–189
Web 2.0 paradigm, 5, 87, 89–90
  disruptive nature for traditional telecom carriers, 90
Web 3.0 paradigm, 5
Web page size, growth of average, 127
Web services, 84
  intrinsic reliance on, 84
Weights, use in 802.11n standard, 45
Westbound interfaces, 88
Wi-Fi Alliance (WFA), 50
Widgets, for mashup applications, 96
WiFi, 15
Wikis, 83
Window Scale Option, 107
Wired networks, TCP performance in, 100
Wireless access networks, TCP-impacting characteristics, 101–106
Wireless backhaul, 239, 240
  BPL for, 202
Wireless Boosted Session Transport (WBST), 112
Wireless clouds, 15
Wireless Identification and Sensing Platform (WISP), 9. *See also* WISP platform
Wireless LANs, 802.11n standard enhancements for, 27–33
Wireless Local Area Networks (WLANs), TCP-impacting characteristics, 105–106

Wireless network operators, femtocell benefits for, 130
Wireless technologies, enhancing to improve TCP, 113
Wireline-based smart grid networks, 227
Wireline core networks, TCP performance enhancement in, 99–100
WISP platform, 9, 23

X

xDSL, 167, 169, 184
XM Radio, 158

Y

YouTube™, 82, 83